To Robert From Karl

A Century of
Electrical Engineering and
Computer Science
at MIT, 1882–1982

This book was set in Univers Light by the MIT Press Computer-graphics Department and printed and bound by Murray Printing Company in the United States of America.

Library of Congress Cataloging in Publication Data

Wildes, Karl.
A century of electrical engineering and computer science at MIT, 1882–1982.

Bibliography: p.
Includes index.
1. Massachusetts Institute of Technology—History. 2. Electric engineering—Massachusetts—Cambridge—History. 3. Computer engineering—Massachusetts—Cambridge—History. I. Lindgren, Nilo. II. Title.
TK210.M3W55   1985   621.3'07'117444   84-19420
ISBN 0-262-23119-0

Karl L. Wildes and Nilo A. Lindgren

with best wishes and appreciation

Karl L. Wildes

A Century of
Electrical Engineering and
Computer Science
at MIT, 1882–1982

The MIT Press, Cambridge, Massachusetts, and
London, England

Contents

**Part I
Electrical Engineering:
An Emerging Profession**
1

**Part II
Setting New Directions**
79

Foreword by Paul Gray
vi

Preface
viii

Acknowledgments
x

Introduction
The Beginnings of
Electrical Engineering
2

Chapter 1
The Rise of Useful Learning
6

Chapter 2
Electrical Engineering:
Offspring of Physics
16

Chapter 3
Growth of the Department,
1902–1952
32

Introduction
Some Who Lead
80

Chapter 4
Machines for Solving
Problems: Vannevar Bush
82

Chapter 5
From Educational Research
to Application:
Harold Hazen and the
Network Analyzer
96

Chapter 6
The Building of
Communications:
Edward L. Bowles
106

Chapter 7
The Industrial Cooperative
Program: William H. Timbie
and Course VI-A
124

Chapter 8
Frozen Motion:
Harold Edgerton and
the Stroboscope
138

Chapter 9
Network Analysis and
Synthesis:
Ernst A. Guillemin
154

Chapter 10
The High-Voltage Research
Laboratory: John G. Trump
160

Chapter 11
The Laboratory for
Insulation Research:
Arthur R. von Hippel
166

**Part III
The War Years**
179

**Part IV
The Turning Point**
237

**Part V
Into the Computer Age**
327

Introduction
Threshold of Major Change
180

Chapter 12
Organizing for
War Research
182

Chapter 13
The Radiation Laboratory
190

Chapter 14
Servomechanisms: The
Bridge to a New Period
210

Chapter 15
The Origins of Whirlwind:
Wartime Digital Computer
Research
228

Introduction
A Time of Reassessment
238

Chapter 16
The Research Laboratory of
Electronics
242

Chapter 17
From Whirlwind to SAGE
280

Chapter 18
Growth of the Acoustical
Sciences: The Acoustics
Laboratory
302

Chapter 19
Innovations in
Engineering Education:
The Gordon Brown Era
310

Introduction
A New Identity
328

Chapter 20
The Computer and
Changing Perceptions
330

Chapter 21
The Computation Center
334

Chapter 22
The Evolution of
Time Sharing
342

Chapter 23
The Department of
Electrical Engineering and
Computer Science:
Seeking a New Paradigm in
Engineering
354

Appendix: Chronology,
1846–1980
374

Notes
410

Index
416

Foreword

Paul E. Gray, President
Massachusetts Institute
of Technology

This book traces the central forces and prevailing goals that have shaped the evolution of an extraordinary—and extraordinarily successful—academic enterprise. The manifold activities in electrical engineering at the Massachusetts Institute of Technology over many decades have had an enormous influence on the field—on those who practice and teach in this interesting, complex, and vigorous intellectual domain, and on those who have carried its disciplines and insights into professional fields once far removed. How that happened, and how education and research have been continually infusing and revitalizing one another, are core themes of this volume, which covers the first century of electrical engineering at MIT. It is a story of revolutionary changes affecting our society and the world, seen within the context of a particular institution. At another level, it is the story of how outstanding individuals have influenced other outstanding individuals—the interaction of students and teachers over time. The book reflects the MIT spirit—the underlying ethos that has remained more or less constant for generations of students and teachers despite continual change in the foreground curriculum matter. For these and other reasons, this volume is important both to the profession and to MIT, I believe, for it illuminates a significant history, providing guides for a future whose form will also be continually changing.

My personal association with the Department of Electrical Engineering and Computer Science—as student, as faculty member, and, since 1970, as institutional colleague—spans more than thirty years. During this time I have come to prize what, for me, are the cardinal principles that have shaped the development of the department:

• Steadfast attention to the importance of stimulating teaching and of broadening and deepening the students' horizons. Few other departments have known so many uncommonly gifted teachers. This is no accident. In its collective concern for quality and effectiveness in teaching and learning, the fact that people are the principal objects of this department's efforts is consistently evident.

• A willingness, indeed an insistence, not only to connect with real problems and important educational and technical issues of today—the problems of industry, the needs of society, the priorities of government—but to anticipate the needs of tomorrow through imaginative research programs carried out by student-professor partnerships.

• A complete acceptance of the doctrine that teaching and research play mutually reinforcing roles. The department's curricula have been steadily transformed and strengthened by the regular infusion of knowledge and insights gained from its research activities.

• An overriding concern for quality in all that is done. The influence and strength of the department result in large measure from its insistence that every enterprise, every new development, be a best effort.

These qualities, it seems to me, are hallmarks not only of the department but also of MIT, and they set a high standard for our future.

For those who know this department from their own time here, this book may recall events concurrent with their personal involvement and sharpen their view of how they were part of a larger history. For future students this volume will prove to be more than a formal chronology; it will provide the information and the insights, and the sense of things past, that will allow them to make critical appraisals and to draw inspiration from this remarkable enterprise spanning many generations.

Preface

Karl L. Wildes
Nilo A. Lindgren

In the century that has passed since the field of electrical engineering became formalized, one salient feature has distinguished its growth. Curiously enough, this single feature—namely, the capacity of electrical engineering to change its identity continually—stands almost as a contradiction. The many disciplines embraced today by this field seem a far cry from the telegraph, telephone, electric lamps, motors, and generators in which the field has its roots. Yet things electrical have been discovered to have such a profound universality that it has been possible to build a unified core, though not without a struggle.

Thus, in this protean profession, a student may embark in one direction but within a few years find himself or herself engaged in something totally unexpected. Moreover, perhaps surprisingly, he or she may find himself or herself not only prepared to deal with that unexpected thing but in no small way to have participated in bringing it about.

How does this happen? It is not an accident. What it really means is that the student's education in electrical engineering has been sound. This is no trivial achievement, especially as things electrical have continued to ramify. It means that somehow or other the school has managed to keep to the mainstream, to anticipate future important developments, to participate in those developments, and to distinguish between the merely technical and the fundamental. What is the "somehow or other"? What are the keys to the outstanding success of this school, which in the course of a century has gained recognition as one of the best electrical engineering schools in the world?

How to keep education sound in an environment of continual change brought about by scientific and technological progress is one of the themes of this volume. It began many years ago as an effort to document the growth of the Department of Electrical Engineering at MIT and to describe the work and influence of some of the department's outstanding educators and alumni. It began, moreover, as an effort on the part of the senior coauthor to record and reflect in one volume the collective experience of educators and students at MIT, spanning in personal contact nearly the entire century. The archival version of this story of the MIT Department of Electrical Engineering and Computer Science, which is available in the MIT Library, has long since outgrown the possibility of being contained in one volume. The present volume, therefore, which has been winnowed from the archival documentation, highlights the major stages of the department's evolution. In the process of winnowing it has become clear that, despite the ever-widening diversity manifest in this field, there has been a basic continuity in the approach to what in the beginning was called ''alternate education''—as opposed to what was provided in the classically oriented great universities. The continuity arose not just out of selecting the best and the brightest and letting them have their go. It arose out of an institutional framework, a set of principles, a mutually reinforcing interaction between education and research, and a basic cooperation with the momentum of industry. MIT's motto, *Mens et Manus*, ''mind and hand,'' expresses this approach as well as anything—without, however, defining how it works in practice.

How has it worked in practice? Among the many scientific and engineering developments marking this past century, we have selected the events and the people—students and teachers—that have proved to have a lasting significance in the electrical engineering field and in the growth of the department at MIT. What we see is how at each stage of change professors and students worked together and in solving problems together opened up new fields, which is the key to the way research periodically but consistently unfolds the unknown and reinvigorates the educational function. Over the course of a century we see a steadily growing stream of students coming into the field, working with teachers, becoming teachers, reshaping old fields and opening up new ones. Though we do not draw the genealogical line sharply, it is there—how Cross worked with Bell and nurtured the fledgling Course VI, how Cross brought along Clifford, how Clifford linked to Jackson, how Jackson nourished and brought up ''Van'' Bush and Bowles, how they worked with students like Hazen and Stratton and Brown, all of whom would be instrumental in the post–World War II growth of the department, and so on, down to the present day. In fact, there is no single genealogical line, but many, which have proliferated now into a thousand branches.

In the process there have been sharp struggles. It has been observed that it is these struggles—the ''fights''—that have helped give the department its strength. Periodically, as the field has grown and become too diversified, it has been necessary to hammer out new underlying unities, new core curricula, new paradigms to keep the field together and to keep pace with change. The participants in these struggles were concerned with how to keep the field open-ended and the educational process vital and relevant; they were concerned with how to educate students for the long term, since students and educators alike are always in danger of being outpaced by events. The theme of the centennial celebration of the department in 1982—namely, lifelong cooperative engineering education—testified to this concern.

Among the recent struggles has been one—sketched in our final chapter—that could be the prelude to the evolution of a new paradigm for the Department of Electrical Engineering and Computer Science. We do not pretend to be prognosticators, but shall wait and see how things go. In any case, it is clear that the dynamic interaction of research and teaching—the creation of new knowledge and the infusion of that knowledge into the ever-evolving curriculum—and the interconnections with industry are as vigorous as ever.

Acknowledgments

K. L. W., MIT '22
N. A. L., MIT '48

In preparing this volume, we were helped by many individuals who gave us their insights and shared their experiences and knowledge, who read earlier drafts and gave us their time, their intelligence, and their care. We cannot possibly name them all—at least two hundred friends and critics were in one way or another involved over the many years—but there were a number who were crucially supportive, and these should not go unnamed.

Special thanks go to Gordon Brown, who encouraged Karl Wildes to undertake this work initially and who has unstintingly helped us in many important ways, even down to the final stages of proofreading and illustration selection; to Richard Adler, who generously took upon himself the responsibility of anchorman in the EECS Department, helping us solve both logistical and philosophical problems; to Julius Stratton, who gave us gentle, wise counsel over the years; and to Patricia Burns Lindgren, who gave loving support and encouragement in a project that extended far beyond our original expectations.

For the creation of the chronology at the end of this volume, which highlights world events concurrent with the events in our narrative, and for invaluable advice and research on photographs, we owe thanks to Warren Seamans, director of the MIT Museum, and to his staff; nearly all the illustrations for this history were provided by the museum. For manuscript production, we are indebted to Gianna Sabella, Edna Strom, Aina Sils, Barbara Ivers, and Patricia Nettles.

Although the two authors are responsible for the final content and interpretations in this history, we are indebted, for critical guidance, for informative discussions, verbal and written, and for invaluable research, to the following: Richard Adler, Jonathan Allen, Wilmer Barrow, Edward Bowles, Gordon Brown, Fernando Corbató, Wilbur Davenport, Mildred Dresselhaus, Harold Edgerton, Peter Elias, Robert Fano, Lawrence Frishkopf, Paul Gray, Truman Gray, Cecil Green, John Hewitt, Howard Johnson, James Killian, John Linvill, Loretta Mannix, Marvin Minsky, Philip Morse, Joel Moses, Frank Reintjes, Walter Rosenblith, Arthur Samuel, Warren Seamans, Helen Slotkin, Louis Smullin, Julius Stratton, Frederick Terman, John Trump, John Tucker, Arthur von Hippel, John Ward, Jerome Wiesner, Gerald Wilson, and Henry Zimmermann.

Considering the fact that we represent different generations at MIT, we would like to record the deep pleasure we have discovered in a true collaboration. It is based in part, we believe, on a continuity at "our school" that was not wholly obvious to us when we started, but that bound us together, along with the many others who participated at different stages—even as fellow students—and to whom we express our gratitude.

A Century of
Electrical Engineering and
Computer Science
at MIT, 1882–1982

The Rogers Building, the first MIT building, in the Back Bay area of Boston, was begun in 1863 next to the Boston Society of Natural History (*completed at right*).

Part I

Electrical Engineering:
An Emerging Profession

Introduction
The Beginnings of
Electrical Engineering

When the first electrical engineering courses were being established at MIT in 1882, the state of knowledge about things electrical was fast changing, spurred largely by the activities of individual inventor-entrepreneurs. Researches in electrical phenomena had been going on for many decades—for instance, electrical arc lighting had been demonstrated by Sir Humphry Davy as early as 1809, and the fundamental principle for the electric generator had been discovered by Michael Faraday in 1831—but it is not until the emergence of practical telegraphy in the mid-1800s that one observes a great upsurge in electrical inventions. By 1882 the basic elements of first-generation telephone, electric light, and electric power systems had been invented and were being developed at a rapid pace in cities in the United States and Europe. Within a decade second-generation systems based on new principles were already beginning to supplant the first.

Edison's famous Pearl Street Station in New York City, based on direct-current principles, went into operation in 1882, the same year that MIT physicist Charles R. Cross was setting up the first formal course in electrical engineering. In that same year another inventor from England, Frank J. Sprague, was taking out his first patents on electrically driven trolley systems. Electrical manufacturing industries were in the process of being formed, spearheaded in the United States by the inventor-entrepreneurs Thomas Edison, Elihu Thomson, George Westinghouse, and others, and their growing companies would create a greater and greater need for electrical engineers, a need that educators by then could clearly foresee. Professional societies of electrical engineers were being founded in the

United States and abroad, new journals devoted to electrical development were being established, and these efforts were contributing to the rapid growth of the field. The development of second-generation electric power systems based on alternating current, as opposed to direct current, introduced a new level of complexity in electrical engineering and made professional training virtually mandatory.

As things turned out, it was the professionally trained men in engineering, in electrical theory, in organization, in finance, who created a new plateau of the electrical century. Though Edison, for instance, continued to rise in American mythology as a great hero, he was being supplanted and outdistanced in actual electrical progress by those who had a solid theoretical as well as practical footing. Though companies like General Electric and Westinghouse were working hard to train their young would-be electrical engineers in this new field, by the 1890s schools like MIT had become the chief suppliers of electrical engineers.

How the electrical engineering department at MIT was to grow so successfully over a long period of time and acquire international renown should be of no little interest in a world that depends so deeply on scientific and technological progress in general and on electrical engineering in particular. Was it location, luck, policy, particular people, critical mass, some unusual synergistic combination of timing and world events? Was it the continuous evolution of the electrical engineering curriculum to keep pace with the changing field, was it the way its professors were recruited, was it the insistence on research integrated with education? These are some of the issues examined in this volume covering the highlights of the various epochs in the growth of electrical engineering.

Before focusing on the growth of electrical engineering and on the evolution of the MIT EE Department, it is important to look briefly at the founding of MIT itself and at the broader forces that stimulated the idea of an MIT. More than may be generally realized, the principles underlying the founding of this school have continued to influence its growth down to the present day. The struggles that have attended the building of the EE Department reflect the efforts of many different kinds of people to adhere to the founding principles.

An Act to Incorporate the Massachusetts Institute of Technology, and to Grant Aid to Said Institute and to the Boston Society of Natural History

April 10, 1861

Be it enacted by the Senate and House of Representatives in General Court assembled, and by the authority of the same, as follows:

William B. Rogers, James M. Beebe, E. S. Tobey, S. H. Gookin, E. B. Bigelow, M. D. Ross, J. D. Philbrick, F. H. Storer, J. D. Runkle, C. H. Dalton, J. B. Francis, I. C. Hoadley, M. P. Wilder, C. L. Flint, Thomas Rice, John Chase, J. P. Robinson, F. W. Lincoln, Jr., Thomas Aspinwall, J. A. Dupee, E. C. Cabot, their associates and successors, are hereby made a body corporate by the name of the Massachusetts Institute of Technology, for the purpose of instituting and maintaining a society of arts, a museum of arts, and a school of industrial science, and aiding generally, by suitable means, the advancement, development and practical application of science in connection with arts, agriculture, manufactures and commerce; with all the powers and privileges, and subject to all the duties, restrictions and liabilities, set forth in the sixty-eighth chapter of the General Statutes.

Said corporation, for the purposes aforesaid, shall have authority to hold real and personal estate to an amount not exceeding two hundred thousand dollars. . . .

This act shall be null and void, unless its provisions shall be accepted within one year, by the Massachusetts Institute of Technology, and the Boston Society of Natural History, so far as they apply to those societies respectively.

Acceptance by Massachusetts Institute of Technology

April 9, 1862

To His Excellency the Governor and the Honorable the Council of the Commonwealth of Massachusetts:

At a meeting of the Massachusetts Institute of Technology held on Tuesday, April 8, 1862, at the rooms of the Board of Trade in this city, the Institute having been permanently organized and a government elected, the following Resolve was adopted:

"Resolved that the Massachusetts Institute of Technology accept the Charter and other provisions of the Act of April 10, 1861, relating to said Institute."

William B. Rogers,
President, Massachusetts Institute of Technology

Chapter 1

The Rise of Useful Learning

The Founding of MIT

It is clear from the early writings of William Barton Rogers, the founder of MIT, that he felt an absolute assurance that the scientific and technical institute he envisioned would grow great. In 1846, fifteen years before he would succeed in getting MIT chartered in Boston, he wrote:

A scheme of this kind begun with two professors in the scientific department, two subordinate instructors in the other (the ''practical department''), under the direction of the former, would, I am certain, prove so signally successful as ultimately to require its expansion into a polytechnic college on the most ample scale, in which, along with all the subjects above referred to, would be embraced full courses in elementary mathematics and instruction, perhaps, in the French and German languages. In a word, I doubt not that such a nucleus-school would, with the growth of this active and knowledge-seeking community [Boston], finally expand into a great institution comprehending the whole field of physical science and the arts with the auxiliary branches of mathematics and modern languages, and would soon overtop the universities of the land in the accuracy and the extent of its teachings in all branches of positive knowledge.

It is probable that Rogers, even with his prescience, did not foresee just how far his ''Polytechnic Institute'' would succeed, though he watched and nurtured its growth and independence until his death in 1882, the year the first formal electrical engineering courses were being taught within the MIT Physics Department.

Since the emergence of electrical engineering as a distinct field from physics, there has been a kind of ebb and flow of relationship between EE and pure science. As a scientist and educator at a number of schools, Rogers became convinced that training in science and engineering would best be forwarded by a school strongly independent of the traditional universities. Thus, he worked for many years toward the establishment of such a special school, where science would be a foundation stone interlocked with practical applications in industry and business.

When Rogers wrote out his scheme in 1846 in "A Plan for a Polytechnic School in Boston," he visualized a tripartite institute, embracing a Society of Arts, a School of Industrial Science, and a Museum of Technology. The first of these referred not to the fine arts but to the practical arts; its chief function would be to present lectures by people from many fields about their special knowledge and inventions, lectures aimed at educating a wider public about science, technology, and industry. Although the Society of Arts kept up its activities for many years, by the turn of the twentieth century its role had progressively diminished with respect to the activities of the School of Industrial Science and as the numbers of more highly specialized sciences and technological disciplines increased. Publications of the Society of Arts were merged in 1890 with the *Technology Quarterly*. By the time of the First World War the Society had nearly faded away altogether.

Rogers's Museum of Technology did not become a functioning entity in that early epoch. Some artifacts were collected, but they remained scattered throughout the Institute. Actually, with the stupendous growth of science and technology over the past century, there are both more matter for and more interest in such museums today. At MIT the establishment of a new museum in 1971 represents a realization of part of Rogers's proposal.

William Barton Rogers, founder of MIT.

To the Physical Lecture Room
at MIT there came many older
"students," men and women,
attempting to do experiments
and to grasp the rudiments of
the scientific method.

Of the three entities it was the School of Industrial Science that attracted more and more students. Its courses both widened and became more specialized, and gradually evolved into the core of the MIT curriculum as generations of students came to know it. The focus of this school was on the ways in which the knowledge of science could lead to applications that would be of benefit to industry and business and society generally. In contrast, the popular trade schools taught the *how* without examining the *why*. Implicit in Rogers's philosophy of education, in his vision of "industrial intelligence," was the concept of having lectures supported by "experiments, models, and diagrams." Out of such an approach would flow projects and laboratories and a continually evolving concept of research. Though it would take many decades to articulate and realize in the educational environment, it was the vital combination of new research and a continually evolving curriculum fed by that research that became the key to success in MIT's scientific and engineering training.

The Growth of Useful Learning

At the time Rogers began to work seriously toward the realization of his Polytechnic Institute, the seeds for "useful learning" had been germinating in American soil for nearly a century. The renowned native American inventor, Benjamin Franklin, the prototype for many later inventors, who was to bring prestige to those practicing the useful and inventive arts, was instrumental in establishing in 1751 the first real school aimed at providing a pragmatic education. Called the Philadelphia Academy, it became a model for other such academies that attempted to respond to the needs of everyday life. Although the earliest American schools, such as Harvard, had introduced mathematics and natural philosophy (science) in the early 1700s, such offerings did not much alter the classical flavor of the curriculum, nor did they stress the practical application that was being demanded by the colonies' emerging businesses and trades. (But these innovations at Harvard, such as they were, were enough to drive some Harvard-trained scholars to establish Yale in New Haven in 1701 along strict classical lines.)

Boston was from the beginning an important center for education. Barely had the "Great Migration" begun with the Puritans in the 1630s when schools were established. Harvard, the first American college, founded in 1636, was modeled on Oxford and Cambridge, which in turn had been modeled on the great archetype, the University of Paris, founded about 1170. The famous Boston Latin School, the first private high school, was set up in 1635. Though Thomas Jefferson struggled in Virginia to establish schools supported by the public as early as 1779, again it was in Boston that the first such schools appeared. The first "common school," called the English Classical School but later renamed the English High School, was set up in Boston in 1821. In the nineteenth century Massachusetts was recognized for the liberal attitude of its legislature toward education, inspired by the forward-looking and progressive efforts of Horace Mann; in 1837, for instance, Massachusetts was to establish a Board of Education. Moreover, Boston was at the hub of business and industrial enterprise in that period and had become the mecca for young would-be entrepreneurs and inventors.

Thus, it was on Boston that Rogers began to pin his hopes in the mid-1800s for the establishment of this new kind of school, one based on the sciences and the arts, but dedicated to the support of practical applications that were to be of benefit to society generally. His dream was something far greater than a trade school; it was a university of a new breed, equal in importance to the classical universities, yet offering an alternative path. Rogers wrote:

From the mill, the farm, the machine shop, the laboratory, the shipyard, from the desk of the engineer and architect, the chair of science, the workman's bench, the merchant's counting room and all the other scenes where educated industry is at work, we may claim and expect the aid of vigorous thought and cooperative labour; and knowing how mighty is the momentum of the industrial intelligence around us, we can scarcely doubt that the efforts of the Institute will prove of substantial public interest and advantage.

Looking back, one might ask why it was necessary to establish such an independent technical school at all. Could not the traditional universities, and numerous trade schools, have equally well met the needs of growing industry in an America that in the 1860s and 1870s had reached an economic and industrial take-off stage? It is worth remembering that the Industrial Revolution (so named by Friedrich Engels in 1844) had already been rumbling along for many decades without the benefit of highly specialized technological institutes. True, there were some schools emerging on the Continent, particularly in Germany and Great Britain, where the effects of industrialism were most deeply experienced, but these were still exceptional. Making a comparative study of European schools in 1864, Rogers was to find that the Polytechnic Institute of Karlsruhe in Germany was more like his concept of MIT than any other.

Today, when schools of high science and technology are the rule rather than the exception, and when they have proved their worth to industry and government, it is difficult to imagine the conditions in which there would be considerable hesitancy—as there was in the Massachusetts legislature, despite its progressive tone—in accepting and supporting the proposal by Rogers and many other well-known and respected scientists and industrialists for a true alternative form of education. It was probably difficult to distinguish the

Boston, October 1860: This first aerial photograph, taken from a balloon at about 1,000 feet, shows the Mercantile Building at lower right, where MIT's first classes were held in February 1865.

Because the new MIT building on Boylston Street, begun in 1863, was slow in being completed, Rogers held his first classes in the Mercantile Library, with an initial enrollment of 15 students.

Earliest known photograph, taken in 1872, of the Rogers Building, the neoclassic brick structure in which "Boston Tech" made its home.

concept of the Institute from the trade schools already in existence. Indeed, even today there are probably some who still see such institutes as MIT as little more than narrow trade schools.

The linking of the school with an organized research laboratory was also a novel idea, at least in the United States. There were individual inventors and researchers who built and maintained their own laboratories, but these were idiosyncratic efforts, removed from both academic and industrial settings. Only in the distant future would such research become a core element in scientific and technological education, as well as in industry. The public benefits of these endeavors, then, were not at all obvious to most people.

Thus, when Rogers moved to Boston, seeing it as an ideally suited place for the kind of polytechnic institute he was promoting, he had to struggle for 15 years to realize the formation of his school. He wrote the "Plan for a Polytechnic School in Boston" on March 13, 1846; it took until April 10, 1861, for the Massachusetts legislature to pass the Act to Incorporate the Massachusetts Institute of Technology. And then the American Civil War was to delay the effective opening and functioning of MIT for several more years.

The fostering and development of the new Institute "(Boston Tech")" in downtown Boston near Copley Square has been carefully documented in a history by Samuel C. Prescott, covering the years from 1861 to 1916.[1] Prescott himself was at MIT from his student days (Class of 1894) until his retirement from the faculty in 1962, so that in addition to documents, he was able to draw upon the personal reminiscences of the MIT old-timers who were then still alive. The chronology at the end of our history (see Appendix) highlights some of the technical, social, and political events of this early epoch when MIT's survival as an independent school was most precarious. Only during Francis Amasa Walker's presidency (1881–1897) did MIT find itself on a secure independent basis.

Two views of the Rogers Laboratory of Physics, established soon after MIT was founded. In this laboratory Alexander Graham Bell carried out experiments in telephony leading to the invention that would become one of the major foundations of electrical engineering. (The photograph dates from 1876.)

The Battle for Alternative Education

The new education—the alternative way—did not win its place easily. In addition to attracting unusual types of people, alternative education possessed some unusual characteristics. It took its matter from the world around it, from the ongoing efforts of industry, of business, of isolated inventors, of scientists who then knew all the science there was to know. It was frankly secular in a time when theology was dominant in the traditional universities. It championed a here-and-now approach to what was newest in industrial intelligence rather than remaining fixed on the immutable in Aristotle or on the moral shaping of young people. It had to be flexible and creative; it had to work hand-in-hand with industrial application; it had to seize on new theory wherever it appeared, test it, and apply it if possible; it had to struggle with the unknown in the hope of application—if not now, sometime. It had to set up its own laboratories and do its own research, and then try to promote its results in practical industry. Alternative education was the antithesis of what the traditional universities stood for and practiced; that is why it could not nest well in these institutions—it might have been swallowed up, its growth stunted, or, misunderstood wholly, it might have been relegated to the trade schools where it didn't belong either.

Enthusiasm for this form of education was strongest where the Industrial Revolution was strongest—in Western Europe, especially Germany and Great Britain, where the Industrial Revolution had incubated and burst forth in the 1700s. In America it grew ever stronger through the latter half of the nineteenth century; by the turn of the twentieth century, MIT had become a model for many other schools. Yet there were numbers of times during its first half-century when the new institute seemed on the point of extinction, or about to lose its independent path through being absorbed by its outstanding and well-endowed neighbor, Harvard.

These early vicissitudes are not widely known today. Many people assume that the Institute began to acquire its reputation for doing remarkable things and for turning out remarkable people during its activities in World War II. There is some truth in this perception, for it was then that the Institute really began its profound involvement with government and with society; the onset of World War II pushed MIT research and development activities to a level that previously would have seemed impossible. As one looks into the early history of MIT, however, one sees that these recent epochs were plateaus that could not have been scaled without the work of the preceding generations, of those who transformed a local school, "Boston Tech," into an institute of international standing and scope.

Just as MIT itself can be seen as a representation of America's search for an alternative way in education, so can it be seen as a representation of the nation's search for an alternative way to the future, one in which science and engineering, at first ancillary or service activities, like springs and gushers in the old rural setting, have progressively become the mainstream of the nation's growth. In fact, the major share of the nation's Gross National Product comes from the inventions and innovations of scientists and engineers. This underscores the need for a continuing strong investment in scientific and technical training. During the century with which we are most concerned in this story, invention has changed from an art practiced by a dedicated and passionate few into an organized process in which teams of highly trained professionals take part. The experimental method has become central to the workings of society. Research laboratories have grown from a few into thousands; a grasp of science and technology has become requisite in many professions; and science and engineering, long separate, have become irrevocably interlocked. Scientific training and the economic-industrial system have turned out to be completely interdependent.

Electrical Engineering

Of the many new fields that have been created in the past century under the aegis of alternative education, electrical engineering has demonstrated the most remarkable growth and vitality. In fact, electrical engineering, since its origins in the late nineteenth century, has exhibited certain features that make it a unique subject in relation to the ongoing technological momentum. More than other fields, it has been willing to change its identity (that is, the people in it have been willing to change their identities) and it has been willing to take up new challenges; it has displayed a willingness to take on any task, provided only that issues electrical were somehow embedded in the task. That electrical engineers in recent years could be occupied with, for instance, computational linguistics or theories of human cognition, whereas the electrical engineers of the last century were occupied with the dynamics of electrical power systems, suggests the magnitude of the changes in this protean profession.

How such changes came about, step by step, is part of the story sketched in this volume. This is not to pretend that everything of importance in electrical engineering took place at MIT or emanated from MIT—far from it. But enough did take place, and the ebb and flow of general technological developments were closely enough tracked at the Institute, to give us a grasp of how the study of things electrical grew to be what it is today.

Chapter 2

Electrical Engineering: Offspring of Physics

In accounting for the vitality and growth of electrical engineering, one must see, among other things, its close connection with physics on the one hand and with "external" practical electrical problems on the other. At MIT the electrical engineering course started in the Physics Department and was part of that department for two decades (1882–1902). The first teachers of electrical engineering were physicists, and much of the early staff was recruited from the physics graduates.

As the electrical engineering curriculum evolved, and as the profession grew, the close connection with physics sometimes waned, but at critical periods—as in the early formative years, in the early 1920s, and again in the post–World War II years—the affinity was reestablished, and the theoretical foundations of electrical engineering became ever deeper. From the very beginning, then, the connection between physics and electrical engineering was far more influential than was, say, the connection between physics and civil or mechanical engineering; in some schools, however, electrical engineering emerged from mechanical engineering—owing, no doubt, to the field's involvement with heavy power machinery. From time to time at MIT there were close collaborations between electrical and mechanical engineering, but rarely were these as far-reaching as the interactions with physics or, indeed, with mathematics.

MIT's early EE connections with external developments grew out of the interests of the teachers in the Physics Department, notably Edward C. Pickering and Charles R. Cross. It was Cross particularly who was to encourage a working alliance with Alexander Graham Bell as he was groping his way toward the invention of the telephone. Their relationship, and the involvement of MIT in Bell's invention, are depicted in this chapter; the continuing involvement of members of MIT's electrical engineering staff with outside practical developments, and the influence of such involvements on curriculum revision and reforms, will be traced throughout this volume. Though such ties to developments beyond the campus were not the sole secret of the department's success, they come as close as any other single factor to explaining the department's continuing vitality over the course of many decades.

The reputation of being a tough place also contributed to the élan of the school as a whole; the students, who were beginning to come from far and near, took a stubborn pride in being able to survive the "grind"—a characteristic that has prevailed to the present day. Some of the problems of the early MIT, which helped to shape its unique character, have been described by Samuel B. Prescott in his book *When MIT Was "Boston Tech"*:

When Walker came in 1881, nearly two-thirds of the 302 students lived in Boston or in the cities and towns within commuting distance. From nearby suburbs they walked or came by horsecar daily, and from the outlying towns by train to one of the half dozen stations in the city, whence they generally walked to the school. Students came daily from as far as Worcester, Newburyport, and Lawrence, or even farther. To make a nine o'clock class they had to rise with the lark, snatch a hasty breakfast, and hurry to the depot to get an early train. At the end of the day they got home in time for a late supper, with work for the next day still to be done. Whatever diversions of social life these students had were likely to be in their home town rather than at "the Tech."

Since there were no dormitories, students from more distant points had to find quarters in South End rooming houses or private homes throughout the city, for the vicinity of the Institute was sparsely settled in the

For twenty years the courses in electrical engineering were conducted within the MIT Physics Department on the Boston campus. The scene is a physics lecture room in the "new building" (later named after President Walker of MIT), erected in 1883 as the student body and laboratories outgrew the space in the original Rogers Building.

Edward C. Pickering gave the first lectures and demonstrations in electrical science at MIT in 1866, and conducted experiments in telephony well before Alexander G. Bell arrived on the Boston intellectual scene.

Charles R. Cross, a student of Pickering, was to become the founder of the first formal course in electrical engineering in 1882, the year of William Barton Rogers's death, and the year that the "wizard" of electricity, Thomas A. Edison, began providing electrical service from his Pearl Street Station in New York city.

early years. The full class schedules and the time spent in travel left very little opportunity for diversion or organized recreation. Furthermore many students were "specials," often older men primarily interested in pursuing the work in some specialized field in which they had already had some experience, and not candidates for a degree. Many of the most successful and loyal of MIT's alumni were in this group. In the seventies less than half of the students were taking the full schedule of subjects, and as late as the middle nineties only sixty-one per cent of all students were "regulars." With the emphasis so largely on a full-time program of classes and laboratory work, and with so many commuting and special students, "college spirit" developed slowly. It was generally assumed that Tech students were quite different from the students in the older or "liberal" colleges. They came largely from families of limited means and no traditional college loyalties, and they came with serious and immediate objectives. For many of them association with fellow students was confined to class and laboratory exercises and military drill, with occasional meetings between classes or at the noon period. There was no lunch room for students until 1876, when a small one was provided at one end of the drill shed, and there were never any chapel exercises, either voluntary or mandatory. Yet somehow a general acquaintanceship developed, despite slow trains and long days, and from the beginning the hard-working students developed pride in being Technology men and acquired something of the spirit that had made MIT possible. Even those who remained for only a year or who fell out of the ranks before graduation absorbed a feeling of loyalty which was often evident in later years.

A century later, though MIT's location, resources, size, and personnel have changed, the description is familiar enough to anyone who has been part of the Institute at any period.

Charles R. Cross

What kind of man was Charles Cross, who established the first course in electrical engineering? The records are not always lucid on such matters. His early photographs show him to be a serious if not stern figure, a very formal man, though the "electricals," as they then began calling themselves, probably felt a sneaking affection as well as respect for him. He was clearly a very competent person, an innovator, and an appreciator of technological innovation, as revealed, for instance, by his throwing open the physics laboratories for use by Alexander Graham Bell. Moreover, he was a dedicated educator, noted for his lectures and for his students, who would eventually form the nucleus of a new department of electrical engineering.

The Early EE Courses

The course in electrical engineering at MIT has been generally recognized as the first in the United States.[1] The curriculum was published in the MIT Catalogue for 1882–83 and included, in the third and fourth years, many subjects already standard in mechanical engineering, such as applied mechanics, theory of steam and other engines, and a large amount of mechanical engineering laboratory. Subjects in telegraphy, telephony, electric lighting, and transmission of electric power were listed, but were yet to be developed by the physics staff.

In the fall of 1882, six students were registered in the second year of this course; it was called Course VIII-B to distinguish it from the regular physics course, Course VIII. Professor Charles R. Cross, then head of the Physics Department, had given a series of optional lectures on the practical applications of electricity in the spring term of 1882; he founded and administered the new course.

By 1890 the "electricals" at MIT had assumed a jaunty self-identity as well as taking potshots at the founding professor of their courses, "Charlie" Cross.

"A bunch of Electricals,'90." The name "Electrical Engineer" had begun to be prominent by 1884, but most men who were making practical applications of electricity still preferred to be called "electricians." At a National Conference of Electricians held in 1884, the presiding chairman, Henry A. Rowland, a noted professor of physics at Johns Hopkins University, reflected the spirit of the time: "Let physical laboratories arise," he said; "let technical schools also be founded. . . . It is not telegraph operators but electrical engineers that the future demands."

Cross's lectures were by no means the first appearance of electrical science at MIT. Lectures and demonstrations in electrical science had been introduced by Edward C. Pickering, professor of physics at MIT, as early as 1866. He continued these lectures until 1877, when he left the Institute to become director of the Harvard Astronomical Observatory. Pickering's specialty was acoustics, but he advanced the teaching of electrical science in America with equipment he had purchased in Europe.

Pickering was, in today's terminology, more of a "research type," whereas Cross, one of Pickering's outstanding students in physics, enjoyed lecturing and writing clearly on scientific and technical ideas developed by others. Cross received his degree in 1870 in Science and Literature (the so-called Course VI of that day, there being no established course in physics until Course VIII was set up in 1873–74) and immediately upon graduation was offered an instructorship. He was promoted in 1871 to an assistant professorship. His rise in the department was rapid. In 1875 he was promoted to a full professorship, there being no grade of associate professor until 1884.

In reviewing the laboratory and demonstration equipment in 1873, Cross noted that the department had an "exceedingly full" supply of acoustics apparatus, with almost all the more important instruments made by Koenig of Paris. A set of organ pipes had been given recently by Hook and Hastings, organ builders in suburban Boston. Somewhat complainingly, Cross remarked that a considerable sum would have to be spent on equipment relating to electricity and magnetism "to make it a fair representation of the present state of electrical science." The following year a Holtz machine and an induction coil were acquired, and Pickering announced that a large magneto-electric machine had been loaned to the Institute by its maker, Moses G. Farmer, a well-known inventor of that period. It was Pickering's hope that this machine could be secured permanently for the Institute.

Acoustics Laboratory. Between Pickering and Cross, there was a shift in emphasis in the Physics Department; the apparatus of acoustical science began to go on display, while the electrical laboratories became crowded with new and exciting equipment. In the twentieth century acoustics and electrical engineering would once again establish close research connections.

Physicists in other educational institutions were also teaching the practical applications of electrical science as these applications appeared on the American scene. Professor William A. Anthony, who for many years was the sole member of the physics faculty at Cornell University, constructed, with the aid of his chief assistant George S. Moler, a Gramme-type dynamo, which was displayed at the United States Centennial Exhibition in 1876.[2] The 1876 dynamo was used subsequently to supply electricity on the Cornell campus. The Cornell University Register for 1883–84 announced the new Course in Electrical Engineering to receive registrants in September 1883. Cornell also claims the nation's first doctorate in electrical engineering, awarded to James G. White in 1885.

At the University of Pennsylvania, George F. Barker, professor of physics, was intensely interested in Edison's experiments and invited Edison to join an expedition to Wyoming in the summer of 1878 to view an eclipse of the sun. It was on this trip and during a subsequent visit to William Wallace's dynamo factory in Ansonia, Connecticut, arranged by Barker, that Edison's system of parallel-connected incandescent lamps began to take shape.[3]

At the University of Missouri, Professor Benjamin E. Thomas introduced practical applications of electricity in his physics courses, and in 1883 he received, as a gift from Thomas Edison, a small generator and a few incandescent lamps. In the centennial history of the University of Missouri published in 1939, Professor Mendell P. Weinbach mistakenly stated, "The oldest department of electrical engineering in the country is that at the Massachusetts Institute of Technology, established in 1882." In fact, EE was not yet a department, the course in electrical engineering being administered until 1902 by the Physics Department. Weinbach also states that "a department of electrical engineering leading to the degree Electrical Engineer was established [at Missouri] in 1886."[4] If this statement is correct, the University of Missouri appears to have had the first department of electrical engineering in the United States. Professor Dugald C. Jackson, who was to move to MIT in 1907, believed that the department of electrical engineering that he organized in 1891 at the University of Wisconsin was the nation's first. His claim seems not to have been challenged during his lifetime.

In Europe, also, early courses had been established. Pioneers in electrical science in Germany, France, England, Italy, Austria, and Russia were teaching the practical applications of electricity before 1882. The Germans recognize Professor Erasmus Kittler as founder of the first faculty of electrical engineering in the German-speaking countries, authorized to administer doctoral programs at the Technische Hochschule Darmstadt in November 1882. At the Technische Hochschule Berlin-Charlottenburg Professor Adolf Slaby was the first holder of the chair of electrical engineering in 1883.

MIT and the Invention of the Telephone

MIT's connection with the future inventor of the telephone, Alexander Graham Bell, began in 1871, soon after the young Bell arrived in Boston for the first time. Bell had come to the intellectual and scientific center of the United States already thoroughly trained in the acoustic arts; his father had developed a phonetic alphabet called Visible Speech, which he hoped would prove to be the "universal alphabet" that phoneticians of the day were seeking. As well as teaching speech to deaf pupils, Bell was promoting the Visible Speech work of his father and pursuing his own growing passion for experiments in telegraphy and acoustics. Being well introduced in Boston, he was soon led to the acoustical work of Pickering and Cross.[5]

Five years earlier, in 1866, while still in England, Bell had performed some experiments in acoustics with tuning forks, driven by electromagnets, for the synthesis of human vowel sounds; but he had shortly thereafter learned that he had, in effect, repeated in a partial way earlier experiments of the famed German scientist Helmholtz. Now, in 1871, on his very first day in Boston, Bell was told that MIT had a complete set of Helmholtz's experimental apparatus; he would soon have a chance of participating in a repetition of those initial experiments.

Bell's background and training were such that, though he lacked a formal college degree, he was offered a professorship in 1873 at the new Boston University (which had been chartered in 1869, not long after MIT was founded). With the added prestige of being Professor of Vocal Physiology and Elocution in BU's School of Oratory, Bell's contacts with the Boston scientific community steadily widened, and his acquaintance with Cross and Pickering at MIT ripened. According to a biography by Robert Bruce, Bell went to MIT in March 1874 "to see Charles Cross demonstrate Helmholtz's experiments in a lecture . . . and afterward talked with Cross about tuning forks and sympathetic vibrations."[6] Although Cross did not bequeath to the MIT Archives an account of his work with Bell, some of their common interests are revealed in theses supervised by Cross, as well as in his Society of Arts lectures. These contacts led to an invitation to Bell to address the MIT Society of Arts on "Visible Speech, or the Science of Universal Alphabetics." Apparently it was that lecture, on April 9, 1874, that prompted Cross to offer Bell "free access to the Institute of Technology." This offer included "the use of the Institute's apparatus and laboratories." Bell excitedly accepted the offer, and immediately began doing experiments at MIT that would lead him within the next two years to invent the telephone.

Among the devices on which Bell worked at MIT was a "phonautograph"; a stylus in the device traced on a piece of smoked glass a waveform of voice vibrations. A student of Cross, Charles A. Morey, had modified and improved the device, and lectured on it to the MIT Society of Arts on May 28, 1874. Bell, whose work had centered around the teaching of speech to the deaf and dumb, saw the phonautograph as a possible avenue to making speech visible if characteristic and unique shapes could be developed for each speech sound. According to Bruce, these experiments almost led Bell to the concept of the phonograph before Edison, but he veered off, became more involved in things electrical, and soon was earnestly seeking the way to transmit speech over telegraph lines.

Recent books have traced the successive ideas and experiments leading to Bell's invention of the telephone in 1876.[7] It seems clear from these that Bell consulted quite closely with Cross, who had, in fact, worked earlier with Pickering at MIT on a kind of "tin-box receiver" for picking up electrically the vibrations made by tuning forks. Biographer Bruce writes, "As so often has happened in the history of technology, men's minds were converging on a single point. . . . To get there first would take a special set of talents, interests, temperament, and environment. Bell was born with the right talents and temperament. His upbringing gave him the special interests. Chance brought him to Boston, which perhaps more than any other city in the world provided the proper intellectual, technical, and economic environment for the invention of the telephone."

It is noteworthy that just a few years earlier, in 1868, Thomas A. Edison had also come to Boston, where his own inventive style began to mature. (In fact, Boston at that time had the second highest per capita invention rate in the United States; by 1880 it had moved to first place.)[8] Edison had worked at a shop on Court Street in Boston, run by Charles Williams, which attracted many inventors. Bell too found his way there and in 1875 enlisted the aid of a young assistant, Thomas A. Watson, who worked for Williams. Watson would become famous along with Bell. The first "mass" production of telephones was carried out in Williams's shop.

Bell's first public demonstration of the telephone took place at his MIT Society of Arts lecture on May 25, 1876, entitled "Telephony, or the Telegraphing of Musical Sounds." A year later he gave a sketch of the history of telephony and described his simplified receiver with its permanent magnet and coil.

At the corner of Osborn and Main streets in Cambridge is a bronze tablet that reads, "October 9, 1876, the first two-way long-distance telephone conversation was carried on for three hours. From here in Cambridgeport Thomas A. Watson spoke over a telegraph wire to Alexander Graham Bell at the office of the Walworth Manufacturing Company, 69 Kilby Street, Boston, Massachusetts." Many years later, at the Golden Jubilee Banquet of the MIT Alumni Association in 1916, Bell paid tribute to Professor Cross, "who had made many advances in the telephone itself and had inspired many students to go forth from the Institute to perfect the work."

Building the Nucleus of EE

Three 1876 graduates in physics, Silas W. Holman, John B. Henck, Jr., and William W. Jacques, were to play major roles in the early years of electrical engineering at MIT. Holman and Henck became assistants in physics immediately upon graduation. Holman was a research type somewhat like Pickering, and was soon in charge of the physics laboratories. He was especially helpful to Cross when the course in electrical engineering was launched in the fall of 1882. Cross

did most of the lecturing in the new field, while Holman, with his outgoing personality, worked closely with the students, helping them to strengthen their understanding of the principles and practical applications of mechanics, heat, electricity, and magnetism.

Cross had a dry sense of humor. During his lecture demonstration of static electricity, he would rub an ebonite rod with a catskin, calling forth a chorus of meows from the back of the room. Then, when he rubbed the rod with a piece of silk to induce the opposite charge, he would say without cracking a smile, "Unfortunately, gentlemen, the silkworm does not have a characteristic cry." Personal acquaintance with Cross was not easy, but behind his formality were broad interests: he had traveled to Europe and visited the great galleries; after the Boston Symphony Orchestra was formed in 1881, he rarely missed a concert.

By contrast, Holman was easier to know and love. Even after he was confined to a wheelchair by a crippling disease, he continued his active involvement with students. After the "new building" (later named Walker) became available in 1883, he originated a large laboratory for electrical measurements and prepared laboratory manuals for all the physics laboratories, including the one for dynamo-electric machinery. He developed pioneer subjects, notably Precision of Measurements, with illustrations largely from physics and electrical engineering. The last of his several books, a philosophical discussion entitled *Matter, Energy, Force, and Work*, was completed at home in 1898 with the help of his wife.[9]

Henck retained his staff position in the Physics Department until 1880 and then became employed full-time as "electrician" with the American Bell Telephone Company, but he retained his close connection with MIT, becoming Secretary of the Institute in 1890–91. Jacques became a Fellow of Johns Hopkins University

Silas W. Holman.

In the new Walker Building (*at left*), built nearly twenty years after the original Rogers Building (*right*), were housed the first large electrical laboratories.

The Laboratory of Electrical Measurements, started by Silas W. Holman.

The Electrical Machinery Laboratory, 1885. In the center of the laboratory is an Edison bipolar dynamo, a gift of Thomas A. Edison.

in 1876, with brief studies in Berlin, Leipzig, Vienna, and Göttingen, receiving the Ph.D. degree in 1879. He then joined the American Bell Telephone Company as electrician along with Henck, becoming a part-time instructor in telegraphy at MIT in 1883. In 1884 he gave a Society of Arts lecture on underground telephone cables. Professor Cross presided at the meeting, commented on tests he had made on Bell Telephone lines, and referred to a paper on the subject he had written in 1882.

As a measure of the state of knowledge about electrical phenomena at that early period, it is interesting to note the misconceptions about the character of electrical inductance. Cross and Jacques still believed, for instance, that line inductance should be reduced to achieve optimum transmission. Jacques pointed out that an attempt had been made to reduce the self-inductance of a pair of conductors by placing the two conductors very close together, but the idea had been abandoned because, unfortunately, this also increased the capacitance. Heaviside, the famous British scientist and engineer, was shortly to enlighten the telegraph and telephone electricians as to the beneficial effect of inductance.

Another important member of the early electrical group was William L. Puffer, who joined the physics staff as assistant upon his graduation in 1884; he became interested in dynamo-electric machinery, eventually equipping a separate laboratory for electrical machinery in the "new building."

How Did Electrical Engineering Get to Be Course VI?

In MIT's realignment of courses in 1873–74, ten courses were listed:

I. Civil and Topographical Engineering
II. Mechanical Engineering
III. Geology and Mining Engineering
IV. Building and Architecture
V. Chemistry
VI. Metallurgy
VII. Natural History
VIII. Physics
IX. Science and Literature
X. Philosophy

Up to 1873, Course VI had been Science and Literature. In 1884 Metallurgy was combined with Mining to form Course III, Mining and Metallurgy, and Electrical Engineering became Course VI, still administered by Cross and within the Physics Department. The first EE graduates in 1885 were therefore called Course VI men, a name that has remained unchanged to the present day.

Parenthetically, the MIT Alumni Association, founded March 17, 1875, held its first regular meeting on January 27, 1876, and elected Robert H. Richards, '68, president and Charles R. Cross, '70, secretary for two years. The first number of *Technology Review*, the MIT alumni magazine, appeared in 1899, under the leadership of Arthur D. Little, '85, C. Francis Allen, '72, and James P. Monroe, '82. The MIT student newspaper, *The Tech*, began publication in November 1881 under a board of directors of which H. Ward Leonard (later famous for his system of motor control) was president and Isaac W. Litchfield (member of the first Course VIII-B class) was treasurer. Arthur D. Little was on the editorial board.

The Tech of March 7, 1883, mentioned a long article in the *Electrical Review* on the new MIT course in electrical engineering. The same issue reported that a new course in electrical engineering had also been established at Tufts College under "Professor Amos E. Dolbear, one of the pioneers of the telephone." Dolbear, in fact, was one of those who had contended with Bell, claiming that Bell had swindled him. Though the case was "settled" in 1879, Dolbear tried to start his own telephone company in 1881, but it was successfully enjoined by the Bell Company. Dolbear, who taught at Tufts for thirty years, was still proclaiming to his students as late as 1895 that he was the inventor of the first telephone.

In the fall of 1883, soon after the Thomson-Houston Company had moved from New Britain, Connecticut, to Lynn, Massachusetts, Cross invited Professor Elihu Thomson to lecture to the MIT students in electrical engineering, and thus began Thomson's close and influential relations with MIT. Not only did he become acting president of MIT in 1920, but perhaps even more influential in the development of the MIT EE Department was the merger of his Thomson-Houston Company with the Edison Companies in 1892 to form the giant General Electric Corporation. Its proximity to MIT, as well as the large scope of its business, led to many vital interactions between the two institutions. For instance, when the cooperative course in electrical engineering was formed in the early twentieth century, it was at General Electric that the students got their first contact with practical electrical engineering.

Besides the EE teaching done by Cross, Holman, Puffer, and William H. Pickering in the fall of 1884, lectures were given by George W. Blodgett, '73, electrician of the Boston and Albany Railroad, on the application of electricity to railway signaling, and by Anthony C. White, '82, Horace B. Gale, '83, and E. H. Hewins of the New England Weston Electric Light Company, on electric lighting installations. A complete set of electric railway signals was given to MIT by the Union Electric Switch and Signal Company, and a dynamo-electric machine with lamps of several patterns was loaned by the Brush Electric Company. The following year, a Weston 60-incandescent-light dynamo and a Gramme machine were added.

There were ten Course VI graduates in 1886, of whom Harry E. Clifford and Dana P. Bartlett did a joint

William L. Puffer, an 1884 graduate in physics, was influential in the field of dynamo-electric machinery at MIT.

The "Great Battle" between direct-current and alternating-current systems waged between Thomas Edison and George Westinghouse in the late 1880s and early 1890s found its reflection in the MIT physics faculty. Harry E. Clifford, a member of the "new generation," became the department's scholar of electromagnetic wave behavior by the end of the century.

thesis, "The Electrical Transmission of Power." Clifford became an assistant in physics and Bartlett in mathematics. Clifford rapidly developed into a scholar of unusual depth in the theory of electricity and magnetism. Alternating-current technology was well on its way when Clifford joined the staff in 1886.

Another strong addition to the staff in electrical engineering came in 1889 as Frank A. Laws graduated from Course VI. He began to take over the laboratory for electrical measurements, inaugurated and built up by William Pickering, Cross, Holman, and Clifford. Louis Derr, an 1892 Course VI graduate, became proficient in teaching dynamo design and telegraphy.

Without doubt the competent and dedicated staff of teachers and investigators built up by Cross accounts in large measure for the success and popularity of Course VI, reaching a high point of 141 registrants in 1893. The Civil and Mechanical Engineering Departments recovered their ascendancy in 1897, when Course VI registration decreased to 100. Later, however, in the 1920s (about the time of Cross's death), EE would become the Institute's largest department.

During the era when Cross was fostering electrical engineering (1882–1902), not only was the nucleus of the future department built up, but among the graduates who would "go out into industry" were a considerable number of outstanding people whose subsequent careers and continuing involvement with MIT influenced the growth of the department. To name a few: Frank A. Pickernell, '85, who became chief engineer of American Telephone and Telegraph; Charles A. Stone and Edwin S. Webster, '88, who did a joint thesis, "Efficiency of Alternating-Current Transformers," and who formed the firm Stone and Webster, which in 1916 built MIT's new home in Cambridge; Alfred P. Sloan, '95, who became president of General Motors, and who in 1937 established the Sloan Foundation to promote a better understanding of economic problems; Gerard Swope, '95, who became president of the General Electric Company, proposed honors groups at MIT in electrical engineering and physics, and funded fellowships; William D. Coolidge, '96, who became a renowned researcher at General Electric's Schenectady Research Laboratory, begun in 1900 by Willis R. Whitney, '90, a former teacher at MIT. As the first strong non-academic research laboratory, the GE facility was to become a model for many other industrial research laboratories over the years. It came to represent, as well, an alternative and prestigious career path for many scientists and engineers who might otherwise have remained in academic environments.

Alternating Current: Impulse to EE Education

Beyond the founding of the field itself, a development of major importance for electrical engineering education in the closing years of the nineteenth century was the triumph of alternating-current systems over direct-current. These systems brought in a new level of complexity, demanding deeper theoretical and mathematical tools, and thus marked a turning point for the individual amateur inventors who had been able to work with only a modicum of theory. Engineers trained in electrical theory, who were versatile with mathematical concepts, were needed by the burgeoning electrical industries.

At MIT, one of the leading theoretical electrical engineers, Harry E. Clifford, who would become important in the building of the department, cut his teeth as it were on AC. Stimulated by such achievements as the great AC power station at Niagara Falls, which went into operation in 1895, Clifford was lecturing on the propagation of electromagnetic waves by the turn of the century.

Chapter 3

Growth of the Department 1902–1952

During its first half-century as a full-fledged department (in contrast to being a "course" administered within Physics) the Electrical Engineering Department was headed by five very different men: Louis Duncan (1902–1904), Harry E. Clifford (1904–1907), Dugald C. Jackson (1907–1935), Edward L. Moreland (1935–1938), and Harold L. Hazen (1938–1952). During the long tenures of two of these men, Jackson and Hazen, the department went through the greatest change. Under Jackson, it was essentially built to "critical mass," and shaped as a strong educational entity with a growing and influential graduate program and a growing emphasis on research. Under Hazen, it made external commitments on such a scale as to set the stage for its own transformation, while achieving truly international stature.

This chapter highlights the changes in electrical engineering and in the EE Department at MIT during the latter's first half-century; it focuses on the Jackson era, when the department really took off. Building initially on a strength in electric power engineering, the department, under Jackson's pushing and encouragement of research, also became strong in the new fields of communications, computational machinery, and electronics. Its curriculum changes and the textbooks published by the EE staff members were at the same time influencing the development of other EE departments around the country.

Jackson, a keen, sometimes choleric man, reigned for twenty-eight years, longer than anyone else, and became known and respected as the great "department builder." What Cross had started and nurtured, Jackson consolidated, expanded, and carried forward through momentous changes in electrical engineering and in science and engineering generally. The chronology at the end of this volume (see Appendix) suggests some of the changes in the world during the

Jackson period. Electrical engineering grew steadily in its influence and role in industry. Electric power systems spread rapidly in the United States, their grids gradually beginning to interconnect at state boundaries, moving toward a national network.

As the power lines grew in length and connectivity, transmission voltages rose steadily; engineers sought greater efficiencies, electrical generators grew larger and larger, and all electrical components developed apace. Higher voltages demanded better insulators, better transformers, better switches. Transmission line theory developed, but the growing complexity of the interconnected grids pushed ahead of theory, and in the Jackson period the first important simulation systems were developed at MIT to assist in the solution of power system instabilities. This work in turn helped to stimulate the development of the first analog computers at MIT. The thesis work carried out by students at MIT during these years of steady power growth reflects the problems and the changes of these years.

During this period MIT relocated its campus from Boston to Cambridge, and went from being a local school of high reputation to an institute of national renown. Franklin D. Roosevelt, assistant secretary of the Navy, was there in 1916 as an official representative to celebrate the move to the Cambridge campus overlooking the Charles River. Years later, just before World War II, as president of the United States, FDR would appoint an MIT man, Vannevar Bush, to mobilize the nation's scientists and engineers in the war effort.

In science, during the Jackson period, the classical viewpoint was shattered by the exciting emergence of nuclear physics in the micro-domain and by the cosmological Theory of Relativity in the macro-domain. A reflection of these changes in scientific viewpoint is

seen at MIT in the appointment of physicist Karl T. Compton as president in 1930. Part of his mission was to build a stronger scientific basis for the Institute—to make it a strong scientific school as well as a strong engineering school—and to bring about a greater interaction between physics and engineering. Numerous examples of that new interaction are cited in later chapters. MIT might not have been as prepared for later challenges had not Compton been brought in at this period. In a sense, his activities were a return to the presence and style of William Barton Rogers, except that the physics was radically new. Compton's close work with Bush, described later, is representative of the movement of events during the 1930s and 1940s.

In the arts, during the same period, the world saw Cubism burst out of Impressionism, dally with Dadaism, and move into pure abstraction; in music there were experiments with atonal composition. In the political sphere, there was the horrible butchery of World War I, and besides the great communist experiment in Russia, there was the steady rise of nazism and fascism in the 1920s and 1930s; in the United States the New Deal struggled to rise above the Great Depression.

In this period, also, the unique Tennessee Valley Authority project was begun in 1932 when Roosevelt became president of the United States. Though TVA started as a multi-faceted flood-control, river-navigation, land-conservation project (what today is called regional development), its dams were also used for the generation of electric power—cheap electric power—and as time went by, people began to see it principally as an electric power project on the grandest scale. With more than thirty dams, many beautifully designed, the project came to symbolize America's hopes in modern technology and especially electric technology. In the depths of the Depression, such symbols were desperately needed.

TVA had its critics. The private electric power companies saw it as a threat and they fought TVA as a piece of ideological foreign matter in the body of the country—"creeping Socialism" they called it. Indeed, electric power prices dropped in the regions around the Tennessee Valley. Prices were initially pegged at a penny a kilowatt-hour, the lowest in the nation, and FDR touted it as a yardstick to measure the efficiencies of all electric systems. There was a serious question, however, whether TVA could truly act as a yardstick, since it was government supported, so that it would be almost impossible to compare costs and efficiencies point by point. Dugald Jackson was one of the critics of TVA on this and other grounds; he and other MIT professors lectured against it.

This was all symptomatic of just how far things electrical had entered the national bloodstream, changing nearly everyone's ways and means (especially as rural electrification grew) and reaching into questions that were at the heart of the nation's social and institutional structure.

At the very end of Jackson's tenure, the nation was focused on the holding company scandals that revealed intricate financial pyramids among the electric utilities. These scandals not only tarnished the reputation of the electric utilities; they undoubtedly diminished the prestige of the electric power field as well. Out of the holding company debacle there emerged a greater government regulation of the electric utilities; the new view was that these utilities were for everyone, that they performed a public service. In just a few decades electricity, in the public consciousness, had grown from an exotic commodity into a social necessity.

The other long-term department head in the first half-century, Harold L. Hazen (1938–1952), served in a tumultuous period when electrical engineering became deeply involved in national defense, war research and development, and the post–World War II reconstruction period. His tenure saw the culmination of a great period of electric power engineering; it also saw schisms in the field, and new directions for research. Not only was the engagement with government growing, but vast new funds and resources were flowing into the electrical engineers' world. The EE Department tripled in size, both in student enrollment (reaching a peak of 1,215 in 1947–48) and in faculty numbers, many of them dedicated to war research. The presence of the Radiation Laboratory on the MIT campus, with its scientific and engineering staffs numbering about 4,000, broke the seams. "Temporary" wooden buildings were put up along the back of the campus to serve as research laboratories. Research threatened to overwhelm the educational function, but instead became corporate with engineering education in a way that had never been experienced before on such a scale, or with such success.

To manage a burgeoning department amid such forces required diversity, breadth, resilience, and reason. It required a person who could encourage many individuals to express themselves without dictating to them, a person who could make sense out of potential chaos, while remaining calm in the center. That was the role that fell on Hazen. During his tenure all science and engineering went through profound changes. It would take years to sort it all out, but everyone recognized that electrical engineering could never be the same again. By the time Hazen retired, finishing his term in 1952, the field of electrical engineering had expanded and diversified to such an extent that it was becoming evident that some new synthesis of engineering education would be needed.

After an exuberant and uncontrolled growth of exotic specialties, somewhat at the expense of traditional power engineering, a consolidation and unification was in order.

To appreciate the scope of the changes that World War II brought about, it is necessary, however, to examine in detail many of the engineering and scientific projects undertaken in the period 1902–1952 and to show how they influenced the educational curriculum. No less important were the struggles that went on as individuals sought to clarify their direction and to sense where the forefront of their field was moving. Everyone who has had any acquaintance with an educational institution (or with any large organization for that matter) is aware of the infighting that goes on, sometimes dark and brooding, sometimes scandalously delightful, depending on the characters of the people involved. In the kinds of people who have been attracted to MIT over the decades, there has been enough variety to keep things lively and to provide the Human Comedy with some new episodes.

The role of a department leader is somewhat like that of a symphony orchestra conductor, who must guide and compose the efforts of diversely creative individuals, manage their advancement, and keep things vitally integrated. The styles of conductors are diverse—authoritarian, democratic and participative, brilliant and pedantic, romantic and passionate, idiosyncratic and fervent, serene, or even, sometimes, downright dull. Among the men who headed up the Electrical Engineering Department from 1902, when it was first formed, to 1952, when revolutionary changes began to be introduced, when the old electric machinery was thrown out, at the beginning of what is still regarded with some awe as "The Gordon Brown Era," we find a number of interesting types of conductors.

Louis Duncan (1902–1904)

The first of these, Dr. Louis Duncan, didn't last very long. Duncan was the first "outsider" to be brought in by President Pritchett, who had come to MIT in 1900 apparently somewhat persuaded that the Institute was too inbred, having drawn too heavily from its own students for new faculty.

At the turn of the century Professor Cross was, in fact, urging President Pritchett and the Executive Committee of MIT to form a Department of Electrical Engineering under direction other than his own. The issue came to a head in the spring of 1902, when it became evident that classroom and laboratory space would be inadequate for the increased registration in the fall term. Fortunately, gifts from members of the Lowell family, long known for its support of scientific and technical education, allowed the construction of a new building dedicated to facilities for electrical engineering. The building was completed that very summer. Said President Pritchett in his report of December 1902, "The building is, for the purposes of the Department of Electrical Engineering, almost ideal, and, when the equipment which has been purchased is in place, we shall have in Boston one of the most perfect and at the same time one of the most practical electrical laboratories in the world."

The question was, who would head the new department? Associate Professor Clifford at that time seemed to be the logical choice. His 16 years under Cross were marked by excellent teaching, the development of new subject matter, and visits to Europe, where he kept in touch with the latest foreign advances in the field. He was also emerging as an organizer and leader in the philosophy of education. He was strongly favored and supported by the EE staff. Among these were Associate Professor Puffer, who had become a specialist in dynamo-electric machinery and laboratory testing, and Assistant Professor Laws

Louis Duncan.

This early photo of the dynamo room in the Lowell Building Electrical Engineering Laboratory shows that the new department was off to a vigorous start in terms of its research capabilities.

(known in later years as "Pop Laws"), who had created a laboratory for electrical measurements, standardizing, and testing. Ralph R. Lawrence, '95, and Harrison W. Smith, '97, were instructors and Harry E. Dart, '01, was an assistant. Elihu Thomson, one of the original inventor-entrepreneurs of the early electrical era and now scientist and inventor at the Lynn works of the General Electric Company, was at this time (1902) appointed Nonresident Professor of Applied Electricity.

Despite the support of these men, President Pritchett was unwilling to appoint Clifford. He saw two fundamental reasons for not selecting the new department head from the incumbent group. First, they were all (including the outside lecturers for 1902) graduates of MIT; they would not be open, he argued, to broadening influences from other institutions. Second, none of the MIT men was a practicing engineer of renown, a qualification that seemed to be requisite for the building of a really great electrical engineering department.

Pritchett concluded, after a careful search, that Dr. Louis Duncan was just the right man. Duncan had been chief engineer in charge of construction of the Third Avenue transit system in New York City and consulting engineer for the city railways in Baltimore and Washington, and had organized the firm of Sprague, Duncan, and Hutchinson, designers and builders in 1892–93 of the first large electric locomotive (60 tons, 1,000 horsepower). He was chairman of the Board of Judges at the Philadelphia Electrical Exhibition of 1884 and a member of the Board of the World's Columbian Exposition, 1892, in Chicago. Twice elected president of the American Institute of Electrical Engineers (1895 and 1896), he held membership in several professional organizations, including the American Philosophical Society, the American Electrochemical Society, and the Mathematical and

Elihu Thomson, member of the original EE faculty. A founder of the large Thomson-Houston electrical manufacturing firm, which merged with the Thomas Edison companies to form General Electric, Thomson remained a prolific inventor throughout his career. This photo was taken in 1886, about the time of his invention of electric welding.

Physical Societies of France. A graduate of the United States Naval Academy in 1880, Duncan received a doctorate in 1885 from Johns Hopkins University with a dissertation on the determination of the absolute unit of electrical resistance. In 1886 he started a course in electrical engineering for a few physics students at Johns Hopkins, and was reported to have inspired much original work in his students. In short, his credentials were outstanding.

Duncan arrived at MIT in the fall of 1902, after the staff had planned and carried out the move from the overcrowded Walker Building into the new Lowell Building. Cross and Derr remained in the Physics Department, but continued to give the new EE Department head their consultation and to give instruction in telephony, telegraphy, and dynamo design. In the fall term, 1902, Course VI had 38 seniors, 37 juniors, and 43 sophomores. A few students, mostly special students, were enrolled in graduate studies. With its new leader, plenty of eager students, and commodious quarters, the department seemed off to a good start.

Duncan's chief contribution to the students was a series of lectures to seniors on his specialty, electric railway engineering. His first public lecture at MIT, "Long-Distance Electric Railroading," was read before the Society of Arts on October 9, 1902. In this lecture he reviewed the design of trolley cars for urban transit, pointing up some of the failures and breakdowns of the early models. He then went into the development of interurban lines with speeds up to 60 miles an hour and finally dealt with problems of heavy traction for passengers and freight. His enthusiasm for electric railway engineering was apparently unbounded, and he continued a considerable consulting practice while at MIT.

The Institute's educational atmosphere, however, appeared not to please him. His experience at Hopkins had been with small groups of physics students with whom he worked in close personal contact and with meager equipment. The lively MIT community of staff and students, striving to keep up with the latest in science and with all the new motors, transformers, and other practical devices arriving almost daily to equip the new Augustus Lowell Laboratory, seemed to baffle him. In a number of ways, Duncan apparently didn't mesh well with the established, and wholly MIT-trained, EE staff. For their part, the staff may not have taken kindly to cooperating with an interloper and may have joined ranks against him; Duncan, in his turn, was critical of the Institute.

Duncan's views were expressed in an article, "Technical Education," in the *Engineering Magazine* of November 1903. He wrote: "The fault with our technical education in America is that it is too technical and that there is too much instruction. The end of a technical education should be not to obtain information but to apply it." He then made the point that technical courses treat idealizations of the real world piece by piece without studying their interrelations: "Problems cannot be worked out as engineering abstractions, but must take into account financial conditions, social questions, and all the complex considerations that make up real life rather than ideal life. . . . I am afraid our largest and best-equipped institutions do not show as fair a percentage of leading engineers as some of the smaller, poorly equipped colleges. . . . The students in the smaller colleges are more directly in contact with the instructors and do not have the facilities for work that the more ambitious colleges have; that is, when an experiment is to be tried, the ingenuity of the student, and not his patience, is exercised." He expressed the opinion that at least half of the instruction at MIT had done harm rather than good. A thoroughgoing study of the Course VI curriculum was

The plethora of electrical equipment purchased to furnish the Lowell Laboratory gave the new EE Department one of the best engineering laboratories in the nation.

already under way, but Duncan brought about no significant curriculum changes during his two-year term. Without having taken an effective hold on the MIT educational problem, he returned full-time to his consulting practice in the fall of 1904.

Graduate Education

Postgraduate studies had been carried on by the engineering departments from the Institute's earliest days, but no graduate engineering programs had been established. (One master of science degree had been awarded in civil engineering in 1894, and subsequently five in mechanical engineering and three in chemical engineering.) To encourage engineering research, the Graduate School of Engineering Research was announced in 1902, to be inaugurated in the fall of 1903. In his report of December 1902, President Pritchett wrote:

The world has awakened to the fact that education and training in the end outstrip natural ability and untrained initiative, and the civilized nations of the world are bending their energies, in proportion as they are alert, to the problems of technical education. We in America must keep step with the needs of our own country and the efforts of other countries in this direction, and the time has now come when the American engineer must be capable, not only of the most modern practice, but also of conducting investigation and research. To show how clearly our work in this direction is watched and followed, I venture to quote from the inaugural address of the present Rector of the great technical school at Charlottenburg. "The German need fear in the industrial world neither the Englishman nor the Frenchman, only the American; and to compete with the American engineer, we must strive constantly to improve and extend our engineering courses." The time has come when the Institute must be not only a teaching body, but it must as well lay the foundations for a school of investigation in the physical sciences.

Referring to the Graduate School of Engineering Research again in December 1903, Pritchett said,

The plan contemplates provision for research in applied science and in engineering directions for a very small number of well-equipped men who are capable of executing pieces of research of real value and importance, and who shall have such freedom of work that they may throw themselves with devotion into what they undertake. This is the first effort of any technical school in the country to offer research work distinctive from that of the college, and directed toward engineering subjects. While this school will offer an opportunity for earning the degree of Doctor of Engineering, it is intended that the conferring of the degree shall be a minor feature of the work, and that the few men who are admitted to this work shall be men who are aiming at research and not degrees. For the present year [1903–4] the only candidate who has been admitted is Professor Harold B. Smith, head of the Department of Electrical Engineering in the Worcester Polytechnic Institute.

The requirements for the doctor of engineering degree included satisfactory resident work of not less than two years.

The first master of science degrees in electrical engineering were awarded in 1904 to Herbert M. Morley for a thesis entitled "An Investigation of the Operation of a Double-Current Generator," supervised by Puffer, and to Waldemar R. Kremer for his thesis, "The Change of Permeability in Norway Iron and English Steel due to a High Fluid Compressive Stress," supervised by Clifford. Morley was a 1903 graduate of Course VI, and Kremer did his undergraduate work at the Technische Hochschule in Munich, Germany.

Harry E. Clifford (1904–1907)

Upon the resignation of Duncan in the fall of 1904, Harry E. Clifford was placed in active charge of the department. The promotion of Ralph R. Lawrence to an assistant professorship and an increase in graduate assistants from one to four constituted the change in the teaching staff. Puffer, who had developed the laboratory instruction in dynamo-electric machinery and who in 1901 had visited the laboratories of European technical schools in preparation for his responsibilities in planning the Augustus Lowell Laboratories, was still in charge of this work; he also offered a graduate course in power station design. Laws was well on the way in his notable work in electrical standardization and measurements, and Lawrence was already publishing in the field of electrical machinery.

Having carried major responsibilities in the setting up of the new department in 1901–2, and having heavily contributed to its management during the Duncan administration, Clifford was able to give immediate and competent leadership to department affairs. He looked upon the lecture method of teaching undergraduates as inferior to the use of good text material with class discussions of difficult concepts and the working of problems under the supervision of instructors. All lectures in department undergraduate subjects were therefore replaced with recitation sections taught by faculty-grade instructors, "a change which has given most satisfactory results." To quote further from his report of January 1906, "The matter of the course scheme has been considered by the Department with great care and at considerable length. The conclusions reached are that a radical simplification is most desirable, in fact is absolutely necessary, since the students are working under pressure on too many subjects. The time for the digestion and assimilation of what is presented to them for clear and careful thinking must be very considerably increased." He observed that employers desired men who could think clearly, "the knowledge of a great number of details

Harry E. Clifford.

being of secondary importance." A new system of laboratory instruction with preliminary reports and conferences aided the students in gaining independence and understanding.

In one of his reports Clifford pointed out a chronic problem of that early period, namely, the need for greater continuity in the corps of assistants and instructors. "It is regrettable that the Institute cannot hold out to such of these men as show exceptional capacity inducements in salary and rapidity of advancement which will permanently enlist their services, to the end that there may be in the laboratory a body of instructors possessing experience, in addition to enthusiasm." A tutorial system of staff conferences with individual students was put into effect, and this more intimate contact produced "a marked improvement in the student attitude toward the regular work."

Clifford was a clear and effective expositor, having given great care to the preparation and logical arrangement of his material. His manner was pleasant but businesslike; according to one of his later students, the intense expression in his eyes assured students and fellow faculty members that he meant "no fooling around," even though he had a ready sense of humor.

The academic year 1905–6 found the department well established in its new building, with laboratories splendidly equipped for the task of teaching electrical engineering. Without exception, the five members of the electrical engineering faculty were devoted to the education of their students and to their own researches. Organizationally, the department was running smoothly, except that it was overloaded. Perhaps one of the contributory reasons for the effectiveness of this group was the fact that they were all graduates

of MIT, as was the instructor, Charles H. Porter; five out of six of the assistants were Course VI graduates of the previous June.

From the administrative viewpoint, however, the department was decidedly ingrown, and President Pritchett sought to broaden its educational base by looking outside MIT for new staff members. As part of this quest he asked Clifford to visit the University of Wisconsin, where Professor Dugald C. Jackson had launched a successful department of electrical engineering in 1891. Clifford's favorable report led to negotiations with Jackson to become head of MIT's Department of Electrical Engineering and with George C. Shaad, one of Jackson's lieutenants, to accept an assistant professorship to begin in the fall of 1906. Because of Jackson's previous commitments, he was unable to take up his duties at MIT until February 1907. During the fall semester, 1906–7, Clifford continued to administer the department.

Though Clifford's talents and executive abilities had twice been bypassed by MIT's president, he did not go unrecognized elsewhere. In 1909 he went up the Charles River to accept a prestigious chair at Harvard, the first Gordon McKay Professorship of Electrical Engineering. (In 1905 Harvard had made its fourth serious attempt to merge with MIT and to make MIT its own engineering school, a move that Pritchett approved, but that MIT's faculty and alumni resisted.)

The Dugald Jackson Era (1907–1935)

To break successfully into the snug group of five MIT alumni comprising the department faculty would require much diplomacy and understanding on Jackson's part. The adjustment was aided by the resignation of Puffer in the fall of 1906 and the assumption of his work in the dynamo laboratory by Jackson's Wisconsin associate, George Shaad. A strong advisory committee for the department had been set up at Clifford's suggestion in 1906; it consisted of Elihu Thomson (MIT corporation and General Electric Company), Charles L. Edgar (president, Edison Electric Illuminating Company of Boston), Hammond V. Hayes (chief engineer, American Telephone and Telegraph Company of Boston), Louis A. Ferguson (vice-president, Chicago Edison Company), and Charles F. Scott (consulting engineer at Westinghouse Electric and Manufacturing Company of Pittsburgh). Jackson lost no time in making full use of this council of eminent engineers. A thorough revision of the curriculum was made and carried into effect with the sophomore class of 1907–8. Jackson's ideas on engineering education were fairly close to the educational objectives of MIT's founder, William Barton Rogers, and to the traditions that had developed during the first 40 years of MIT's history.

In a paper presented to a joint session of the American Institute of Electrical Engineers (AIEE) and the Society for the Promotion of Engineering Education (SPEE) in 1903, Jackson had characterized the electrical engineer in industry as a man ''competent to conceive, organize and direct extended industrial enterprises of broadly varied character. Such a man must be a keen straight-forward thinker who sees things as they are and must have an extended, and even profound, knowledge of natural laws with their useful applications. Moreover, he must know men and the affairs of men—which is sociology; and he must be acquainted with business methods and the affairs of the business world.''[1]

First MIT Doctorates

Significant, although not directly connected with Course VI, was the stimulation to graduate work produced by Professor Arthur A. Noyes, '86, and his Research Laboratory of Physical Chemistry. Established in 1903, this laboratory produced the first three MIT

Dugald C. Jackson, who during his long tenure built the department into one of world standing.

doctorates in 1907. It became the training ground for such men as Willis R. Whitney, '90, who in 1900 became the founding director of the General Electric Research Laboratory in Schenectady, and William D. Coolidge, Course VI '96, who, having achieved many inventions, including hard X rays and ductile tungsten, succeeded Dr. Whitney in 1932 as director of the GE laboratory. The research work of Noyes, Whitney, Coolidge, and G. N. Lewis and their students gave MIT an international reputation in research and graduate education.

In contrast, the engineering departments were slow in getting doctoral-level research under way, and in January 1905 Dr. Sedgwick, secretary of the Council of the Graduate School of Engineering Research, had reported, ''No one has this year been registered as a member of the School or as a candidate for the degree of Doctor of Engineering.'' Professor Smith of Worcester had discontinued his doctoral program, and new graduates who could have done successful doctoral research were in great demand by the industries. Jackson was firm in his belief, however, that the capable engineering graduate should pursue further studies in advanced applications of electricity, physics, mathematics, economics, and other subjects that would serve as foundation material for a creative engineering career. The advanced thesis experience would launch the graduate student into the world of investigation.

Jackson's first-year objective was to study the existing department and institutional situation, to find his own place in the MIT community, and to plan his strategy for the administration of the department. He judged the teaching staff to be well balanced, but agreed with Clifford that the lower levels of instructorships and assistantships needed to be given greater continuity and supervision by the faculty members. Jackson's most recent teaching at Wisconsin had been in the fields of alternating current and alternating-current machinery, but, finding that Clifford had the teaching of these branches well in hand, he turned his attention to a fourth-year elective on the executive and administrative aspects of industrial concerns, which later evolved into a three-hour weekly session entitled ''The Organization and Administration of Public Service Companies.''

Jackson's Philosophy and Style

Late in his career, Jackson articulated in a series of essays and reports some of the principles that had guided him as he built up the Electrical Engineering Department at MIT. ''Engineering,'' he wrote, ''is fundamentally a structure which bridges the gulf between the impersonal exact sciences and the more human and personal affairs of economics and sociology.'' For electrical engineering to perform the role he described, it was necessary always to push ahead to solve the next problem, and the next. This meant research. There was no possibility, in his view, for engineering to stand still, to depend on cut-and-try or on rote textbook formulas. In his management of engineering education, in his consultation work, and in his observation of many engineering schools, Jackson noted that the faculties in the weaker schools failed ''to recognize that the proper use of research vitalizes all levels of engineering education, from the sophomore undergraduate level to the most advanced levels.'' It was during Jackson's period, in the 1920s and 1930s, that research truly became a vital force in electrical engineering education at MIT.

As an administrator Jackson had a remarkable array of abilities, insights, and principles. His broad and intensive reading and his world travels gave him an understanding of diverse cultures that sharpened his assessments of politics, economics, science, engineering, the arts, and philosophy as components of society. Within this breadth of interests he concentrated on the details of his specialized engineering field in such a way as to develop a clear philosophy of engineering practice, engineering research, and engineering education.

Not highly inventive himself, Jackson's strength lay in the organization of ideas, materials, and persons into an effective, smoothly running department. He was quick and accurate in his evaluation of persons as to character and potential, and multiplied his administrative powers through the placing of projects in the hands of trusted colleagues with the intent and assurance that they would succeed. He expected personal growth and achievement in each department member and was not lacking in the power to stimulate such growth. He could be tough and realistic. On one occasion, for instance, when Edward Bowles (about whom we shall hear more later) was given a particular assignment, he asked, "Professor Jackson, am I responsible for this project or is the responsibility shared with others?" To which the reply was, "Bowles, if the project fails, *you* are responsible; if it succeeds, we are all responsible."

No member of the staff was spared an occasional chastisement, which the chief administered deliberately to strengthen a man's ability to receive criticism and defend himself with reason and vigor. Sometimes, however, Jackson would blow up at a staff member or a student without much provocation. He recognized this as a defect and called it his "cantankerousness," sometimes the result of overwork or faulty digestion; it was not part of his character, however, to apologize for such incidents.

During his first two years at MIT Jackson formulated the following principles and procedures:

- The educational process is not an exact science but must be guided and improved through "balance of evidence" judgments and through experimentation with new and promising methods.

- Educational objectives should be clearly delineated and the department should be staffed and equipped to carry out these objectives.

- Each faculty member should have an important and absorbing field of effort, which he is making his own.

- Staff members should be encouraged and recognized through merited promotions in rank and increases in salary.

- Staff members who can increase their effectiveness and realize their personal ambitions through the acceptance of calls to service outside the department should be free to leave after obligations are fulfilled. In fact, "it seems desirable to go to some inconvenience to release in the middle of the year younger men of the staff who may have calls to greater responsibilities in teaching or scientific research elsewhere. This is on the theory that our duty lies in aiding the interests of engineering teaching and of scientific research anywhere in the country, in addition to carrying on our own teaching and research in the most effective manner practicable."

- Staff members should be encouraged to visit other institutions and industries, both in the United States and abroad, in order to keep abreast of the latest developments, and MIT should provide financial assistance for such visits.

- It ought to be a matter of fixed policy that some staff members should attend the annual meetings of the professional engineering societies, such as AIEE. "Such a policy would unquestionably add fertility to the teaching of our undergraduate classes, and it may be considered one of our proper contributions to the formal efforts that are being made for the improvement of electrical engineering practice."

- Some members of our staff should attend meetings of SPEE (now the American Society for Engineering Education—ASEE) for the purpose of comparing experience with engineering teachers from all parts of the country.

- It is desirable that a fund be made available to the department from which to pay the expenses of delegates to suitable conventions.

- Advanced study should be further extended and developed, and the abler men of each graduating class should be encouraged to continue their engineering studies to embrace one or two or, in exceptional cases, three years of graduate work.
- The department should rely to a great extent on other departments to teach such collateral subjects as mathematics, physics, mechanics, and thermodynamics in order to put our students in contact with men of competence and vision in these fields.
- Occasions and facilities for sociability among students and between students and staff must be encouraged and provided.

This stress on greater sociability was not merely an effort to bring more fun into the hard grind for which MIT had become both famous and notorious, but reflected a deeper long-standing problem in the building of the Institute. For instance, in earlier years, while it was Boston Tech, MIT was still very much a local school drawing the larger share of its students from nearby areas; most of them lived at home and commuted by horsecar and train. Also, until it moved to its large campus in Cambridge in 1916–17, MIT was continually struggling for space and sufficient buildings to keep up with the growing student body and need for classrooms, laboratories, and the like. It had no dormitories for students. Thus, special efforts had to be made to bridge the gulfs and create a sense of social cohesion.

In his search for ways of building greater cohesion in the department, Jackson turned to the MIT Electrical Engineering Society. Made up of third- and fourth-year students, it was accustomed to meet monthly at the Tech Union. The programs included guest speakers and student reports on technical subjects or on unusual events, such as trips abroad. Jackson saw this organization as a means for sociability among students and staff; he raised funds to provide, without

cost to the students, three dinners per year that would bring engineers of national stature to the Society. The guests for the academic year 1907–8, for instance, were H. G. Stott, president of AIEE, Charles P. Steinmetz of the General Electric Company, and Elihu Thomson of the General Electric Company and MIT.

With his vision and personal influence, Jackson continued to build a department of individualistic stars, yet having such cohesion and cooperation as to establish a standard of excellence in engineering education for the country. He set a personal example, not only through his activities at MIT but through his participation in the broader councils on the national scene. He was president of SPEE in 1906–7 and president of AIEE in 1910–11. He was on the juries of award at the Chicago World's Fair in 1893 and the Buffalo Exposition in 1901. He was a working member of many local and national committees in engineering and engineering education. At the national meeting of SPEE in July 1907 he proposed a joint committee on engineering education to comprise two members from each of the five major engineering societies—AIEE, the American Society of Civil Engineers, the American Institute of Mining Engineers, the American Society of Mechanical Engineers, and the American Chemical Society—and three from SPEE. His proposal was passed and he was charged with setting up this organization, a task he had completed by the time of SPEE's 1908 meeting.[2]

The summer of 1909 brought about some significant changes in the department faculty as Clifford and Shaad left for other institutions. Clifford went to Harvard, as mentioned earlier, and Shaad, who was an associate professor at MIT, resigned to become head of the Electrical Engineering Department at the University of Kansas. Jackson maintained department strength through the appointment of Harold Pender to replace Clifford and Assistant Professor William E. Wickenden to replace Shaad.

At the time of his appointment to the MIT EE Department in 1909, Dr. Pender was already well known, having been invited to England and France to demonstrate his doctoral research (Johns Hopkins, 1901) and subsequent developments on the magnetic effects of moving electrostatic charges. After his return to the United States from the University of Paris, he was employed in power transformer engineering with the Westinghouse Electric and Manufacturing Company in Pittsburgh; from 1904 to 1909 he worked with the New York Central Railroad and engaged in other consulting projects in New York City. Pender's title and duties at MIT were initially the same as those of Clifford. Having charge of the introductory subjects for Course VI juniors, he immediately set about writing a textbook, which was later published under the name *Principles of Electrical Engineering*.[3] He was already at work on the *American Handbook for Electrical Engineers*, to be known for many years as "Pender."[4]

Wickenden, who replaced Shaad, was a graduate of Denison University, where he specialized in mathematics and physics. From 1906 to 1909 he was an instructor in electrical engineering at the University of Wisconsin, where he gained a reputation as an effective and magnetic teacher. At MIT he offered a senior elective in photometry and illumination, subjects in which he already had a book in the process of publication. Professor Ralph R. Lawrence took charge of the senior instruction in alternating-current machinery and Instructor Waldo V. Lyon became his particular assistant, with responsibility for the home problem assignments for the third and fourth year students.

Building the Graduate Program

When Pender and Wickenden joined the department in 1909, Jackson expected them to give special attention to graduate education and department research. The manufacturing companies and public utilities were recruiting heavily at the bachelor's level and did not encourage students to study for higher degrees. Most of the graduates who stayed on as laboratory assistants were enticed into industry before completing a master of science program. Jackson was on the lookout for an able student who would carry his studies to the doctorate and in 1908 found a promising candidate in Harold S. Osborne. Jackson was evidently eager to get the doctoral program going with a strong start, and Osborne had the potential. He was awarded the Henry Saltonstall scholarship for graduate study in 1908–9 and an Austin Research Fellowship in 1909–10. In June 1910 Harold Osborne received the first doctoral degree to be awarded for work in the engineering departments at MIT. His thesis, "Potential Stresses in Dielectrics," proved excellent in every respect and set a standard for the thousands of later doctoral theses in engineering. Some of Osborne's data were taken at the factory of Simplex Electrical Company (forerunner of Simplex Wire and Cable Company), where cables were being designed with "graded insulation," a new development discussed in the thesis. Osborne presented a paper on this subject at the AIEE meeting in New York in October 1910, with several prominent engineers, including Dr. Charles P. Steinmetz, taking part.

A second doctor of engineering degree (Eng.D.) was awarded in 1911 to Reginald L. Jones, whose thesis was entitled "The Effects of Heat and Magnetization on the Magnetic Properties of Iron."

It should be noted that over the decades, as it became more common practice for Americans to stay in the United States to do their graduate work, the honor and prestige associated with graduate degrees went through subtle changes. In the last century, and the beginning of this, it was a virtual necessity to go to Europe to do one's graduate work. But in the first two decades of this century, that began slowly to change through efforts like these described at MIT.

Though in the 1930s America's best physicists, by and large, were still going to Europe to learn the new quantum theory and recent mathematics, that avenue was beginning to close with the rise of Hitler, as well as with the growth of American physics. The tide in engineering began to turn earlier, but there was still a difference between a doctor of engineering and a doctor of science. The onus on being "only an engineer" persisted through World War II and after. Vannevar Bush, for instance, reports that on one occasion during World War II, when his personal prestige and power were at a peak, he received a chilly response from a British counterpart on being introduced as an engineer. By that time, Bush had believed, the distinction had evaporated, but the incident proved to him that it had not.

In any event, the popularity of the doctor of science degree grew steadily, once it began to be offered to American-trained engineers; later, in the 1950s, the Ph.D. began to be preferred. Such changes, though they may seem superficial considerations, do reflect the shifting relationships between science and engineering and, in turn, the relation between them and the popular perception. At the Institute, the doctor of engineering degree was discontinued in 1918, and MIT subsequently offered its recipients an exchange for the doctor of science degree. Of the six in electrical engineering who had received the Eng.D. degree up to that time, Reginald L. Jones, Robert J. Wiseman, and Tsuenzo Hada made the exchange, but Harold S. Osborne, Roy D. Huxley, and Vannevar Bush decided to retain their original degrees.

The early department research program did not, in retrospect, investigate electrical phenomena in a highly creative way. Bachelor's theses were limited to such subjects as measuring armature reaction or air-gap flux density in rotating machines, the feasibility of using motor drives in factories or of installing a privately owned generator rather than buying power from the central station. There was also a preponderance of theses and other studies relating to financial considerations, which seems to have been a reflection of Jackson's interests. (His initial delay in coming to MIT in 1907 came about because he was conducting a telephone rate case in Chicago.)

The continuing involvement of Jackson and other members of the department in such business problems, which required engineering methods in their solution, prompted Pender in 1912 to suggest that a department of business engineering be established at the Institute. After discussion and concurrence by the faculty, President Maclaurin announced a new course to be called Course XV, Engineering Administration, with three options, civil engineering, mechanical and electrical engineering, and chemical engineering. In the fall of 1914 the new course was inaugurated with a second-year registration of 57 students. A new department was not formed, but the course was placed under the direction of Professor Davis R. Dewey, head of the Department of Economics. The researches in engineering economy were consequently not removed from the program of the Electrical Engineering Department.

In the summer of 1913 a Division of Electrical Engineering Research was set up within the department, with Pender as director. Research funds on a continuing basis were provided by contributions from various companies, including the General Electric Company, the Public Service Railway Company, Stone and Webster, and the American Telephone and Telegraph Company. By the fall of 1914 a part of the Vail Library had been catalogued and became available to students and staff. This library of 30,000 titles was a gift of AT&T in 1913 through its president, Theodore N. Vail, who became at this time a life member of the MIT corporation. This collection was said to contain every important book in electrical engineering published since 1860; Vail also provided a continuing fund of $5,000 a year to keep the collection up to date.

The "New Technology": The Move to Cambridge

It was during the Dugald Jackson period that MIT solved a problem of long standing, its chronic need for more space—a problem that also involved the issue of MIT's independence. It had already been clear as early as 1904–5, when Pritchett was president of MIT, that the Copley Square area in Boston could not provide land for the expansion of the school facilities and that a new site needed to be found. In 1904 President Eliot of Harvard University proposed that MIT merge its industrial educational program with that of Harvard. This was President Eliot's fourth attempt at a Harvard-MIT merger, and seemed likely to succeed this time because it offered a solution for a new MIT site—the present site of the Harvard Business School—and because a large bequest had been made to Harvard from the estate of Gordon McKay, announced in November 1903 and applicable to the promotion of applied science. President Pritchett supported this merger strongly because he and Eliot believed it would solve many problems faced by the two institutions; however, the plan was strongly opposed by the alumni at their reunion in June 1904, and later by a vote of 1,351 graduates against and 451 for the plan. The MIT faculty also opposed the merger at its meeting on May 5, 1905, by a vote of 56 to 7. In spite of this opposition, the MIT corporation kept the issue alive until the Supreme Judicial Court of Massachusetts decided on September 6, 1905, that MIT did not have "title in fee simple" to its land on Boylston Street and was not allowed to sell it. MIT was thus stuck in Copley Square.

By the time Richard C. Maclaurin became president of MIT in 1909, the space problem had become even more severe, and Maclaurin knew that a solution had to take first priority on his agenda. The story of how Maclaurin was able to pull all the pieces together to move MIT to a new campus is a fascinating one; as it has been fully described by Prescott,[5] we shall only skim over some of the highlights here. In the end Maclaurin had what he wanted, which made him a hero to the school's governing body, faculty, students, and alumni: he had the guaranteed support of large sums of money and he had the site he wanted. In a letter to MIT's most generous benefactor in the early part of this century, a man who would be known for many years only as "Mr. Smith," Maclaurin described the place as "a tract of land with a frontage of a third of a mile on the Charles River Basin. This site is ideal for the Institute's purposes—near to the heart of things, wonderfully accessible from all points of the city and surrounding country, and occupying a position that commands the public view and *must command it for all time.*" For Maclaurin (a man described as possessing "the intuition of a canny Scot"), the unobstructable view of the "New Technology" on the river front would announce the permanence, power, and prestige of scientific and technical education beyond dispute.

In the view of Howard Johnson, chairman of the MIT corporation (until June 30, 1983) and past president, it was the move to Cambridge that was decisive in transforming the Institute from a local school into an internationally recognized center of learning. The move, much celebrated in its time, was thus more than a mile jump from Boston to Cambridge; it was truly a jump into the larger world. Many factors conspired to make it so: the fact of scientific and technical research coming into its own, the serious engagement of MIT people in World War I weapons

In 1912 MIT acquired a large land-fill area on the Cambridge side of the Charles River, which would become its permanent home a few years later. This drawing from that period shows one of the dredges used to pump the mud forming the land area. (Courtesy Boston Public Library.)

research and development; the Congress of Technology, held in 1911; the growing numbers and influence of the school's alumni; a growing body of students attracted from nations all around the world; and, as much as anything, the incredible technological momentum that had been achieved in the United States, in which electrical power was becoming an enormous stimulant to the nation's productivity. All in all, for MIT this period proved to be a threshold of change almost as great as that of the World War II experience later.

By 1912 almost 50 acres of land in Cambridge, bounded by Massachusetts Avenue, Vassar Street, Ames Street, and the Esplanade, had been purchased at a price of $775,000, $500,000 of which had been given by alumnus T. Coleman du Pont, '84. The "mysterious Mr. Smith," who was eventually revealed as George Eastman of Eastman Kodak Company, had pledged $2,500,000 for the construction of the buildings. In early 1913 Welles Bosworth, a graduate of MIT's Department of Architecture in 1889, was selected as architect. By fall the plans were announced and the Stone and Webster Engineering Corporation secured to carry out the construction. (Charles A. Stone and Edwin S. Webster were fellow Electrical Engineering alumni, Class of 1888.) The first educational laboratory—in fact, the first permanent construction on the new site—was the Electrical Engineering Department's 500-foot power-transmission span complete with towers and 200,000-volt insulators, given and erected by Stone and Webster in the fall of 1913.

MIT's President Maclaurin and President Lowell of Harvard, supported by their corporate organizations, now envisioned a new level of service to the engineering profession and to the city of Cambridge, the Commonwealth of Massachusetts, and the nation through a sharing of resources by the two educational institutions. The principal provisions of the agreement finally reached between the two schools in early 1914 were as follows:

• Each institution was to maintain its own name, organization, and property.

• The two institutions were to cooperate in the conduct of courses leading to degrees in mechanical, electrical, civil, and sanitary engineering, mining, and metallurgy and in the promotion of research in these branches of applied science, the major activities to be carried out in the buildings of the Institute.

• Most of the equipment related to these fields and owned by the two institutions was to be located at the Institute site, and most of the funds held or acquired for work in these fields, including not less than three-fifths of the net income from Harvard's Gordon McKay endowment, were to be spent on the combined program. New buildings were to be erected from only the share of funds supplied by the Institute.

• The president or acting president of the Institute was to be the executive head of all the work carried on under this agreement.

• The faculty of the Institute was to be enlarged through the addition thereto of the Harvard professors, associate professors, and assistant professors of the stated fields, but faculty members were to be appointed or removed by the corporation that paid their salaries. All faculty members in the specified fields were to hold the same titles in both institutions.

• Students would pay the Institute's tuition rate and would be advised in their studies and recommended for degrees by the combined faculty, but degrees would be conferred by the Institute and by Harvard acting separately. This provision would usually result in two separate degrees for each graduate.

Although the agreement was not to be effective officially until MIT occupied its new buildings in the fall of 1916, President Maclaurin was able to say in his report of January 1915, "During the present year, the professors of engineering at Harvard are taking part in the regular work of the Institute's faculty, and it has been most gratifying to observe how smoothly the joint effort is working and how easily difficulties have been overcome that to the vision of some seemed in prospect to be formidable. The readiness of all concerned to make the joint effort eminently successful is a tribute to the breadth of spirit of the teaching profession. I do not think there can now be the slightest doubt that the arrangement is workable, and that immense benefit will accrue to the community from the combination of effort on which we have agreed."

In electrical engineering, equivalent classes in the two institutions were combined, and the crowding in the Boston buildings was considerably relieved through the use of office, classroom, and laboratory space at Harvard. Harvard professors Arthur E. Kennelly, Comfort A. Adams, and Harry E. Clifford were added to the MIT electrical engineering faculty. Kennelly took over the direction of the Research Division, replacing Pender, who resigned to become head of the Electrical Engineering Department at the University of Pennsylvania. Adams strengthened the instruction in alternating-current machinery.

The department's final year, 1915–16, in the Boston buildings was a period of expanding activities in both teaching and research, in spite of the additional planning necessary for the move into the new buildings. The research program was enlarged under the leadership of Kennelly. Instructor Otto R. Schurig, '11, became secretary of the Research Division, and Philip L. Alger, '15, who would later become known for his lifelong work at General Electric on induction motors, was one of the new assistants. Research meetings were held once a month, with regular attendance and participation by the graduate students and a considerable attendance by seniors. Three candidates for the doctorate were included among the postgraduate group. The number of undergraduates was increasing at such a rate that the space allocated to the department in the new Cambridge buildings looked as though it would be entirely filled by the fall of 1916. To maintain the effectiveness of teaching the larger number of juniors taking Principles of Electrical Engineering, an experiment was tried in which the students were grouped in sections according to aptitude.

The prospect of the move during the summer of 1916 brought to MIT alumni a mixture of nostalgia for the Boston campus and expectations for the "New Technology" in Cambridge. Commemorative events were held during the reunion of June 1916, and "old home week" was set up for June 12, 13, and 14, culminating in the dedication of the new buildings. On Monday morning, June 12, James P. Munroe, '82, was the orator in a "Farewell to Rogers" in the building's Huntington Hall. In the afternoon the cornerstone of Walker Memorial was laid in a ceremony shared by President Maclaurin and Charles A. Stone, then president of the Alumni Association. Among the celebrants, watching power squadron maneuvers on the Charles River, was Franklin D. Roosevelt, then assistant secretary of the Navy.

Tuesday, June 13, was commencement day for the 360 recipients of degrees, the exercises being held for the last time in the old Rogers Building. This was followed by a pageant in which the official seal was transported across the Charles River Basin, borne on a replica of the white Venetian barge *Bucentaur,* its oars moving in unison, and captained by Henry A. Morss, Course VI '93. Dedicatory addresses were made by President Maclaurin, Governor Samuel W. McCall of Massachusetts, President A. Lawrence Lowell of Harvard, and Senator Henry Cabot Lodge.

President Maclaurin and Charles A. Stone, president of the Alumni Association, doff their silk hats and shake hands at the laying of the cornerstone for Walker Memorial in 1916.

Franklin D. Roosevelt, assistant secretary of the Navy, reviews a power squadron on the Charles River. Designed by alumni as submarine chasers in anticipation of American entrance into the war, the boats joined the flotilla for the reunion. At right is Edwin S. Webster.

The *Bucentaur,* used for the ceremonial crossing of the Charles River. A hawser had damaged the plaster on the port bow.

The Golden Jubilee Banquet Wednesday evening was celebrated at Symphony Hall in Boston. The evening provided an early demonstration of teleconferencing, inasmuch as the American Telephone and Telegraph Company set up connections to alumni gatherings in 34 cities in different parts of the United States. The chief engineer of AT&T, J. J. Carty, remarked that it was especially fitting that this demonstration should be an MIT occasion, since the telephone owed more to MIT than to any other institution. The technical address of the evening was given by Professor Michael I. Pupin of Columbia University, whose loading coil had made long-distance telephony possible. Alexander Graham Bell spoke briefly, paying tribute to Professor Charles R. Cross, who had made many advances in the telephone itself and had inspired many students to perfect the work.

The fall of 1916 found the Electrical Engineering Department in its new Cambridge quarters in Buildings 4 and 10, which were destined to be its home, with suitable give and take, for more than half a century. In the basement of Building 10 was a giant space that housed the showplace of the department, the Dynamo Electric Machinery Laboratory, with its substation, instrument room, and large machines; the smaller machines and transformers were set up on a second-floor balcony. A ten-ton electric traveling crane served the two-story part of the main laboratory. The three motor-generator sets of the substation received power from a 2,300-volt three-phase supply and delivered 187.5 kVA of 60-Hz three-phase pure-sine-wave power, 50 kVA of 25-Hz three-phase or 60-Hz four-phase power, and direct current. All generators were equipped with voltage regulators.

EE Curriculum Changes

Just before the academic year 1916–17, the Course VI curriculum was revised according to the recommendations of a committee composed of Wickenden, Laws, and Adams. The revised agenda was predomi-

A banquet in Symphony Hall
wound up the 1916
celebration.

The new Massachusetts Institute of Technology.

The Dynamo Laboratory in Building 10.

nantly aimed at the problems of electrical power systems, with some illumination and telephony. The "upstarts" radio and electronics were not yet visible in the formal curriculum.

The Principles of Electrical Engineering (PEE) series of subjects, required of electrical engineering students, are listed in table 3.1; tables 3.2 and 3.3 list the fourth-year and graduate electives.

The undergraduate mathematics courses of this period covered two- and three-dimensional analytic geometry and calculus. A special third-year subject (M35) covering differential equations was required of Course VI students. The physics requirement was a one-year course of theory (801) and laboratory (807) and a ten-week, one-hour-a-week course, Precision of Measurement (803), all in the second year.

The Impact of World War I
In recent years we have come to expect a close connection between scientific and technological institutions and defense and war efforts. It was not always so, of course, but the fact that in the United States a significant percentage of the Gross National Product is now devoted to technological research and development for war preparedness has perverted the perspective of just how far these things have evolved. What was a droplet at the time of the Spanish-American War, a bare trickle in World War I, became a river in World War II. Many of the young scientists and engineers who had their initiation in trying to work with the government and the military in the First World War were able in their mature years during World War II to build on that experience and attain much greater effectiveness.

The growing impact of the European War (1914–18) on the American economy initially involved MIT in the government's program for self-sufficiency. A program of research to supply our own nitrogen products, nitric acid and ammonia, in the event that imports of saltpeter from Chile might be cut off, for instance, was proposed by Professor Noyes. He was eventually made chairman of a committee to carry out this work on a national scale.

By 1916–17 the entrance of the United States into the war became highly probable, and MIT converted a major portion of its activities to projects that would assure its effectiveness in the defense of the country and the expected conduct of the war. Numbers of MIT staff members became involved in specific government assignments. For example, Lieutenant Jerome C. Hunsaker, who had established MIT's course—the first in the country—in aeronautical engineering, withdrew to take up responsibilities in the design of aircraft for the Navy.

The declaration of war in the spring of 1917 upset the normal educational program. Both staff and students were inclined to enlist immediately in the military services. Students in the senior class who enlisted before their prospective graduation in June 1917 were awarded their degrees. The National Defense Act created the Reserve Officers Training Corps in 1916, and during the academic year 1917–18, MIT established Coast Artillery and Signal Corps (limited to EE students) units, with plans for Ordnance and Engineering Corps units to follow. Enlistments were accelerated by the announcement of the selective draft and the consequent closing of enlistments by December 15, 1917.

During this period it appeared that the United States might make the mistake England and France had made in putting manpower indiscriminately into the trenches, thus depleting the supply of technical personnel so important to modern warfare. MIT's President Maclaurin advised students below the senior class to delay enlistment until the draft situation could

Table 3.1
Principles of Electrical Engineering series

Subject number	Material covered	Year	Term	Hours per week	In charge
601	Fundamental concepts of electric and magnetic circuits	2	2	2	Wickenden
602	DC machinery	3	1	4	Wickenden
603	Electrostatics, variable and alternating currents	3	2	4	Lawrence, Lyon
604	AC machinery	4	1	6	Lawrence, Adams
605	AC machinery, transmission	4	2	6	Jackson, Lawrence, Lyon

Laboratory subjects 671, 672, 673, and 674 were required in the third and fourth years and were under the charge of Professor Laws (Technical Electrical Measurements) and Professor Green (Dynamo-Electric Machinery).

Table 3.2
Fourth-year electives

Subject number	Material covered	In charge
632	Transmission equipment	Wickenden
635	Industrial applications of electric power	Adams
637	Central stations	Hudson
642	Electric railways	Clifford
645 & 646	Principles of dynamo design	Adams
655	Illumination	Wickenden
658	Telephone engineering	Kennelly

Table 3.3
Elective graduate subjects

Subject number	Material covered	In charge
625	AC machinery	Lyon
627	Power and telephone transmission	Kennelly
634	Organization and adminstration of public service companies	Jackson
639	Power stations and distribution systems	Wickenden
643	Electric railways	Clifford

Advanced laboratory subjects were also available.

be clarified. Originally students were placed in Class I (subject to immediate draft), but upon reconsideration in Washington, most of the students in good standing were allowed to continue their education under an accelerated curriculum. Regular, or in many cases modified, subjects were offered in the summer of 1917, permitting students to pursue their studies at a faster pace and allowing a large number of the class of 1918 to finish in January rather than June. In the fall of 1917, 90 percent of the regular students were in attendance, and several schools for war training had been established on the campus.

The Navy Aviation School, the only one of its kind in the country, was set up in the summer of 1917 at the request of the secretary of the Navy. It was placed under the command of a naval lieutenant, with Professor H. W. Smith of the Electrical Engineering Department as dean. Walker Memorial, the new student center, was nearing completion and was turned over to the Navy Aviation School, which by November 1917 had more than 400 registrants.

Relief from the ambiguities of military conscription, and a conviction that technical training was of paramount importance to the war effort, stopped the disruptions that had been occurring and brought renewed serious and effective study on the part of regular MIT students. The administration tried to preserve the freedom of the students' extracurricular time and encouraged a reasonable level of social activities. On the whole, the year 1917–18 was marked throughout the country by efficient though war-oriented study and increased production in American industries.

By summer 1918 the misfortunes of the war were bringing the Allies to disaster. American forces had fought victoriously at Chateau-Thierry, Bois de Belleau,

and the Marne, but with staggering losses. More Americans were being trained and sent over as rapidly as possible. A plan was already under way to form a Students' Army Training Corps to be administered by the Committee on Education and Training of the War Department.

On July 1, 1918, President Maclaurin was appointed by the secretary of war to organize and direct the Students' Army Training Corps (SATC) for the whole country. A plan was worked out whereby each college was to become a training center for officer material; students over 18 years of age were to be voluntarily inducted into the service and then placed on furlough to complete their regular educational program plus a heavy dose of military training. This plan was to go into effect at the beginning of the fall term, but on August 31 Congress passed the Man-Power Bill, which reduced the draft age to 18, eliminated the furlough arrangement, and required every student to be placed on active duty. All costs of education, housing, and subsistence were covered by the government, including full pay for students. All activities were under military command; classes were conducted by the corps of MIT instructors.

Conversion to SATC required the hasty construction of barracks and a mess hall, and although these were ready on the announced opening day of the fall term, September 28, a severe influenza epidemic swept the nation, forcing a postponement of the opening for three weeks. The Phi Beta Epsilon house at 400 Charles River Road became the first SATC infirmary in the nation, fraternity houses on other campuses taking over this function as the need arose. The registration in SATC numbered 850 men and in the Naval Detachment (SNTC) 350.

All of these disruptions in facilities, together with a severe depletion of staff—Jackson went to France in May 1918 as a major in the Engineering Reserve,

The Navy Aviation School occupied the new Walker Memorial
Building during the war. Here
aviators and engineers received
special training.

and Wickenden resigned in August 1918 to manage personnel problems for the Western Electric Company—brought the normal department activities almost to a standstill. From the declaration of hostilities on April 6, 1917, to October 1918, the department furnished from its staff a total of 25 men for enlistment in the military services or for associated government work. A number of MIT scientists and engineers were involved in war research, such as the development of antisubmarine devices.

While in France, Jackson was chief engineer of the Technical Board of the General Purchasing Agent, American Expeditionary Forces. He was charged with procurement of power requirements for the great port developments, power plants, hospitals, and storehouses. Equipment was obtained not only in France, but in Great Britain, Italy, Switzerland, Spain, and Portugal. Plans were under way for large supplies from the Scandinavian countries when the armistice was signed. His post-armistice duties involved the inspection of industrial plants in various cities and towns along the Hindenburg line so as to assess the war damages Germany would have to pay to the Allies. One of his aides was his consulting firm partner, Edward L. Moreland.

The armistice of November 11, 1918, came as a surprise; expectations and plans were for a long war. The MIT community celebrated the termination of hostilities and then settled down, during the remainder of the fall term, to a renewed interest in studies and a revival of social activities.

The second term started right after Christmas vacation. Class officers were elected; in addition to *The Tech*, which had continued reporting throughout the war period, *Voo Doo* (a humor magazine), Technique, crew, Tech Show, and musical clubs were reorganized. All the professional societies became active, the Electricals (EE Society and Wireless Society) holding frequent meetings and conducting trips to local power plants, manufacturing establishments, and Harvard's Cruft Laboratory, where Professor George W. Pierce described the wartime developments in radio. The naval aviators terminated their occupation of Walker Memorial, and on January 31, 1919, the building was formally turned over to the students. Dugald Jackson was back at his desk on May 7, a year to the day from his departure for France.

Postwar: A New Direction

Several problems faced the MIT administration in the spring and summer of 1919. The campus had to be put in shape for the opening of the fall term with the prospect of a greatly increased enrollment. The temporary wooden war buildings were to be torn down, except for the aircraft hangars, which could presumably be used for aerodynamic laboratory equipment and for storage of service trucks and tractors. Of far greater importance, however, was a new financial crisis, whose resolution would have a long-term effect on MIT's educational programs and on its research relations with American industry.

The crisis arose from the dissolution of the Harvard-MIT collaborative agreement. Although Harvard and MIT had combined their resources for engineering education in 1914 and had successfully worked together, the agreement involved the expenditure at MIT of three-fifths of the Gordon McKay fund (which was expected to grow to $22 to $30 million by 1956). The trustees of the will opposed the Harvard-MIT agreement, because they believed that the testator desired the development of a strong and efficient school of engineering at Harvard under the exclusive jurisdiction of the university. The case was taken to court in 1915, and on November 27, 1917, the Supreme Judicial Court of Massachusetts decided that Harvard and MIT could not carry out the terms of their 1914 agreement. The cooperation continued during the war but was terminated in June 1919, when Adams and Clifford returned to Harvard with its newly organized Engineering School. Kennelly accepted a half-time faculty appointment at each school. Another element in the financial crisis was the refusal of the Massachusetts legislature to renew the state funds that had been granted in 1911 at $100,000 a year for a period of ten years and would therefore cease in 1921.

President Maclaurin met these financial reversals with a plan to establish an endowment fund. Working intensively during the summer, he turned to his friend "Mr. Smith," who responded by offering an additional gift of $4 million on the condition that MIT raise a like amount by January 1920. In the belief that this amount could not be received solely in personal gifts of MIT alumni and friends, Maclaurin, with the assistance of alumni and faculty, devised the "Technology Plan," whereby American industries would be offered library and laboratory facilities, together with the services of certain members of the MIT staff, for the solution of industrial problems. Each company would make a financial contribution to MIT in proportion to benefits received. Professor William H. Walker and Arthur R. Stubbs, '14, were appointed to investigate such a service. Safeguards had to be written into the contracts to ensure that the new plan would neither interfere with personal consulting services, such as Jackson and Walker were carrying on, nor encroach upon the educational responsibilities of faculty members.

The Technology Plan was envisioned to be a realization of founder Rogers's dream of an institution that would not only teach industrial science but also aid "generally by suitable means the advancement, development, and practical application of science in connection with arts, manufactures, and commerce." This project was organized as the Division of Industrial Cooperation and Research, with William Walker as its director.

At the Alumni Dinner on the evening of January 10, 1920, announcement was made that the Technology Plan had brought contracts in the amount of over $1 million. That night Merton L. Emerson, '04, alumni director of the educational endowment drive, was able to announce that the $8 million mark had been passed at 4:55 p.m. the day before. The sacrificial effort of the MIT community and its friends had dispelled the financial gloom in spite of the breakdown of the Harvard-MIT collaboration.

In the long run, it is now agreed, the dissolution of the Harvard-MIT agreement has proved beneficial to the growth of the Institute. Had MIT been allowed, in effect, to "coast" on the McKay endowment, it would not have been forced to turn full-face to American industry and to become deeply engaged with its real problems of development. It was this very engagement that strengthened the educational program and brought research and development issues into the minds and hands of faculty and students.

Another remarkable school of electrical engineering, which was built up at Stanford University by Frederick Terman, followed a somewhat similar course. Terman, who in 1924 received the eighth doctor of science degree in electrical engineering awarded in the new graduate program, almost stayed at MIT, where he was offered a position on the faculty. But illness intervened, and Terman went back to Stanford; he had grown up there and his father had had a long teaching career at the university. Upon his recovery Terman decided to stay at Stanford. Over the years he built up strong ties between his department members and local high-technology industries. Rather than setting up a specific program to do so, however, he gently encouraged these ties at all possible times; over the years, also, he recruited many people from MIT— men like Skilling, Tuttle, McCarthy, the Linvills, Angell,

The EE staff, 1920.*Front row, left to right:* Professors Vannevar Bush, Theodore H. Dillon, Frank A. Laws, Dugald C. Jackson, Arthur E. Kennelly, Ralph R. Lawrence, Waldo V. Lyon. *Second row:* Instructors Harold W. Bibber, Clifford E. Lansil, Hartley B. Gardner, Arthur L. Nelson; Professor Ralph G. Hudson; Instructors Frederick S. Dellenbaugh, Claire W. Ricker; Professor William H. Timbie; Instructor Carlton E. Tucker. *Third row:* Assistant James A. Carr; Instructor Frederick B. Philbrick; Assistant Clifford E. Hentz; Instructor John W. B. Kennard; unidentified.

Widrow, and others—who were familiar with the MIT industrial program, including the special cooperative program that the EE Department had established in 1917. Today these two schools stand recognized as the foremost in the country in electrical engineering.

Postwar Departmental Buildup

Jackson's chief tasks during the summer of 1919 were to fill the depleted staff ranks and to reestablish the EE cooperative program with industry, known as Course VI-A. Colonel Theodore H. Dillon of the Army Engineering Corps was appointed professor; he was to give his attention to power transmission and electric railway development. Dr. Vannevar Bush, engineer at the American Radio and Research Corporation, was made an associate professor and took charge of the introductory course in electrical engineering formerly conducted by Wickenden, as well as devoting considerable time to the department's research program, since Kennelly was available only half-time. William H Timbie was appointed associate professor to direct the affairs of Course VI-A, revived in the fall of 1919 (see chapter 7). Many new instructors were added to the staff.

During this postwar period, as Jackson was rebuilding the department and laying the groundwork for the emergence of new, important elements in electrical engineering—including electronics, communications, and machine computation—the entire MIT community went through the readjustment owing to the death of President Maclaurin in January 1920.

The Search for a New President, 1920–1923

The morning of January 15, 1920, brought the news of President Maclaurin's death. Maclaurin had developed symptoms of grippe on Friday morning, January 9, and his physician expected that a day of bed rest

Students studying the action of direct and alternating currents under different conditions, 1919. This kind of laboratory work, repeating earlier experiments, was still leagues away from the kind of participatory teacher-student research on new problems that would begin to evolve in the next decade.

would enable him to attend the alumni banquet the next evening and to deliver the address he had prepared for the occasion. By Saturday his malady was found to be pneumonia, and in his enforced absence his address was read to the gathering of alumni by William T. Sedgwick, head of the Department of Biology and Public Health. After paying tribute to founder William Barton Rogers and to General T. Coleman du Pont (president of the Alumni Association and donor of half a million dollars toward the purchase of the Cambridge site), Dr. Maclaurin had intended to reveal the identity of the "mysterious Mr. Smith," whose financial generosity had made the new MIT buildings possible. Preferring anonymity, but realizing that the MIT treasurer's report would have to list the Eastman Kodak stock that constituted the most recent $4 million gift, "Mr. Smith" had consented to the disclosure of his name on this occasion. At the end of the prepared address General du Pont revealed him to be George Eastman of Rochester, New York.[6]

Among the many tributes to Maclaurin, Robert E. Rogers, editor of the *Technology Review*, wrote, "Caesar Augustus found Rome brick and left it marble. Richard Cockburn Maclaurin, ten years ago, found a great institution barred from potential supremacy in its field by poverty. . . . To him, more than to any other single man, the world owes the new Institute, which is the old Institute, for the first time capable of its opportunities."

Immediately following Maclaurin's death, an administrative committee was established to take charge of most of the duties previously performed by the president; on March 10, Elihu Thomson was elected acting president. MIT was headed by Thomson and the administrative committee for a period of almost three years. Although Dr. Ernest Fox Nichols, former professor of physics and colleague of Maclaurin at Columbia University, was elected president on March 30,

1921, and inaugurated on June 8, he soon became ill and was unable to take up his new office, his resignation being announced on November 20, 1921.

Thomson, the veteran inventor-entrepreneur, whose early work in the last quarter of the nineteenth century had helped launch the new electrical age and who continued to work actively in electrical research at his laboratories in Lynn, Massachusetts, was the first electrical engineer to head the Institute. He would not be the last. Thomson's appointment reflects the growing maturity of the electrical field. Until Thomson, with the exception of General Walker, an economist, the heads of the Institute from 1865 to 1920 were noted scientists. Thomson was the first engineer. In the late 1940s and especially after World War II, the number of engineers who rose to head major industries grew enormously. In 1980 MIT honored 41 of its graduates who headed Fortune 500 companies; among these 41, nine were electrical engineers, the grandchildren, as it were, of the Jackson period, when the department was beginning to establish itself as a world force.

The quest for a new president finally led to Dr. Samuel W. Stratton, director of the National Bureau of Standards in Washington, D. C. Stratton, professor of physics at the University of Chicago, had studied the need for a government standardizing agency, had formulated and steered through Congress the legislation to authorize the Bureau of Standards, and then had been named by President McKinley as its first director. In his 21 years at the bureau Stratton had built an organization of a thousand scientists and scientific technicians. His appointment as president of MIT was ratified by the corporation on October 11, 1922, to be effective January 1, 1923. On June 11, at Symphony Hall in Boston, Dr. Stratton was inaugurated the eighth president of MIT.

President Maclaurin's funeral.

Research Takes a Leap

Unprecedented stimulation and excitement began to characterize the EE Department's research as Professor Vannevar Bush became the central figure in these activities. He was quick to grasp the technical significance of a problem, to explore it, and to inspire young men to work on its solution, either as individuals or as a well-coordinated team.

One of the research questions undertaken was the steady-state operation of transmission lines, which had been studied in the laboratory by means of a lumped-constant artificial line designed by Pender in the early days of the Research Division. This research, which was to go through a number of stages,[7] was to lead into Bush's development of the integraphs and differential analyzer (see chapter 4).

By 1922 Jackson was aware that his Research Division was really forging ahead with vigor, volume, and originality. Bush worked cooperatively with Timbie to get the first VI-A master's-degree group set up with interesting research problems in the spring of 1922. The number of master's degrees jumped from 4 in 1921 to 37 in 1922 and to 45 in 1923. The average number per year in the ten years prior to 1922 was 4.3; in the ten years beginning with 1922 the average yearly number was 51.5.

In 1925 Arthur E. Kennelly retired from the active faculty; he too had been influential in building up the research tenor of this period.[8] A decade and a half after Kennelly's retirement, Bush, who had done his own thesis under the elder scientist in 1915, observed of him: "He was at his best at a scientific meeting, where his geniality and ready wit enlivened many a discussion. In international gatherings in particular, his precision of language, his unfailing courtesy, and his wide acquaintance aided greatly in bringing about understanding and good will."[9]

In contrast to Kennelly's practice of publishing jointly with students whose theses he had supervised, Bush, the representative of the next generation of researchers, more often encouraged his students to publish solely under their own names, even though Bush's contributions may have been included. Outstanding examples of Bush's stimulating and inventive activities in teaching and research in the period are examined in later chapters; his interests and influence ranged over all research within the department.[10]

Parry Moon was another of those whose long careers in research and teaching, especially in the field of illumination, began in the Jackson period. Realizing that important aspects of illumination included the physics of radiation and the psychology and physiology of vision, color, and aesthetics, Jackson and the department set up in 1934 under Moon's leadership an elective curriculum suitable for students who wished to emphasize illumination. This was probably the first time the EE Department became involved in fields like psychology, which still seemed very distant from engineering. A curriculum for Course VI-B, Illuminating Engineering, was published in the MIT Catalogue for 1935–36. Moon's book *The Scientific Basis of Illuminating Engineering* was brought out in 1936.[11]

Moon coauthored many books and journal articles with Domina E. Spencer, an MIT Ph.D. in mathematics.[12] Probably the most prolific writer in the department, Moon authored or coauthored 11 books and over 200 journal articles. The most recent book by Moon and Spencer, *Photic Field Theory*, was published in 1981. The Illuminating Engineering Society (IES) Gold Medal was presented to Moon and Spencer at the New Orleans meeting on July 14, 1974.

Bush acted as department head during the academic year 1929–30 while Jackson took sabbatical leave for a round-the-world tour, including attendance as United States delegate to the 1930 World Engineering Congress in Tokyo. A handwritten footnote by

Jackson on Bush's assignment sheet for September 1930 says, "Research in the Department rests on your stimulating charge." Not only were Bush's talents as a research facilitator, driver, and organizer clear to all those about him, he himself was keenly aware that his capabilities were continually growing and their full potential barely tapped. Once, in 1931, in a moment of frustration, he confided that he felt stymied in his career, that he could not foresee when the next step up would come. It came soon, however, as described later.

Another dynamic and highly talented researcher and organizer of that same generation—and throughout their careers, a passionate competitor with Bush—was Edward L. Bowles. The teaching and research in communications were under his immediate charge. Bowles had a rather special status, partly because of the personal interest in his work on the part of President S. W. Stratton, and partly because of the financial support of the Round Hill research facility by Colonel Edward H. R. Green (see chapter 6).

The forceful leadership of Bush in directing the department's research was recognized in 1931 by President Compton and by Jackson. They envisioned the establishment of a research professorship in honor of Elihu Thomson, who would be 80 years old in 1933; they intended that Bush should be the first holder of this new professorship. A preliminary survey was made to raise the necessary funds, but the Great Depression made the years 1931–1933 an inopportune time to seek this funding. As a matter of organizational policy Jackson provided a "runner-up" for each responsible member of the department staff. He considered Bush his own deputy and successor and had placed him in charge of the department on various occasions.

When Compton became president of MIT in 1930, the administrative responsibilities had become too heavy to be carried by the president, and the office of chairman of the corporation was set up, to be occupied by former president S. W. Stratton. The death of Chairman Stratton on October 18, 1931, brought about a reconsideration of MIT's administrative structure, and in the spring of 1932 Bush was made vice-president of MIT and ex officio member of the corporation, the position of chairman of the corporation remaining unfilled. A new plan of MIT organization was announced, in which the Institute comprised three "schools" and two "divisions." The schools were the School of Engineering, the School of Science, and School of Architecture. The divisions were the Division of Industrial Cooperation and the Division of Humanities. In addition to his vice-presidency Bush was made dean of engineering to head the new School of Engineering. Bush's promotion out of the EE Department necessitated a reorganization of department activities.

During his last few years as department head, Jackson leaned more and more on the assistance of Professor Gustav Dahl and, in effect, groomed him as his successor. However, somewhat inexplicably, upon Jackson's retirement in 1935 Dahl was bypassed, and Jackson's partner in his consulting firm, Edward L. Moreland (who had also been Jackson's student in 1907) received the appointment. Moreland's tenure, as things turned out, was rather brief; but it went smoothly, with Dahl serving as Moreland's chief aide.

An Assessment of the Jackson Era
Professor Jackson, in his 28 years of tenure, was certainly the department builder, not only in the design of research programs and curricula, but perhaps chiefly in his stimulation of young staff members in their personal growth and professional development.[13] He urged his staff members to select special fields of interest and to become masters of these fields. During

his career he fostered many new fields and helped shape the careers of many hundreds of engineers. It was during his era, for instance, that men like Fred Terman, Vannevar Bush, Edward Bowles, Gordon Brown, Julius Stratton, Harold Hazen, Harold Edgerton—and others whom we shall hear more about in part II of this volume—began their prominent careers. Although Jackson was undoubtedly an outstanding administrator, he was always on the lookout for young people who could become strong intellectual leaders in electrical engineering, who could bring an added creative stimulus to research, especially on the graduate level. Through his choice of personnel, his keen technical judgment, and his wide contacts, Jackson built a department of electrical engineering that was second to none in the world.

As a consulting engineer, Jackson had been involved in some of the largest power and railway electrification projects of his time, including the Conowingo hydroelectric development on the Susquehanna River, the Fifteen Miles Falls project on the Connecticut River, the heavy electrification on the Cascade Mountain Division of the Great Northern Railway, and the New Jersey suburban service of the Lackawanna Railroad. Most of his consulting was done in connection with his firm, Jackson and Moreland, organized at the close of World War I.

By the end of Jackson's career, electrical engineering had changed its complexion; the death knell was sounding for the predominance of electric power engineering, and young students were turning to the excitements and mysteries of electronics and communications. Engineering in general was undergoing great changes, and MIT itself was rising into international prominence. The research projects established under Jackson, as well as the people running them, would prove later to have a significant impact in the larger world. Thus, in assessing the

Vannevar Bush with President Karl T. Compton (*right*), who came to MIT in 1930 with the mission of making MIT as strong in fundamental science as it was in engineering.

EE faculty, 1929, near the end of Jackson's tenure as department head. *Front row, left to right:* Murray F. Gardner, (partially visible), Charles Kingsley, Samuel H. Caldwell, Chester Peterson, Harold L. Hazen, Wilmer L. Barrow, Harold E. Edgerton, Lloyd A. Bingham, Austin S. Norcross, Frank M. Gager. *Second row:* Ralph G. Hudson, William H. Timbie, Ralph R. Lawrence, Frank A. Laws, Dugald C. Jackson, Vannevar Bush, Herbert B. Dwight, Joseph W. Barker, Louis F. Woodruff, Otto G. C. Dahl, Ernst A. Guillemin.

Third row: Julius A. Stratton, (not visible), Richard D. Fay, Clifford E. Hentz, John B. Russell, Ernst G. Bangratz, Eugene V. B. Van Pelt, Arthur L. Russell, Murrice O. Porter, Marvin H. Dixon, James L. Entwistle, Karl L. Wildes, Clifford E. Lansil. *Fourth row:* Carlton E. Tucker, Lyman M. Dawes, James E. Mulligan, Arthur F. Morash, William L. Sullivan, Richard H. Frazier, Henry M. Lane, John B. Coleman, Edward L. Bowles, Robert E. Quinlan, Parry H. Moon.

achievements in formal electrical engineering education, we could say without reservation that this fostering of research, which has grown continually stronger ever since, was one of Jackson's major contributions.

Some of the seminal research projects that got under way while Jackson was heading the department, and some of EE's most influential teachers and researchers who grew up under him, are described in the succeeding chapters. They were to set the style and many of the directions of the EE Department down to the present day. The great problems of electric power transmission and distribution were intensively studied, both practically and theoretically. Power-system stability problems, which preoccupied the electric power people in the 1920s and 1930s, were essentially mastered during this time; and computational machinery, of an analog type, developed under the leadership of Bush, along with communications (under Bowles) and electronics (under Truman Gray), were brought to a fair beginning. But more than anything else, the tight linkage between electrical engineering education and research was formed irreversibly during this period, especially at the graduate level.

In summarizing his own accomplishments in 1935, Jackson observed that three-quarters of the 2,590 bachelor's degrees, substantially all of the 759 master's degrees, and all of the 28 doctor's degrees in electrical engineering up to that time had been conferred during his tenure. He stressed the importance of carefully organized research as an instrument of undergraduate and graduate education, in addition to its service as a means for adding new knowledge. He considered the following to be his principal contributions to engineering education: Course VI-A in cooperation with industry; the Honors Groups for exceptionally able students; and the teaching of

Course VI subjects in several sections, graded according to the mental speed of the students. He recommended that the annual exchange of an assistant professor for a corresponding professor from another institution be continued. In 1934–35 Harold L. Hazen, a future head of the EE Department, was at Ohio State University in exchange for John F. Byrne; in 1935–36 Richard H. Frazier went to the University of Kansas in exchange for Robert W. Warner.

It would be a mistake to leave the impression that there were no frictions, competitions, or factions among the staff, but these often lent zest to relations that were, on the whole, friendly and conducive to the solidarity and achievement of the department. When, in June 1935, Jackson retired at age 70, the VI-A annual, *Sparks*, carried this dedication: "To Dugald Caleb Jackson—Distinguished as head of the Electrical Engineering Department since 1907; renowned for his brilliance of mind and vigorous leadership; distinctive in America and abroad for his faithful service and boundless energy as engineer and as educator; beloved by all for his untiring efforts as advisor and loyal friend."

The Moreland Administration (1935–1938)
In his final year as department head, Jackson had taken great care to see that administrative matters were in the best possible condition for smooth transfer to his successor. Moreland's future aide Gustav Dahl had worked closely with Jackson after Bush left the department in 1932 to become vice-president and dean of engineering. Dahl was thoroughly acquainted with department policies and operations, and had succeeded Bush in the direction of departmental research. Miss Ednah Blanchard, headquarters secretary since 1919, continued in this role with Moreland as he picked up the many threads of his new responsibility. Her great virtue was her unerring memory for

In 1931 Professor Truman S. Gray established an Engineering Electronics Laboratory in the EE Department. This photo shows students constructing electron tubes.

Edward L. Moreland.

Important for many years for maintaining continuity in the Electrical Engineering Department was Ednah Blanchard, headquarters secretary.

persons and for administrative details; she had also carefully and systematically built up the department files during the latter half of Jackson's tenure, so that Moreland could be briefed on any earlier matters on short notice.

Moreland himself was not a stranger to MIT. He had received a master's degree under Jackson in 1907 and had been for several years active in the MIT Alumni Association, as member of its Executive Committee, as vice-president in 1933, and as president-elect at the time he assumed his new Electrical Engineering Department post. He was also not a stranger to education, his father having been a school principal, professor of physics, and dean, first at Washington and Lee University and later at the Pre-Medical School of the University of Tennessee. His native southern geniality and his alertness in grasping the significance of his new tasks made possible the uninterrupted progress of department activities. As a Fellow of the American Institute of Electrical Engineers and member of the American Society of Civil Engineers and the American Society of Mechanical Engineers, he was well known on the national scene and served as a working member and chairman of various professional committees.

Dahl continued as chief aide to the department head, his office adjoining that of Moreland. The team dealing with graduate student matters was highly experienced, with Dahl in general charge, Gardner as registration officer, Brown in charge of the research laboratories, and Caldwell (differential analyzer), Hazen (cinema integraph), Edgerton (stroboscopic light), and Fitzgerald (network analyzer) working in their individual research areas. The dynamo and measurements laboratories were in the hands of Tucker and Bennett respectively.

The Moreland administration began with 7 professors, 7 associate professors, 9 assistant professors, 11 instructors, 4 research associates, and 4 assistants. Important events during these three years were the arrival of Dr. Arthur R. von Hippel from Europe and outstanding advances in Dr. John G. Trump's high-voltage research (see chapters 10 and 11). An intensive course revision, begun in 1933 under the direction of Edward Bowles, was pursued in these years with the encouragement of Moreland. The succession of class notes eventuated in a series of textbooks that began to appear in 1940 as Technology Press–Wiley publications. Because the books were bound in blue cloth, they became widely known as the "blue books." In their time they had an important influence on American electrical engineering education.

Though things were running smoothly under Moreland, his administration was of short duration owing to a set of circumstances that would prove to be of profound importance for MIT as well as for the nation. In the spring of 1938 Vannevar Bush, who had served for six years as vice-president of MIT and dean of engineering under Karl Compton, left the Institute; he had accepted the presidency of the Carnegie Institution of Washington, D.C. The sense of loss in the MIT community was immediate. Bush had had a deep impact on the EE Department, and his creative stimulation of research and teaching had permeated the entire Institute.

Bush's move to Washington took on a new meaning as events unfolded. Compton, who had been appointed chairman of the National Science Advisory Board in 1933 by Franklin D. Roosevelt, had been, in effect, commuting to Washington and had become deeply involved in scientific affairs at the national level. The same was true for Bush. Already in the 1930s, for instance, he had a special secret contract for developing a computational machine for deciphering coded messages, which was credited later for substantially helping to break the Japanese code during World War II. In the late 1930s Compton, Bush, James Conant (who had become president of Harvard in 1933), and others worried together about the growing Hitler menace; they realized that the scientific and technical communities needed to be mobilized to get valuable lead time in the development of new weaponry for the inevitable war. Bush's move to Carnegie put him in a strategic position for organizing American scientists and engineers. With FDR's "O.K.," the buildings of the Carnegie Institution were turned over to the defense and war effort, at first under the National Defense Research Committee (NDRC) and then as the Office of Scientific Research and Development (OSRD) (see chapter 12). At first Bush divided his time between Boston and Washington, retaining his MIT vice-presidency; he relinquished the MIT position altogether on January 1, 1939.

Moreland replaced Bush as dean of engineering, which meant that a search began once again for a new head of the Electrical Engineering Department. A natural choice at that time might have been Dahl, who was most experienced in matters of department administration, but he had resigned from MIT in 1937 to devote his full time to the engineering problems of the firm of Jackson and Moreland, where he had worked for several years on a consulting basis. There was no dearth of administrative talent in the department; Professors Bennett, Bowles, Tucker, and Timbie were obvious candidates for the headship, but they were all deeply committed to important projects. During the summer of 1938 Harold L. Hazen was elevated from associate professor to professor and appointed head of the Electrical Engineering Department.

Harold L. Hazen (1938–1952) — Tumultuous Years

Hazen's 18 years of association with MIT as student and staff member had given him a broad acquaintance with the Institute's personnel, as well as with its teaching and research activities. In addition to Electrical Engineering Department research (some of it as a student of Bush), Hazen had cooperated in 1934 with Thomas B. Camp, associate professor of sanitary engineering, in the solution of city water distribution problems by setting up an analogous system on the electrical network analyzer. In 1936 and 1937 he worked with Professor Kenneth C. Reynolds of the Civil Engineering Department in the development of a hydraulic model of the Cape Cod Canal. He designed the electrical apparatus that automatically controlled the simulated tide levels at the Buzzard's Bay and Cape Cod Bay ends of the canal, as well as the water-level measuring equipment along the canal. His industrial employers included the General Electric Company, the American Telephone and Telegraph Company, and the Raytheon Manufacturing Company. He was a second lieutenant in the Air Service Reserve from 1924 to 1929 and a lieutenant commander in the Naval Reserve from 1936 to 1949. His year as exchange professor at Ohio State University gave him a view of engineering education outside MIT and an opportunity to evaluate the department's activities in a broader perspective. At MIT his teaching experience ranged from the EE introductory undergraduate course (6.00) to graduate courses in power system stability, in all of which he contributed significantly to the subject matter.

With the realization that he was heading an extremely competent and effective team of researchers and educators, he spelled out his administrative policy as one that would create and maintain a department environment conducive to the optimum accomplishments of

Harold L. Hazen.

individuals and groups. In his Honors Groups counseling he had helped his students to look beyond the mere solving of set problems to the sources and meanings of the problems themselves. Similarly, in his department administration he enjoyed and participated in the insights and ambitions of his staff members as they perceived and tackled the perplexing demands of those stormy years.

Almost the whole span of Hazen's term was heavily involved with national defense and war activities, followed by the construction of a peacetime educational program. Hazen's headship of the department lasted 14 years, second longest of the department administrations. Considering the heavy responsibilities and the achievements of this period, it seems natural to inquire how Hazen's personal qualities and experiences prepared him for his varied sequence of activities. His memoirs, especially the chapters dealing with his MIT student years, 1920–1931, are particularly illuminating. He was a keen observer and sought an understanding of the phenomena that came to his attention. As for early training in things technical, his Sunday-school teacher, Adam Armstrong, in the Presbyterian Church of Three Rivers, Michigan, introduced him to his machine shop, where Hazen acquired experience with elementary machine tools and metal casting. His father also had a well-equipped basement shop where he could work out many of his ideas. For instance, he acquired an automobile-generator armature, for which he was able, with the help of various books on dynamo machines, to fashion a frame, compute the magnetic circuit, and design the field winding.

Entering MIT in the fall of 1920 from Three Rivers High School, Hazen found the encounter with his professors, especially in mathematics and physics, a deeply emotional experience. His entry into the world of science, he testifies, began immediately to reshape his life. He was also imbued with an innate musical sensitivity; he went often to Boston Symphony Orchestra concerts and the productions of visiting opera companies, and joined the Handel and Haydn Society as a singer, participating in some 30 performances. He enjoyed hiking in the White Mountains, and even climbed Mt. Washington, the highest peak in New England. All in all, he was a highly diversified man, and socialized easily, not at all like the stereotypical engineer.

When in 1952 Hazen was appointed dean of the Graduate School to succeed John W. M. Bunker, he left to the new department head, Gordon S. Brown, a strong, well-balanced, and dynamic organization. Hazen had guided the department as its members engaged in wartime research and had established a peacetime program of research and education with heavy financial support from the government. His department staff had grown from 9 professors in 1938 to 16 in 1952, from 10 to 18 associate professors, and from 7 to 21 assistant professors. The corresponding increases in resident junior staff were as follows: instructors, 11 to 19; research associates, 3 to 12; research assistants, 33 to 123; teaching assistants, 8 to 22. The peak postwar student load had come in 1947–48, with a total of 1,215 students as compared to 448 in 1938–39; by 1951–52 the number had settled back to 908.

In the spring of 1949 Hazen set up several working committees to spread the heavy postwar administrative load around the department. As chairman of the Committee on Graduate Students, Hazen saw the need for a study of the graduate program and a clarification of policies regarding graduate activities. In a letter of May 23 he designated a new Committee on Graduate Students with Gordon Brown as chairman

and Professors Gardner, Getting, Guillemin, and Wiesner as members. Retaining himself as an ex officio member, Hazen said, ''I wish, however, to have the chairman take full initiative in running the Committee.'' By 1950 Hazen was finding Brown helpful in solving the knotty administrative problems of the department, and in November 1950 Brown was made associate head of the department, which turned out to be good preparation for his headship, beginning in 1952. The story of Gordon Brown's period—and the beginning of a radically new epoch in electrical engineering, which flowered out of the war and postwar experience—is taken up in chapters 16 through 19.

In part II we shall trace in detail some of the elements of the EE heritage that took initial shape in the first half-century (1902–1952) and that were the basis for the postwar epoch. In the period 1952–1982 it has become clear to nearly everyone that what was once regarded as alternative education has in fact been the mainstream education. But how that happened, we believe, is not clear to everyone, even to some of those who are close students of the history of science and technology.

A major leader of research in the 1920s and 1930s was Vannevar Bush, '16, shown here (*at left*) operating an early computing integraph, with Walter F. Kershaw, Frank G. Kear, Harold L. Hazen, '24, and Murray F. Gardner, '24.

Part II

Setting New Directions

Introduction
Some Who Lead

Just after the beginning of Jackson's tenure, the acting president of MIT, Arthur Amos Noyes, said in an address,

Not enough men come to Technology to prepare for teaching positions. The Institute gets more demands for such positions than it can fill. The idea is somewhat prevalent that a man goes into teaching because he has not energy enough to go into practical work. But the best men enter it in spite of the fact that the financial gain is less, because they feel that the opportunity for service is greater. A man may each year lead a dozen or two dozen or only one or two young people into habits of thought which will determine their whole career. To a man who is naturally a student, the scholarly, intellectual life of teaching is more enticing than practical work.[1]

Noyes was speaking of a situation that within the next decade would begin to change in a fundamental way. There were to arrive on the MIT scene, and in the Electrical Engineering Department, numbers of bright, young, dedicated people who would work especially hard at the unification of the intellectual and the practical life. Spurred by Jackson, they made graduate study and research in engineering an accepted high-priority part of the EE program. At first a few and then an increasing number of teachers and students began to take on substantial research projects. The urge to innovate and to take risks was contagious, and, as some have observed, the spirit of success began to permeate the halls. Here were the beginnings of what would become a key factor, in an unmistakable form, of the success of electrical engineering in its research—namely, the involvement of students, both graduate and undergraduate, with responsibility in authentic engineering programs. This process would transform engineering education and would keep research alive in an open-ended way with a stream of inquiring young minds. By 1920 electrical engineering at MIT had really taken off.

Chapter 3, which traced the growth of the Electrical Engineering Department in this early era, focused principally on the role of Dugald C. Jackson. In the short chapters comprising part II we focus on just a few of the individuals who got started in the Jackson era and who illuminated the path of the department during this seminal period. The men whom we meet here came to MIT during the initial great growth period of electrical engineering, roughly spanning the Jackson years. Vannevar Bush came in 1915, William Timbie in 1919, Edward Bowles and Harold Hazen in 1920, Fred Terman and Ernst Guillemin in 1922, Harold Edgerton in 1926, John Trump and Robert J. Van de Graaff in 1931, Arthur von Hippel in 1936.

Their careers, their special strengths and influence in MIT's EE Department, though different, are cases of a general theme—the growth of research, the involvement of students in research, and the progressive evolution and enrichment of the EE curriculum. Though outstanding among many outstanding electrical engineers, they are representative of what was happening in the country. Between 1902 and 1927 power consumption in the United States increased 25 times; during the same period enrollment in electrical engineering in colleges all over the United States went from 3,000 to more than 19,000. Most college educators kept their eyes on the growing electric power industry and tried to train their students as its prospective employees. MIT educators did that, too, but through the work of the men described here and many others, they also did something different, something remarkably successful. Through the involvement with research and the spirit of thinking things through—teachers and students together—they planted the seeds for continuing education, training their students for the long term. Fred Terman, who came to MIT in 1922 and received his doctorate in 1924 (the eighth in EE to be awarded by the Institute), went to Stanford to do the same thing. Both EE schools today rank at the top. The process evolved by these two schools during the 1920s and 1930s seems to have been the key to their continuing growth and success.

Chapter 4

Machines for Solving Problems: Vannevar Bush

Vannevar Bush, whose early researches at MIT are treated in this chapter, figured prominently in the movement in electrical engineering in which education, research, and practical problems were integrated. Another remarkable electrical engineer of a later period, Jerome B. Wiesner, who would rise to be president of MIT and serve as science advisor to President Kennedy, wrote of Bush: "No American has had greater influence in the growth of science and technology than Vannevar Bush, and the twentieth century may yet not produce his equal. He was an ingenious engineer and an imaginative educator, but above all he was a statesman of integrity and creative ability. He organized and led history's greatest research program during World War II and, with a profound understanding of implications for the future, charted the course of national policy during the years that followed."

Like many engineers Bush published extensively in professional journals as his research interests expanded. In 1917 he followed up his 1916 thesis with an article on oscillating current circuits and another on coupled circuits in which he had introduced a novel mathematical approach; in 1919 he reported work on gimbal stabilization in gyroscopes; in 1920, further mathematical work on circular and hyperbolic functions and development of a simple harmonic analyzer; in 1921, work on a new rectifier. In 1922 he coauthored with William Timbie their well-known *Principles of Electrical Engineering* and research work on gaseous conduction. In 1923 he turned his attention to transmission line transients; in 1924, to operational analysis; in 1925, to further work on power system transients. In 1927 there appeared his first published collaborations on methods of machine computation (reporting on work actually carried out as early as

Vannevar Bush, shown here
with the product integraph, the
"first electrical thinking machine," an early forerunner of
the differential analyzer.

1925) and a second paper, with Harold Hazen, on the integraph solution of differential equations. He reported further work on mechanical solution of engineering problems in 1928; in 1929 he published continuing theoretical work on transient stability and a textbook on operational calculus. Then, in 1931, Bush reported work on the machine for which he acquired early fame, the differential analyzer, his new device for solving differential equations that could not practically be solved "by hand" or by ordinary analytical methods. Though others had attempted such machines before, the MIT differential analyzer was the first practical and useful computational machine; though an analog (not digital) machine, it marked the beginning of the "Second Industrial Revolution," the Information Revolution.

Thereafter in his professional writings Bush turned more and more to policy and organizational questions: in 1937 he wrote on the engineer and his relation to government; in 1939, on the professional spirit in engineering; in 1941, on the base for biological engineering; in 1943, on research and the war effort; and in 1945, reflecting his stature in national affairs, he prepared *Science the Endless Frontier: Report to the President on a Program for Postwar Scientific Research.*

In the intensity of the war research and development effort, in 1940, Bush had taken time out to write a memoir on Arthur Edward Kennelly, who had been his doctoral thesis advisor in 1916. That thesis had brought Kennelly and the young Bush into a head-to-head encounter and had showed Bush's capacity for making an organization serve his drive rather than the other way round.

The problem was, in a sense, a bureaucratic one, but it revolved around several interesting personalities. In those years, when the graduate study program was being carefully built up, there was great concern to admit only top-rank students and to aim at the very best theses, so as to bring to the department recognition of excellence and quality. Only four doctorates of engineering had been awarded by the department prior to Bush; a candidate was expected to labor long and earnestly. Kennelly, who had worked with Thomas A. Edison from 1887 to 1894 and who was a professor of electrical engineering both at Harvard (1902–1930) and at MIT (1913–1924), was one of those who did not expect overnight magic. But Bush managed to beat him out.

Bush's own account of what transpired is intriguing for the light it throws on Dugald C. Jackson, whom Bush regarded as one of the few "able, inspiring" teachers he had encountered. Of Jackson, Bush said,

He was my boss, as head of the Department, when I joined the staff after I had secured my doctorate and the First World War was over. I owed him much, and I was exceedingly grateful to him, although we battled often and vigorously. He was at times a fire-eater. We used to say that, if one wished to visit him in his office, it was well to toss one's hat in first and then, if it stayed in, to follow it. He worked well only with those who traded him blow for blow, and I was one of these, perhaps thanks to a small trace of Irish in my blood. My debt to him goes back to my graduate study days. Having been out of college for several years, and not idle on academic matters during that interregnum, I proposed to finish my work for the doctorate in one year. This was quite necessary, for I had funds for only one year, and there were few loan funds or nice teaching fellowships at that time.[1]

Vannevar Bush did in fact complete his MIT doctoral program in a single year, 1915–16. Before entering this program he had browsed in the works of Oliver Heaviside, where he learned of operational methods applied to the solution of transient problems.[2] In the fall of 1915 he proposed a doctoral research project involving this new kind of mathematics and outlined with his supervisor, Dr. Kennelly, the contents of his thesis. Kennelly agreed, and Bush made a point of getting the agreement in writing, although Kennelly clearly did not believe it would be possible for Bush to succeed in his aim. When in April 1916 Bush had finished his proposed research, Kennelly objected that a doctoral program could not be completed so rapidly and said that the thesis must be extended. Bush appealed to Jackson, who backed up the original Bush-Kennelly agreement, and the 169-page thesis was accepted by the department.

In his thesis Bush worked with Heaviside's "operational calculus," in which differential equations were reduced to algebraic equations by replacing the symbols d/dt and $\int_0^t dt$ by p and l/p respectively. The solution of a class of calculus problems was thereby reduced to the solution of an algebraic problem instead—a much easier task. However, as Bush continued to delve into operational methods and use them in his teaching of power transmission when he returned to MIT as a faculty member in 1919, puzzling difficulties with the methods came up in his classes. In those early years Bush cleared them up in discussions with his students or in consultation with Professor Wiener of the Mathematics Department. Some of his students in the years 1920–1927 recall the exciting discussions that took place as Bush engaged them in mathematical research in the classroom. He sometimes came to class with a complex matter still unresolved, perhaps in spite of a previous evening session with Wiener. Bush refers to this kind of experience in his last book:

I think it is very salutary for a class to reach points where they realize that the professor is working into an area which is even for him a matter of exploration. I think it creates a fine spirit among a class when the professor says, "Well, now, that's a good question, but I can't give you an offhand answer. Let's see what we can work out," and starts to examine the thing in detail right before the class. This attitude is of course at the heart of nearly all good teaching in the graduate school. Much of the way in which classes regard their teacher depends upon whether they think the man is honest, or whether he is just a poser doing some kind of act.[3]

In the fall of 1927 Bush went to Russia for conferences with Russian physicist A. Joffe on breakdown phenomena in thin dielectrics. During his five- or six-week absence, Bush's class in operational calculus was taught by Instructor Murray F. Gardner, who carried out this assignment so competently that Bush encouraged him to continue to teach it and to develop it in his own way. In the fall of 1934 John A. Barnes, VI-A '29, returned from his doctoral study at Princeton to join Gardner in teaching operational calculus. Gardner and Barnes decided to abandon the shaky operational methods of Heaviside and to apply the rigorous functional transformation of Laplace to the set of integrodifferential equations that described a system under study. This new point of view brought about a change in the name of the subject to "Transients in Linear Systems" and the publication in 1942 of a textbook by the same name. At the close of World War II this subject had become one of the most popular in the Graduate School.

Bush's Analog Machines

Vannevar Bush's interest in calculating machines started early. While still an undergraduate student at Tufts College, he conceived, built, and patented a kind of "surveying instrument," which he called a profile tracer.[4] Its concept was at the core of the machines he was to develop later at MIT.

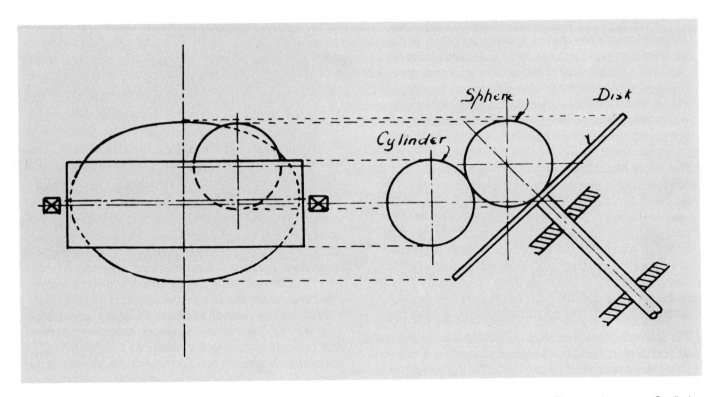

Thomson integrator. Studied but not used in the Bush-Stewart-Gage Integraph.

Hannibal Ford integrator. A practical instrument used in thousands of naval fire-control systems. It had the disk-ball-cylinder elements of the Thomson integrator.

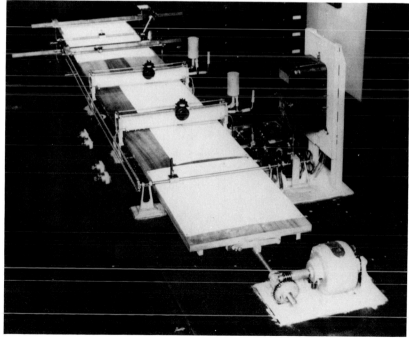

Bush-Stewart-Gage integraph.

The Second Machine

Research assistants King E. Gould and Harold L. Hazen had been intrigued by the power of the Bush-Stewart-Gage machine to solve differential equations, and the next major development came as Hazen suggested a need for two stages of integration to solve an oscillator problem and other second-order problems. According to an unpublished historical note (1948) by Professor Samuel H. Caldwell, "He [Hazen] sketched a design of the additional element and prepared for Bush a note showing how the augmented machine would handle the more difficult problem. As usual, Bush arrived the next morning with a 20-page memorandum scrawled on thin copy paper—orange color. There he had written a carefully argued demonstration of the generality of the step Hazen had taken."

At this point a new machine with two integrators was designed by Hazen and built by machinist Walter F. Kershaw in the department shop.[10] It used the watt-hour-meter integrator of the first machine, but the added integrator was a disk-and-wheel instrument (see diagram). Even though the two-integrator machine was useful, it had many shortcomings, such as the tubular resistors carrying currents of more than two amperes and requiring water cooling, the clumsiness and imprecision of the watt-hour-meter integrator, and backlash in the gear trains. But it provided valuable experience leading to the conception and design of a differential-equation solver of greatly increased precision and power. By the fall of 1928 Bush had secured MIT funds to proceed with the experimental work leading to a radical new design.

The Third Machine: The Differential Analyzer

In these times of "a computer on a chip," it is difficult to imagine the state of the art in the 1920s, when the most effective mechanical aids to computation were the slide rule and the adding machine. The history of arithmetic calculators shows how long it takes before scientific ideas reach full fruition. Several inventive and mechanically minded persons, including Charles Babbage (1791–1871), Sir William Thomson (1824–1907) and his brother James (1822–1892), Herman Hollerith (1860–1929), Alexei N. Krylov (1863–1945), and Hannibal C. Ford (1877–1955), conceived mechanical computing devices that were forerunners of the Bush problem solvers. However, Bush's differential analyzers, though they were analog machines, were the first practical computational machines to be developed.

Bush's and Hazen's new machine, begun in 1928, was to be a rugged but precise instrument with six integrators to handle a sixth-order differential equation system—that is, one differential equation up to the sixth order or three second-order equations. In less than three years the new machine was put into operation and was named (after Waldo V. Lyon's suggestion) the differential analyzer. The complete machine is shown in an accompanying photograph. David D. Terwilliger, VI-A '35, is at one of the input tables. The magnifying glasses were to aid the operator as he followed the curve manually. The eighteen shafts carrying the dependent and independent variables can be seen running lengthwise through the machine. The integrators, input and output tables, and multipliers, located along the sides of the machine, were connected to the longitudinal shafts through cross shafts and gears. Functions were added through differential gears. All the operations were mechanical except the drives and controls.

To keep the wheel of the integrator from slipping as it rested on the disk, Hazen used a Nieman torque amplifier. Its output shaft followed the motion of the input shaft precisely, but with a torque of about 10,000 times the input torque, the extra power being sup-

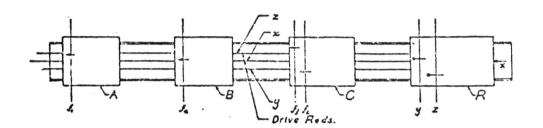

Schematic diagram of the
Bush-Hazen integraph. Func-
tions f_1, f_a, and f_2 were plotted
on platens A, B, and C respectively, and three operators followed these curves with the
pointers of potentiometers. The
result of the first integration y
$= \int f_a (f_1 + f_2) dx$ appeared as a
graph on platen R. A fourth operator kept a second pointer on
f_d (also on platen C), f_d and y
being added through a differential gear. The result of the sec-
ond integration $z = \int (f_d + y) dx$ appeared in graphical form
on platen R.

First differential analyzer.

plied by the motor that drove the drum. A disk-and-wheel integrator with its torque amplifier is clearly seen in the open box at the right of the photo. Another important Hazen development in this machine was the "frontlash unit," which compensated for the backlash at any critical point in the train of shafts and gears. The differential analyzer was able to work to a precision of one part in a thousand in a normal application, as compared to one part in a hundred in the Bush-Hazen integraph.[11]

This model of the differential analyzer proved so useful that similar machines were built at the Ballistics Research Laboratory of the United States Ordnance Department in Aberdeen, Maryland, at the University of Pennsylvania, and at the General Electric plant in Schenectady. Abroad, Douglas Hartree, professor of physics at the University of Manchester, England, cleverly assembled a small differential analyzer, mostly from parts of a children's Meccano set, and exhibited it largely for amusement. This machine gave results within 2 percent, and it was used to solve problems in which its accuracy was sufficient. Following Hartree's lead, there were other European developments, including small machines built by Harrie S. W. Massey and his associates at Queen's University, Belfast, Ireland, and by Professor John E. Lennard-Jones of Cambridge University. A twelve-integrator analyzer with Nieman torque amplifiers was built by Svein Rosseland in Oslo, Norway, in 1939, and a six-integrator machine by Bruk in the Energy Institute of the Leningrad Academy.

Bush's Fourth and Last Machine
Wide acceptance of the differential analyzer and experience in operating it led to dreams of a larger and more versatile machine. The setting up of the differential analyzer for solving a particular problem required the manual interconnection of shafts, which wasted considerable time in changing from one problem to another. In 1935 a program was launched to design and construct a very large machine that would set up a problem quickly and automatically through instructions fed from punched tapes. Three such tapes, called the A-, B-, and C-tapes, supplied information to the machine. The new analyzer was demonstrated informally on December 13, 1941, to a small group of MIT personnel but was not effectively operational until the middle of 1942. Publication was deferred until the close of World War II, because Bush and Caldwell were completely occupied with war activities. This analog machine was largely funded by the Rockefeller Foundation and was called the RDA (Rockefeller Differential Analyzer). It weighed 100 tons and had 2,000 electronic tubes, 200 miles of wire, 150 motors, and several thousand relays. It achieved its purpose of easy operation and gave an additional decimal place beyond the precision of the 1931 machine.

The "Radiation" Integraphs
A new concept in integraphs was suggested by Norbert Wiener of the MIT Mathematics Department and developed in the same time span as the mechanical devices just described. A rectangular plane source of infrared radiation was maintained at a constant temperature and its radiant energy collected by a one-eighth-inch by thirty-six-inch strip of 256 thermocouples connected in series. Each of the two functions to be multiplied and integrated was represented by a metal mask and interposed between the source and the strip of thermocouples. This idea was developed as part of the Bush program, and a model was built and tested by King E. Gould in a doctoral thesis, "Integration of a Functional Product by Measurement of Infra-red Radiation," presented in June 1927.

Professors Phillips, Bush, Hazen, and Caldwell at the Rockefeller Differential Analyzer (RDA) demonstration on December 13, 1941.

Full view of Rockefeller Differential Analyzer, 1941.

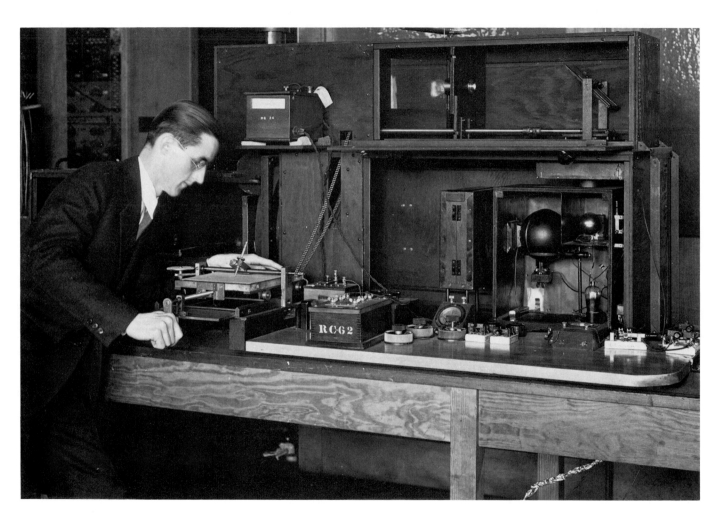

Truman S. Gray and his photoelectric integraph, 1930.
Covers have been removed to
display the optical system, including the integrating sphere
at the right.

The next embodiment of this idea was a kind of inversion of the Gould machine, using a line source of visible light, which was passed through two cardboard function masks; the total transmitted light was measured by photoelectric cells. This machine was called the photoelectric integraph, and its development and testing were the subject of a doctoral thesis by Truman S. Gray, supervised by Bush and submitted in June 1930.[12] These machines were useful in many applications, including the evaluation of Fourier and superposition integrals, and had an accuracy of 2 to 5 percent.

Experience with the photoelectric integraph revealed that a redesign could greatly improve its speed and accuracy. Consequently, a new machine called the cinema integraph was developed by Gordon S. Brown and his associates in the Research Division and became the subject of Brown's doctoral dissertation, presented in June 1938. The name came from the use of motion-picture film instead of cardboard or metal for the function masks. Another doctoral thesis in December 1939, by Walter R. Hedeman, Jr., was entitled "The Numerical Solution of Integral Equations on the Cinema Integraph." At Bush's suggestion the cinema integraph was used by Hazen and Paul T. Nims, S.M. '40, to study the motions of a ship responding to forces encountered in a seaway. There were many other applications.

As mentioned earlier, Bush's interests and activities covered a wide range of fields and subjects. This chapter has dealt only with his development of analog-type problem-solving machines. When he became vice-president of MIT and dean of engineering, he continued to stimulate his friends in electrical engineering by referring to them the various inquiries that continued to come to him in his former fields of interest. His flexible mind and quick insight enabled him to attack the broader problems of MIT with understanding and effectiveness.

During the 1940s, while president of the Carnegie Institution, Bush helped develop a photographic type-composing machine that triggered a revolution in the printing industry. In this 1949 photo Professor Samuel Caldwell looks on as Bush works at the keyboard. (A full discussion of this development appears in the November 1949 *Technology Review*.)

Chapter 5

From Educational Research to Application: Harold Hazen and the Network Analyzer

Superpower and Simulation

During the 1920s and 1930s, while the communications aspects of electrical engineering were being built up at MIT, the specialists in electric power were also tackling significant problems in their field. This chapter depicts the role of the network analyzer in connection with a movement in the 1920s called "superpower."

Like all institutions, engineering schools must debate periodically whether it is really within their means to expand their educational and research activities into new and exotic arenas that may seem promising, but for which there are no guarantees that the promises will materialize. Research of any kind, it has been learned over and over again, is a very high-risk activity, which pays off far less often than is supposed. Educational institutions—whose goals are not economic gain, but primarily teaching and preparing people to take leading roles in new fields—are natural places to conduct high-risk, advanced research (when they have the means) because there is as much learning in the failures as in the successes. By the same token, they are not the places in which to carry out development of products, or the mere delivery of services, either of which could have a negative impact on the primary educational function. However, it is not always clear exactly when the research function has peaked out and a project has phased into the development function. The outcome of the project depicted in this chapter is a case in point.

The project began in the 1920s, not long after a short-lived "energy crisis" that had followed World War I. Utilities began to expand rapidly, pushing electric service out from urban concentrations into rural areas, especially where there existed concentrations of need. For instance, growth in the Southeast of the

Harold L. Hazen's career at MIT spanned nearly 60 years, from the time he came as an undergraduate in 1920 to his service as dean emeritus in the 1970s. He did pioneering research in machine computation and automatic control as an undergraduate, influenced and directed by Vannevar Bush, and eventually served as head of the Electrical Engineering Department from 1938 to 1952. This photo dates from the 1930s, when Hazen was most preoccupied with teaching.

United States was spurred by the power needs of the textile industries, and the "long lines" stretched out from one textile center to another; in the Far West, in California, lines went out to produce power for land irrigation.

In the search for greater efficiencies, the small independent utilities characteristic of the earlier era began to interconnect and to consolidate generation equipment. Moreover, equipment was beginning to be more standardized as the giant electrical manufacturers like GE and Westinghouse began to serve national as opposed to local needs, so that interconnection became easier. The earliest large-scale interconnected grid, for instance, evolved in the Southeast; it was well under way by World War I and became something of a model for the rest of the country, as well as providing the matrix in which TVA would eventually take shape. In an editorial phrase of the times, "the independent networks have naturally touched elbows and then joined hands."

This process of growth was both natural and economically constrained. With longer lines, higher voltages were needed for transmission efficiencies, and with the growing density of electricity usage, larger centralized power plants were economically more effective. Interconnection between utilities was also efficient, since any surplus electricity generated by one utility could be shipped in bulk and sold to a neighboring utility that was facing demands greater than its facilities could meet. (Thus, at an early stage, so-called demand-supply economics became very important in the utility industry.)

The general trend toward interconnection became popularized in the early 1920s by an engineer named William S. Murray. In a book on the subject, entitled *Superpower,* Murray envisioned that eventually the entire nation would be interconnected.[1] The term caught hold in the public consciousness, and many superpower schemes were studied in different parts of the country. In Pennsylvania, for instance, Governor Gifford Pinchot, a noted proponent of the conservation movement, whose name was strongly associated with President Theodore Roosevelt, put forward plans for a superpower system that would generate great amounts of electric power at the mouths of coal mines in the western regions of that state; it was proposed that this cheap electric power be shipped in bulk all across Pennsylvania and to neighboring states. But the plan was beaten down by political opponents—it was too progressive for its time, for it involved a deeper government involvement in electric power services than the private utility sector would tolerate. Eventually, however, the basic ideas involved came to fruition during the Depression years in TVA, with water rather than coal as the power source.

Technologically, the superpower schemes brought new kinds of problems to electric power engineering. In interconnected systems everything was larger—generating plants, transformers, power lines, insulators, and so on—hence more expensive. Moreover, there had been no previous real experience with extensively interconnected systems. Potential instabilities in such systems were not well understood theoretically, and the level of complexity in extensive systems was such that ordinary analytic methods were inadequate. That is, the number of things that could vary in a big system and therefore influence each other in a dynamic fashion entailed calculations that were beyond existing mathematical means. (The great Northeast blackout of 1965 is just one spectacular latter-day example of what can happen when things start going wrong in an interconnected system, but there have been many smaller occurrences as well.) Engineers facing these problems in the 1920s began to look for new kinds of solutions and design tools.

One of the new approaches that would prove effective was *simulation*; that is, by building or devising some kind of model of the proposed system, engineers could "run" the model and then see what would happen under different conditions. In a sense, such models were physical extensions of theoretic models. However, because they were models—requiring assumptions and approximations of various kinds—simulated systems incorporated different degrees of realism. Once the actual full-scale systems were built, their behavior could be observed, measured, and compared to simulation results. In this process of going back and forth between simulations and actual systems, the models generally got better—though they inevitably became more complex—as predictors of actual system behavior. They evolved, in short, from advanced research experiments to proven design tools. Eventually, with the development of highly sophisticated computers, it became possible to simulate systems of much greater complexity. Thus, after two or three decades of intensive use, physical models of electrical networks began to be supplanted by computerized counterparts.

MIT's Network Analyzer

Very early in the 1920s, when the superpower transmission projects were getting under way around the United States, members of the MIT EE Department began to get interested in the problems of such systems, particularly their so-called transient behaviors.[2] One example under study involved a 500-mile, 200-kV three-phase transmission line to deliver 400 MW of power. Engineers in the public utility and manufacturing companies began to devise miniature model systems so that this new problem as well as the more common problems of line loading in power networks could be worked out by simulation in advance of actual system construction. Anticipated changes in the system would be made in the model through the settings of its resistors, inductors, and capacitors, and the electrical conditions at critical points would be observed as readings of ammeters, voltmeters, and wattmeters.

The first attempt at a laboratory simulation of a power system was the DC Short-Circuit Calculating Table,[3] developed at the General Electric Company, in which phase angles were ignored and results for certain kinds of short-circuit studies could be determined within 10 percent accuracy. A more realistic model was conceived in 1919 and assembled by O. R. Schurig,[4] an engineer of the General Electric Company. He used actual three-phase 3.75-kW, 110-V synchronous generators to simulate power stations, but a multimachine system proved to be unstable, hunting itself out of synchronism, because of the small rotor inertias. R. D. Evans and R. C. Bergvall, Westinghouse engineers, overcame the hunting problem to some extent with a test-floor network using larger machines, 625-kVA generators, and 425-kVA synchronous condensers.

At MIT Ralph D. Booth and Vannevar Bush made some intensive analyses of electromechanical transients for Jackson and Moreland, consulting engineers, who were studying a superpower problem. At the winter convention of AIEE in 1924 in Philadelphia, Moreland presented the Booth and Bush point of view as he discussed the six papers on superpower presented at this meeting. As a result of these problems, Bush began to think of a possible power-system model that could be set up in a laboratory.

Bush's idea was to use single-phase phase-shifting transformers for the generators and motors, with slide-wire resistors as loads, and sections of the MIT single-phase artificial line to simulate the actual transmission lines. The design, construction, and testing of such a power-system model was the subject of the joint thesis of Hugh H. Spencer and Harold L. Hazen in the spring of 1924. The resulting design used unit values of 100 watts, 100 volts, and 1 ampere. The setup of three generating stations, 200 miles of line, six loads, and metering equipment required only 50 square feet of table space. A thermionic-tube voltmeter was used to avoid loading the system. The photograph shows a Hazen-Spencer phase-shifting transformer built with frame and other parts supplied by the General Electric Company of Lynn. This thesis led to the publication of a paper by Spencer and Hazen presented at the AIEE convention in February 1925.[5]

Following graduation in 1924, Hazen and Spencer entered employment at the Schenectady plant of the General Electric Company and had the opportunity to work with Robert E. Doherty, who was much concerned with stability studies in power systems. Here considerable doubt was raised as to whether the phase-shifting transformers used as generators could show correctly the power-angle relations among the generators. To settle this question, Hazen returned to MIT for a two- to three-week period and made suitable power-angle runs on the thesis equipment. His tests showed that the laboratory model could indeed adequately represent these power-angle relations.

In the fall of 1925 Hazen returned to MIT as a research assistant and was asked by Bush whether the thesis development could be expanded to represent an actual power system, such as a large urban network or the interconnection of urban systems. Hazen spelled out the two most serious problems. The first was to produce inductance units that would have an adequately high reactance-to-resistance ratio at a frequency of 60 Hz. The second problem was to seek out or design measuring instruments that would not load the system appreciably.

Hazen solved the inductance problem using approaches suggested by two MIT professors—Bowles, then a consultant to the Samson Electric Company, and Dellenbaugh, consultant to the Acme Apparatus Company. In their design of audio-frequency transformers and choke coils these engineers had experimented extensively with similitude and model theory in the design of devices with iron-core configurations of various kinds of iron. They had also tried out many kinds and shapes of copper conductors. After much thought, experimentation, and application of model theory, Hazen reached an optimal design that provided a low and almost constant ratio (Q) for the several tap settings of each coil.

Hazen attacked the instrument problem with the cooperation of the West Lynn works of the General Electric Company, where Isaac F. Kinnard and Myron S. Wilson contributed their expertise in producing instruments of adequate accuracy. Thermocouple ammeters and voltmeters were available, but the wattmeter problem required special consideration. Here Hazen worked with the West Lynn engineers to arrive at the optimum distribution of losses between the voltage and current coils, using a special marine-type suspension.

With these two problems solved, Bush approached Doherty of the General Electric Company with Hazen's plan for the construction of a model in which an existing or contemplated power system could be simulated. The final design and construction of this device were carried out jointly by the General Electric Company and MIT. In May 1929 Professor Murray F. Gardner, then in charge of the Research Laboratory, requested nominations for a name for the new device. Out of a large number of suggested names, including some Greek ones (*dynarkytox*, for example), Professor Ralph G. Hudson's *network analyzer* was adopted. As the manufactured components arrived in the Research Laboratory on the third floor of Building 10, the

Part of a power-system model
was this phase-shifting transformer designed by Hazen and
Spencer in their 1924 bachelor's thesis.

The MIT network analyzer developed by Hazen, used for many years to solve a variety of power system problems. Hazen is seated in the center; to the left is Samuel H. Caldwell, to the right is Sidney E. Caldwell.

analyzer began to take shape under the general direction of Bush and the detailed supervision of Hazen. Among the technicians who assembled the analyzer was Albert M. Grass, at that time a high school graduate attending Franklin Union, who became a Course-VI student (1930–34) and who later founded the Grass Instrument Company of Quincy, Massachusetts.

Under the leadership of Gustav Dahl a group of 58 engineers representing manufacturers and public utility companies met with the department staff and graduate students in June 1929 for a discussion and demonstration of the network analyzer. With its eight phase-shifting transformers (representing synchronous machines), 100 variable line resistors, 100 variable reactors, 32 static capacitors, and 40 load units, systems of considerable magnitude could be represented and a great variety of problems solved. Instructions for use and illustrative problems were prepared by the EE staff and printed by the Technology Press. A paper, "The MIT Network Analyzer," describing the analyzer's design and applications, was presented at the North Eastern District meeting of AIEE in May 1930 by H. L. Hazen, O. R. Schurig, and M. F. Gardner.

Intended primarily as an educational facility, the network analyzer was made available to engineers outside MIT for the solution of commercial problems. A fee was charged for the service, "its amount being such as to meet the fixed and operating expenses, and provide for expansion." During the first ten years the most intensive outside customer was the American Gas and Electric Service Corporation. Next in order were the General Electric Company, the Canadian General Finance Company, Jackson and Moreland, the Illinois Power and Light Company, the Union Gas and Electric Company, the Tennessee Valley Authority, and five or six others.

When in 1939 a proposed federal power grid was suggested for study on the analyzer, Hazen and Harry P. St. Clair, engineer in charge of the analyzer studies for American Gas and Electric, saw that an increased number of simulated generating stations would be needed. Philip Sporn, at that time vice-president of American Gas and Electric in charge of engineering, solicited funds from nine public utility companies involved in the proposed study and provided advance payment to MIT covering the charge for the studies, which at the same time helped to finance the extended facilities.

In the summer of 1940 the analyzer was moved to a larger room and a new section of board was added. A new central metering desk increased the efficiency of data taking, and a large air conditioner kept temperatures nearly constant, with greater comfort for the operating staff. George B. Hoadley, '32 (later head of the Electrical Engineering Department at North Carolina State University), directed the design, construction, and installation of the new equipment.

The Demise of the Network Analyzer as a Research Facility

The use of the analyzer by outside agencies continued at a sustaining level for many years, but because of waning interest among students and staff, its use as an educational facility dropped off markedly in the early 1950s. Shortly after becoming the new head of the EE Department in 1952, Gordon Brown began a series of sweeping reforms in the department programs and facilities. Among other things he recommended termination of the network analyzer activities. A paragraph from Brown's letter of November 27, 1953, to President Killian summarized the situation:

Recent thinking within the Department has indicated that we no longer occupy a unique position within the network analyzer field. While our board is still useful to power companies, it no longer serves as an adequate vehicle for creative first-rate graduate or staff research in the power area. One can say that it has had a glorious past and, as a consequence, many major public utility companies have purchased network analyzers of their own. Because our machine operates at 60 cycles with a voltage base of 100, current base of one ampere, its components are bulky. We encounter many difficulties in attempting to extend its scope and to break into new areas of analog simulation of power systems. It now seems clear that we have made our maximum contribution to the power field along the present lines, and that by continuing to operate our analyzer, we are limiting our capacity to undertake new creative work. Therefore, in line with the principle that we must discontinue doing something in order to provide space and manpower to do new things, I urge its disposal.

Accordingly the analyzer was sold in December 1953 to Jackson and Moreland, to be dismantled and shipped to the Puerto Rico Water Resources Authority in January 1954. In a eulogy many years later, Gordon Brown was to draw a perspective on the conditions under which Hazen and others made their achievements:[6]

Any discussion of Hazen's technical accomplishments during the years 1923–1938 would be incomplete without mention of the environment and circumstances that on the one hand stimulated and nurtured his work and on the other hand served as a deterrent. At MIT he was a contributor to the great intellectual ferment in graduate study and research in the Department of Electrical Engineering. There was a steady but small stream of unusually capable graduate students who were motivated by the personal stimulation of Bush. It is clear that Hazen sensed this stimulation and ferment and responded in ways that added to it. But the facilities available to the students and faculty at that time were primitive by today's standards for graduate education in engineering. The laboratory benches were bare; the electrical measuring instruments were bulky and of only average sensitivity. Oscilloscopes and electronic instruments were in their infancy, mechanical gearing was clumsy and imprecise, and financial support was measured in terms of one to a few hundred dollars per project—picayune by today's standards. It is appropriate to say that it all was better suited to an industrial factory than to a scientific laboratory. But this primitive state of the art makes the achievements of Hazen, Bush, their colleagues, and their students all the more noteworthy. Their work demonstrated in a convincing way that computers that would greatly aid scientific computation could be built. It spurred others to greater accomplishments.

In March 1956 Eric T. B. Gross, at that time professor of power systems engineering at the Illinois Institute of Technology, summarized the locations and characteristics of the first 40 network analyzers in the United States and Canada.[7] Most of these analyzers operated at frequencies of 440 or 480 Hz, permitting the design of smaller reactance units than in the 60-Hz machines. The slow electromechanical swings of synchronous machines encountered in transient stability studies could be computed on the MIT analyzer as a succession of steady states, but several of the later analyzers included swing-curve calculators. About half of the listed analyzers were accompanied by digital computer equipment.

Most of the power system studies at MIT after 1954 were solved on digital computers. Such a solution was first demonstrated by Phyllis A. Fox in her master's thesis of 1949, as programmed on the Whirlwind I digital computer (see chapter 17).

Hazen at work (about 1933)
on a light-sensitive servomechanism, which served as a curve
follower for the differential
analyzer.

Chapter 6

The Building of Communications: Edward L. Bowles

When one encounters Edward L. Bowles, one of the major pioneers of communications engineering at MIT during the 1920s and 1930s, one meets one of the most vital personalities "brought up" in Jackson's period. Like Bush, Bowles was a leader and innovator in the department and something of a special protégé of Jackson's. Bowles came to MIT in 1920 to do graduate work. He studied under Bush, was encouraged by him, and did his master's thesis under Bush's guidance. He went on to become director of MIT's program at Round Hill, where fundamental research in air navigation, radio and light propagation through fog, and fog dissipation laid the groundwork for MIT's role in radar research and development during World War II. This early work led also to a new synthesis in electrical engineering. In 1933 a committee consisting of Bowles, Ernst Guillemin, Julius Stratton, and Wilmer L. Barrow built a new integrated EE curriculum with electronics, electromagnetic theory, and energy conversion as basic common concepts.

One of Bowles's students, Julius Stratton, who did research with Bowles on the early Round Hill projects, went on to become the first head of the Research Laboratory of Electronics (RLE) at MIT in the postwar period. Still later, as MIT's eleventh president (1957–1966), Stratton would introduce broad innovations in the engineering curriculum and launch a major building program. One sees in the building of communications at MIT a few of the key linkages—professor-student-research—that were personal and professional on the one hand and that opened up MIT as an institution on the other. The MIT motto, *Mens et Manus*, continued to be a living force in the educational process, as both the people and the new knowledge they created through their research went back to reinvigorate the curriculum.

The genealogy of all these relationships that would be so influential in the growth of the department—linking Jackson, Bush, Hazen, Terman, Bowles, Stratton, Guillemin, Edgerton, and many others who became part of MIT during the Jackson period, and the students who followed them (Barrow and Lan Jen Chu, to name just two)—is too rich to be wholly deciphered here. However, the point argued in the preface to this volume becomes clear: the process of involving students in ongoing research proves a vitally important key to the continuing success and vigor of engineering and scientific education.

To set some of these linkages in perspective, it is useful to sketch the state of communications when Bowles came on the scene. Jackson, in fact, in the 1920s had come to recognize the essential weakness of the department in being dominated by its electric power machinery aspects, and he kept looking for new young people to redress this. Though MIT had once been strong in communications, especially in its early contacts with Alexander Graham Bell and its subsequent work in telephony, this strength had withered away. Moreover, though there were a few courses in telephone communications, these were largely centered around telephone practice that was well established within the Bell System. Course 6.28, for instance, which was being taught by Carlton Tucker, contained some theory but was mostly applied telephone principles, reflecting Tucker's keen interests in this area of electrical engineering. Research in the field of radio or wireless communication, which was just then emerging, was to become young Bowles's mission.

In the world of engineering, the rapid development of radio after World War I meant more than just the broadcasting of programs, although that was beginning to happen on a broad scale in the United States.

Edward L. Bowles, a key figure in the pioneering communications work of the 1920s and 1930s.

Julius A. Stratton, shown here
as a senior student in electrical
engineering in 1923.

For engineers radio meant the maturation of a theoretical outlook that had had its beginnings in 1865, when a British scientist named James Clerk Maxwell had formulated a coherent set of mathematical equations that seemed to describe completely the behavior and interrelation of electric and magnetic fields. Maxwell's four equations in effect encapsulated much experimental evidence that had been accumulated painstakingly over an earlier epoch when men like Ampere, Ohm, and Faraday, whose names thereafter came to embody various distinguishable electric phenomena, were busy pushing at experimental frontiers. Maxwell not only presented a unifying picture of what happened when electric and magnetic fields were produced and interacted, he also postulated on theoretical grounds that light waves and electromagnetic waves were just different-seeming aspects of the same phenomenon—that in reality there existed a spectrum of waves going from very long to very short.

Maxwell's work became the foundation for an essentially new world view among engineers: they were trained to see waves and fields as not supported by material media; many of their research and development efforts were aimed at exploring and exploiting the characteristics of these electromagnetic waves. With a growing armamentarium of tools and techniques, engineers working in communications and electronics began climbing the spectrum, pushing to new discoveries at the higher frequencies. On the eve of World War II they were tantalized by the very short, high-frequency waves that would make radar (an acronym for *radio detecting and ranging*) possible; still later, high-frequency work would lead to practical television.

However, in the 1920s the spectral tower that would be built on Maxwell was still only at the first story. Moreover, the older engineers who had grown up with electric machines and "closed" circuits were uncomfortable with the open-ended wave-and-field view of the universe. In recognition of this, the brash new radio engineers had organized their own professional society, the Institute of Radio Engineers, Incorporated (IRE), in 1912, and its membership had steadily grown; they tended to look down their noses at the old-timers in blue serge suits with gold watch chains over their vests who had built America's electric utilities. The old-timers, in turn, had little use for "hams" the young radio buffs not only antagonized them on personal grounds, they were also beginning, in places like MIT, to push for a bigger share of EE department budgets. There is nothing better than a limited budget and multiple growing appetites for bringing out peoples' characters. In universities, where limited budgets are a chronic problem, the fighting for shares is endemic; it took all Jackson's skills to keep things together.

The budgetary problems had been aggravated by the decisive and final divergence of MIT and Harvard. The breaking apart in 1919 of the engineering faculties— which had been integrated with high hopes just a few years before—was traumatic; the break left a number of problems, including the search for new sources of funding. Thus, Jackson had to contend with an entrenched and aging power faculty on the one hand and the young radio warriors on the other. Bush's coming along just a few years before had been a breath of fresh air; Bush was moving fast, pursuing a growing range of interests, although his teaching and research were still primarily in the power field. Despite his talents Bush was only one man; others were needed to widen out the department involvements.

In the communications area there was, of course, Kennelly. He had made his name back in 1902 by beating out Heaviside by a matter of weeks in explaining Marconi's theoretically unexpected success in long-range radio transmission. He postulated that there must be a charged layer in the upper atmosphere that refracted the long radio waves, preventing them from going into outer space and instead reflecting them back down to earth at great distances. Hence the famous Kennelly-Heaviside or E layer. Kennelly kept on teaching and publishing papers, but by the time young Edward Bowles arrived, Kennelly had passed his peak as a researcher.

Bowles and Bush

Jackson dropped the communications mantle upon the promising shoulders of Bowles. And, luckily, the EE Department found a sponsor for Bowles's communications work in the person of Colonel Green, whose estate and whose interests in such research appeared at an opportune time. Were it not for Green, and were it not for Bowles's independent drive to get things done, MIT might not have been half as well prepared for the services it would be called upon to render in World War II.

The operating styles of Bush and Bowles—two of the key figures in building an essentially new department within the skin of the old department—were strikingly different. Bush wanted tight organization, clear lines of authority, and control, even while stimulating research. Thus, individuals tended to be subordinated to the interests of the organization. Bowles, on the other hand, preferred operating in a free-wheeling, individualistic way. Their differing styles gradually set Bowles and Bush off at a respectful distance from each other; yet the two men did entertain a mutual respect. Although Bowles kept a wary eye on Bush, he found he "admired the rascal," and he saw Bush's work in the war years in building the organizations of the National Defense Research Committee and the Office of Scientific Research and Development as his "finest hour."

At eighty, in an interview with the authors, Bowles recollected the "paradox" of his and Bush's relation, and how it unfolded. Though they kept a certain distance, they worked with each other, helped each other, and took each other's measure. In the early years at MIT Bush wanted to have Bowles work with him in his consulting practice. And just after the war, when the Department of Defense was being created and it seemed within the realm of possibility that "Van" Bush might be asked to be the first secretary of defense, Bush sounded out Bowles on his possible interest in being tapped as secretary of the Air Force. One wonders how often Jackson must have mused on these two young prizes of his.

But this edge of tension between them had been useful. Concluded Bowles, "Bush was brilliant, I wasn't; it made me work all the harder." Down to the present day, there have been such spirited competitions between young scientists and engineers, whose net effect has been to build a stronger EE Department.

Bowles: Getting into Radio Engineering

Bowles arrived at MIT in the fall of 1920, having taken his bachelor's degree in electrical engineering in June of that year at Washington University, St. Louis. His intention was to take just a year of graduate work and then to enter industrial employment, probably in some collaborative way with his elder brother. But, noting something characteristic of his life, Bowles says, "I always did everything I shouldn't have done, and that I didn't plan to do, so I've enjoyed every bit of it." Instead of a year, Bowles stayed for nearly twenty, to bring new tasks to the department, a role for which Jackson had already singled him out.

Jackson urged him to stay on for a second year of graduate work and in March 1921 wrote to Edward's elder brother, "He is a young man who has the characteristics of initiative and self-reliance and at the same time analytical and synthetical ability, to whom the additional study will be worthwhile. With this in mind I have offered him an instructorship at the salary of $1,400 for the academic year." His brother gave approval to the second year of study, but expressed the hope that eventually the two of them would become associated in their careers.

Among his courses Bowles took Bush's advanced Alternating Currents and Kennelly's Electrical Communication of Intelligence. He not only mastered these subjects but served as assistant to Bush and Kennelly during his first two years at MIT. During the summer of 1921 he began work on a master's thesis under Bush's supervision. His project was to improve the department's bivibrator quadrantal oscillograph so that this instrument could become a practical and powerful tool for studying hysteresis loops of magnetic materials at various frequencies. To do this, he needed to neutralize the resistance of a winding on the test specimen. One approach to the problem was to add a series negative resistance from the dynatron characteristic of a vacuum tube. Not being able to obtain such a tube, he set out to build one, but the Institute did not have at that time the necessary vacuum pumps and other equipment for the construction of this tube. Bush arranged for him to use the facilities at the American Radio and Research Corporation (AM-RAD) plant at Medford Hillside, where he completed several models of the vacuum tube.

Although his goal was not completely achieved, this experience introduced him to vacuum tubes, which were at the heart of radio telephone broadcasting, at that time still in the experimental stage. In fact, American Radio and Research Corporation's station WGI on the Tufts College campus was one of the first on the air, although station KDKA (Pittsburgh) introduced commercial broadcasting on November 2, 1920, with the announcement of returns in the Harding-Cox presidential election. Those early stations helped kick off a national radio craze.

Quite a few scientists and engineers who were to grow up with the electronics boom were getting their feet wet in amateur radio work. Among those who would leave a deep imprint on MIT were, besides

Bowles, Arthur Samuel, who in the course of several careers became famous for the first sophisticated checkers-playing computers and who thereby popularized the artificial-intelligence movement, and Philip M. Morse, who was brought from Princeton to MIT by Karl Compton in 1931 to help build up the Physics Department, which until then, Morse says, had been more a "service department" to the rest of the school than a powerful department of first rank. Morse was also encouraged by Compton to build up strong research links between the electrical engineers and the physicists, which he did.

Morse's introduction to radio started when he was a teenager in Cleveland; he started working in a "radioelectric shop" in 1921 while an undergraduate at Case School of Applied Science (now Case Western Reserve University). Morse wrote of that era, "It was in 1922–23 that the radio craze hit Cleveland. KDKA, the Westinghouse station in Pittsburgh, had started broadcasting news and music the previous year. Other stations were starting up all over the country. Suddenly everyone wanted a receiving set to listen in with. Newspapers started carrying news about programs, as well as technical articles about headphones, crystal sets, vacuum tubes, and loud-speakers. Our store was one of the few with any knowledge of the new developments and any contacts with the sprouting radio industry."

With his master's thesis finished and signed on November 7, 1921, by Bush, Bowles began an intensive exploration of radio engineering and quickly became an authority in this field. By March 1922 he was writing a radio column for an evening newspaper, the *Boston Transcript*. Continuous-wave radio broadcasting was just beginning its popular appeal, stimulated by articles in the daily papers on the operation and construction of radio receiving sets. Inexpensive parts such as coils, capacitors, rheostats, tube sockets, dials, and switches soon became available in the five-and-ten-cent stores. The expensive component was the vacuum tube, at first selling for nine dollars, but suddenly dropping to six dollars and then much lower with the increase in manufacturing volume.

Bowles was soon enmeshed in the teaching of radio at MIT through the route of one of Carlton Tucker's courses. Tucker, a Course VI graduate in June 1918 and an assistant in the Electrical Engineering Department, had been promoted to an instructorship; by 1919 he was teaching an elective subject, Telephone and Telegraph Engineering. The first term of the three-term MIT calendar was devoted to telephone and telegraph transmission, including long lines. The second-term treatment included repeaters and carrier systems, with discussions of central office equipment and visits to local telephone exchanges. The third term dealt with radio, which Bowles took over in 1921–22. This was a very busy term for Bowles, with his interest in radio and his communications teaching. His assignment sheet also carried the note, "If a class in more advanced radio materializes, and it appears that you have time to teach it, you may also have this assigned to you." Moreover, he was still assisting Bush in the laboratories of the advanced subjects Alternating Currents and Electrical Communication of Intelligence.

While Kennelly was in France during the academic year 1921–22, a new curriculum in communications was worked out by Jackson, Bush, Tucker, and Bowles and announced in the catalogue for 1922–23. Thus, when Kennelly returned to MIT in 1922, he found himself in charge of the new "Communications Option." In the year 1922–23 Kennelly and Tucker taught the subjects dealing with telephone and telegraph systems employing wires. Bowles taught the radio subjects and had charge of the communication laboratories. Bowles, limited in space to an

Philip M. Morse.

Manuel S. Vallarta.

undergraduate communication laboratory of 26 by 40 feet, made a thorough study of extended laboratory space for communications teaching and research. His report became Appendix C of one of the most vigorous and comprehensive reports of the Jackson administration, setting forth to the Visiting and Advisory Committees of 1922–23, with charts and diagrams, the policies and needs of the department.

The report to the Visiting and Advisory Committees mentioned their good fortune in having Dr. Frank B. Jewett as a member at this critical time in the establishment of a communications curriculum. The committees agreed with Jackson that "we now have the most adequate staff in this subject of any of the engineering schools, in the persons of Professor Kennelly and Messrs. Tucker and Bowles." Bush and Dellenbaugh could also have been mentioned.

At the time the Communications Option came into being, the MIT Radio Society was especially active. Julius Stratton was secretary of the society in his senior year (1922–23) in Course VI; he wrote an article for the December 1922 *Tech Engineering News*[1] describing the equipment in detail and noting the Institute's recognition of the coming importance of electrical communications through the inauguration of its new option. Stratton's thesis was written in 1923 jointly with James K. Clapp, under Bowles's supervision, and entitled "Absolute Calibration of Wave Meters." The basic idea was to use a high-precision tuning fork as the reference frequency. A wide spectrum of harmonics was supplied by a multivibrator circuit, thus producing a range of standardized frequencies related precisely to the frequency of the fork. Clapp subsequently extended the method of the thesis and developed a crystal-controlled frequency standard that was widely recognized within the young radio industry.

Bowles followed up his 1922 recommendation for increased laboratory space and by 1923–24 had established a new communication laboratory next to the department's research laboratories on the third floor of Building 10. The further development of the Communications Option came largely out of Bowles's appreciation of the scientific needs of communications education. It was at this point that the department deepened its foundations in mathematics and physics. In the fall of 1923 Bowles attended Professor Phillips's class in vector analysis and the following year incorporated this as a third-year requirement in the Communications Option. Also in the fall term of 1923 Bowles arranged for Manuel S. Vallarta to give a course in electromagnetic theory and wave propagation.[2]

Vallarta (1899–1977) was a 1921 graduate of MIT's Course XIV, Electrochemical Engineering, and had done experimental research with the Electrical Engineering Department's Research Division in 1921–22. He had pursued advanced studies in mathematics, physics, and electrical engineering in 1922–23 and had been awarded an honorary research fellowship in physics for 1923–24, receiving the Sc.D. degree in 1924. Based upon his penetrating studies in the works of Heaviside and Maxwell, he was placed in charge of a two-term physics sequence, 8.252, Electromagnetic Theory, and 8.253, Electromagnetic Wave Propagation, included in the Communications Option curriculum in the second and third terms of 1924–25. Vallarta was to progress rapidly through the faculty ranks at MIT and become a world-renowned physicist. Following his career at MIT, he returned in 1946 to his native Mexico, where he gave strong leadership to the cultural and scientific advance of his own and other Latin American countries. He served as president of the United Nations Atomic Energy Commission in 1946. Among his honors was his appointment in 1961 by Pope John XXIII to the Pontifical Academy of Sciences, in which he succeeded to the chair Albert Einstein had held until his death.

When in 1925 Kennelly discontinued his MIT responsibilities, Bowles was made an assistant professor and placed in official charge of the Communications Option, which by that time had come to be known as Course VI-C. Bowles had received an attractive offer to take up research with the Radio Corporation of America, but, as Jackson wrote in March 1925 to President Stratton, ''He prefers to stay with us in the rank of assistant professor, as he agrees with me that the opportunities for enlarging his reputation while on our staff are very great, and he is able to add a material amount to his income through his writings.'' In January 1926 Bowles gave the Society of Arts lecture on recent developments in radio.

Redressing the Power-vs.-Communications Imbalance

Bowles was keenly aware that although Cross and others in the Physics Department had been deeply involved in the early Bell Telephone developments, the Electrical Engineering Department had emphasized power transmission and machinery, with communications in a minor role. Revival of interest in communications was therefore needed on a broad front. To stimulate new interest, the department's new leadership in communications called on their friends in the Bell System for help. The response was quick and energetic. During the academic year 1925–26 Bell Telephone Lab (BTL) staff members provided three colloquia for graduate students and seniors.

In the spring term, 1928, Timothy E. Shea of BTL (VI S.B. '19) came on loan to MIT to give a course on electric wave filters. Bowles had given some instruction on this subject but felt the need to expand the material. Shea had organized an out-of-hours course for BTL employees, and followed the MIT presentation with the publication in 1929 of his book *Transmission*

Colonel Edward Howland Robinson Green, whose interests in radio and aviation led him to provide financial support and the use of his Round Hill estate for research in these fields.

Networks and Wave Filters.[3] It was at this time that Dr. Ernst A. Guillemin, who would become an important influence in network analysis and synthesis in later years at MIT, switched from power to communications engineering (see chapter 9).

The Setting for Important New Communications Research

During the years 1922–1925, while Bowles was building the Communications Option, Course VI-C, he could not have had a premonition of the Round Hill development, which would soon be his major responsibility. The person who introduced this new element into the communications program was Colonel Edward Howland Robinson Green.[4] He entered the MIT picture when President Stratton wrote to him on August 13, 1923, as follows: "This will introduce to you Professor V. Bush of this Institute and one of our experts in radio telegraphy. Mr. Bush is visiting you in response to your courteous invitation of yesterday. . . . I hope that we can establish some arrangement between the Institute and your institution which will be of great use to both of us." It seems that Marconi's early experiments with wireless had captured Green's interest as early as 1896, when he was president of the Texas Midland Railroad. Train dispatching was attempted by installing receivers using metal-filing coherers in two baggage cars and a transmitter in the station at Terrel, Texas. The coherer detectors of that day did not operate satisfactorily aboard the moving trains, but radio telegraphy was established along the right of way. The arrival of continuous-wave broadcasting in the early twenties revived Green's interest in radio, and he established station WMAF, "The Voice from Way Down East," in 1923 at his estate, Round Hill, in South Dartmouth, Massachusetts. The earliest WMAF broadcasts carried programs originating at South Dartmouth, but on July 1, 1923, Green inaugurated an extended broadcast service in which major entertainment programs were transmitted from New York station WEAF over one of the first program-quality telephone lines of the American Telephone and Telegraph Company, and rebroadcast over station WMAF to the New England area. This early chain broadcast was a forerunner of the network broadcasts, inaugurated by the National Broadcasting Company on November 15, 1926, when 24 stations in 21 cities put a single program on the air.

When radio amateurs began to achieve astounding transmissions in the wavelengths below 200 meters, Green encouraged qualified experimenters to work at Round Hill and built new shops and laboratories for their use. It was this amateur activity that led him to approach President Stratton in 1923; soon afterward Bush and Jackson visited Round Hill. On his 277-acre estate, Green was to build facilities for the MIT communications activities described in this chapter. After Green's death in 1936, his sister inherited Round Hill, and in 1948 she gave it to MIT. Lincoln Laboratory, established in 1950, used Round Hill for important research, especially in the development of new reliable methods of "scatter transmission" of microwaves under the leadership of Professors J. B. Wiesner and W. H. Radford. Professor Houghton had also conducted meteorological studies on the way in which smoke and other pollutants are distributed by natural air currents. After Dr. Charles H. Townes and Dr. Ali Javan joined the MIT physics faculty in 1961, they used the wine vault in the basement of the mansion as a site for experiments with optical masers, because of its extremely quiet environment and isolation from extraneous vibrations. The Round Hill Astrophysical Field Station, established under the direction of Professor Roy Gunter of the College of the Holy Cross, was dedicated on April 23, 1966, the whole estate having been sold by MIT in 1964 to the Society of Jesus of New England.

Not much happened beyond the "ham" level until March 1926, when President Stratton wrote the colonel, "I think we will be in a position to carry on the work more actively during the spring." In May Green sent a check for $2,500, "which I trust will help in carrying out your plans for the good cause." Bush wrote the colonel on July 14, 1926, "I am enclosing herewith the tentative schedule for the extension of the plant at Round Hill for the short wavelength research. Mr. J. K. Clapp, one of our instructors who is at present actively in charge of the research, is now at Round Hill where he will be engaged in taking antenna characteristics for two or three days. Will you kindly let me know whether this plan meets with your approval." A four-page list of new equipment and a description of the use of the experimental buildings near the beach, at an estimated cost of $3,300, accompanied this letter. The colonel approved the program promptly, enclosing his check for $3,500.

Under Clapp's immediate supervision, Walter D. Siddall was assigned by Bowles as the first full-time resident at Round Hill; he reported in the November 1926 QST (a radio amateur's magazine) the following items as initial investigations:

- The variation of the cut-off wavelength (the minimum wavelength at which signals are audible for a given distance) between station 1XV (at Round Hill) and station 1BYX (Clapp's station in Auburndale, 60 miles distant) for different times of day and seasons of the year.
- Fading on various wavelengths.
- Different types of antennas.
- Change of signal strength as the plane of the transmitting antenna was rotated. (1XV had a 360-degree orientable antenna mounted on top of a 50-foot telephone pole.)

Colonel Green's mansion and
broadcasting station WMAF at
Round Hill.

In August 1927 President Stratton wrote the colonel, "I have asked Professor W. G. Brown, a member of our aviation staff, to call upon you and give you such assistance as you may desire in the planning of your aviation field and equipment." Charles A. Lindbergh's nonstop flight from New York to Paris in 1927 had stimulated Green to develop an airport on the Round Hill estate. By spring 1928 a runway was in operation, and during the subsequent months this airport was made to satisfy the government regulations for a first-class experimental flying field. President Stratton, Professor Charles H. Chatfield of the Aeronautical Engineering Department, and Bowles went to Round Hill in the spring of 1928 to work out a plan of cooperative research in relation to the flying field.

In 1927 and 1928 extensive studies on the radiation patterns of various antenna arrays were carried out by Clapp and Research Assistant Howard A. Chinn, VI S.B. '27, S.M. '29. Research Assistant Gordon G. Macintosh was resident operator at the station in 1927–28 and published an article in the July 1928 *QST* entitled "IXV-IXAN, An Experimental Station with the World as Its Laboratory."[5] He described the station equipment and mentioned some of the newer research items, such as the making of skip-distance maps for various times of day and seasons of the year involving 38-meter transmissions to points throughout the world. At this time Research Assistants Howard A. Chinn and Lloyd T. Goldsmith were in charge of the laboratory and field investigations. This station also supplied significant public services. It communicated regularly with the MacMillan and Byrd expeditions; it followed the dirigible Graf Zeppelin in its first flight across the Atlantic; and at the time of disastrous floods in Vermont in 1927, a truck-mounted station 1XM, manned by Messrs. Chinn, Wolcott, Clapp, Goldsmith, and Brolly, drove to the vicinity of Ludlow and Bethel to establish the only communications out of the devastated area.

Operating Room of the MIT research station at Round Hill with Gordon G. Macintosh, Resident Operator, who described this station in a 1928 article as having the "world as its laboratory."

The increase in the number of radio channels and the need to calibrate laboratory instruments such as frequency bridges and wave meters called for a frequency standard of high precision. Such a standard was the subject of Chinn's S.M. thesis in 1929. A piezoelectric quartz-crystal oscillator supplied a 100-kHz reference frequency, and the multivibrator circuit provided a range of standard frequencies from 15 Hz to 15 MHz, with deviation from a given value not in excess of 0.001 percent. Part of the device was a synchronous clock, operated from the 1-kHz output, which had a second hand indicating true solar time. In the fall of 1929, operating in conjunction with the American Radio Relay League, station WIXV-WIAXV sent out a signal frequency having a steadiness unrivaled in the field of radio.

The interests and experience of Julius A. Stratton (not a relative of President Samuel W. Stratton) fitted nicely into the Round Hill picture. Prior to his MIT days, he had qualified as a commercial radio operator, first grade, and had served as chief operator on a freighter in a voyage to Manchuria and Japan. Following his S.B. program in Course VI, he had spent a year in France and, at Bowles's invitation, had returned to MIT in the summer of 1924 as a research assistant. His development of a high-frequency (500-kHz) capacitance bridge became the subject of his S.M. thesis, supervised by Bowles and completed in late 1925. His graduate studies in physics under Manuel S. Vallarta, Paul A. Heymans, and Hans Mueller encouraged him to pursue further advanced study in electromagnetic theory at the Eidgenössische Technische Hochschule (ETH) in Zurich, where he completed in late 1927 a dissertation entitled "Streuungs Koeffizient von Wasserstoff nach der Wellenmechanik," published in German in the journal *Helvetica Physica Acta* in 1928.[6] Bowles again persuaded Stratton to return to the Electrical Engineering Department, this time as assistant professor of the theory of electricity and magnetism, with immediate involvement in the supervision of Round Hill research.

The Institute, recognizing that fog was a primary menace to aerial navigation, proposed to Green that an extensive fog research program be set up at Round Hill. With the colonel's support, Professor Stratton and Research Assistant Henry G. Houghton, VI S.M. '27, initiated in the fall of 1928 a study of electromagnetic-wave transmission through fog. An initial theoretical investigation by Stratton showed that energy absorption was negligible for wavelengths greater than five centimeters.[7] Observing that the absorption reached a theoretical maximum value at a wavelength of about two centimeters, Stratton hit upon the idea that the water particles might be evaporatd by means of radiant energy of this wavelength. Houghton and Barrow pointed out, however, that no suitable generator of two-centimeter energy was available at that time. Bush, encouraged by magnetron research at the General Electric Company, worked briefly and unsuccessfully to produce a generator for two-centimeter radiation.

The transmission of visible light through fog was studied theoretically by Houghton and Stratton, with published results in the *Physical Review* for July 1, 1931.[8] A companion paper by Houghton described an experimental procedure in which artificial fog was produced in a long chamber and the light transmission measured.[9] These studies revealed that the transmission was a function of the size, density, and composition of the fog particles, as well as of the frequency of the light. Accordingly, Houghton developed and published in 1932 methods and devices for the determination of particle size and distribution in actual fogs, thus making the Stratton-Houghton formula a practical means of estimating light transmission of different colors under fog conditions in an airport.[10]

Houghton's investigation into the physical and chemical properties of fog particles led him in 1933 to the idea of fog dissipation through the spraying of hygroscopic salts of suitable coarseness into the limited space to be cleared.[11] The embodiment of this idea was worked out by Research Assistants Houghton and William H. Radford, VI S.M. '32, and a much-publicized spectacular outdoor demonstration took place on the afternoon of July 20, 1934, when a heavy fog rolled into the Round Hill estate. The chief practical application was to clear a ground area sufficiently large to enable an airplane to make a landing with visibility after being guided down a radio beam into the airport (see photograph).

A comprehensive report with extensive bibliography, "On the Local Dissipation of Natural Fog," by Houghton and Radford was published by MIT and the Woods Hole Oceanographic Institution in 1938.[12] During these years of fog research Houghton had the advice of Dr. Carl Gustaf Rossby, who, as associate professor of meteorology in the Aeronautics Department, had established in the summer of 1929 a meteorological station on the Round Hill estate. The weather data furnished by this station were of great value to airplanes flying in that vicinity and to the flights of the nonrigid dirigible *Mayflower*, loaned to MIT during the summers of 1929 and 1930 for Bowles to use in mapping the radiation fields of antennas. This dirigible was obtained through the cooperation of Paul W. Litchfield, '96, president of the Goodyear-Zeppelin Corporation and at that time also president of the MIT Alumni Association. Results of these tests were analyzed and published by Stratton and Chinn.[13] A hangar (which had to be called a "dock") for the airship was funded at $30,000 by Colonel Green and built to Goodyear-Zeppelin specifications under MIT direction in June 1929. The radio,

fog research, and meteorological aspects of the program were unified under the general direction of Bowles. In May 1929 Green made $40,000 available for research at Round Hill.

Bowles rated Houghton an experimentalist of great ability, and these precise and careful investigations laid the foundations for Houghton's subsequent entry into the field of meteorology in 1939 and his selection as head of the new Department of Meteorology in 1945.

Stratton's interest in theoretical studies led him to change his affiliation in 1930 to the Physics Department. He soon developed his own version of the physics subjects in electromagnetic theory and wave propagation taught by Vallarta. His transfer to the Physics Department was part of the reorganization of that department under the headship of John C. Slater when Karl T. Compton began his MIT presidency in 1930. Stratton continued to be loyal to Bowles and to the Round Hill project, which helped to forge another link between physics and EE. He was soon teaching advanced mechanics and working on graduate-level electromagnetic theory, his book on this subject being published in 1941, when he was promoted to a full professorship.[14]

The Great Depression, starting in 1929, had an impact on the Round Hill projects, for Colonel Green was unable to increase his financial donations. In view of the large budget that Bowles had built up for the early thirties, it fell on Bowles and his MIT friends then to arrange to finance the Round Hill projects from MIT funds and from such sources as the Humane Society of the Commonwealth of Massachusetts, the Navy Bureau of Aeronautics, the United States Army Air Corps, and the Bureau of Air Commerce.

Wilmer L. Barrow, VI S.M. '29, joined Bowles's Communication Division in the fall of 1931 as an instructor, after completing his doctorate in February at the

Fog dissipation installation at Round Hill, devised by Henry G. Houghton (*top of ladder*) and William H. Radford (*below center*), here standing with Thomas N. Tedesco.

The *Mayflower*, a nonrigid dirigible used in the mapping of radiation fields of antennas, at its dock.

Technische Hochschule of Munich, with advanced study in pure and applied mathematics, thermodynamics, and the structure and properties of solid materials. He studied theoretical physics under Professor A. Sommerfeld and did his research in acoustics under Dr. J. Zenneck. In response to Professor Jackson's inquiry concerning "things you are anxious to carry on," Barrow wrote in May 1931 the following significant words:

A new and especially promising field for research is that of ultrashort radio waves, by which I mean those at least below a meter in wave length. A suitable and efficient generating method is much to be desired. The technique of measurements must be advanced considerably over its present state. The properties and peculiarities of wave propagation, attenuation, reflection, polarization, etc., at these enormous frequencies in the several media must be quite exhaustively studied. Still more interesting is the theoretical aspect of the question, where little or no headway has been made, due to the fact that the ordinary assumptions of circuit theory based on the quasi-stationary conditions no longer hold. Indeed our usual conceptions of capacitance, inductance, etc., have lost their meaning. This is a most fertile field for new and valuable work.

With his characteristic energy Barrow involved himself in the MIT communications research, assisting F. Malcolm Gager in the undergraduate laboratories and, in the second term, taking the major responsibility for the graduate laboratories. He also began his antenna research under the directon of Bowles and Stratton, and by the academic year 1933–34 he was supervising six master's theses (C. J. Alba, F. W. Baumann, R. L. Fossett, Jr., F. T. Hall, Jr., D. B. Smith, and J. R. Sloat), mostly in the ultrahigh-frequency area. Much of this research was carried on in a "radio shack" at the rear of the old brick block that housed the Tech Coop and Walton's Restaurant on Massachusetts Avenue.

Barrow restudied and extended the theory presented by Lord Rayleigh showing that ultrahigh-frequency waves could be transmitted through tubes without a central conductor.[15] Barrow first demonstrated this experimentally through the use of an old galvanized-iron cylindrical air duct about eighteen inches in diameter and sixteen feet long. At the joint meeting of the Institute of Radio Engineers and the Union Radio Scientifique International in Washington, D.C., he announced on May 1, 1936, the principal findings of his classic paper entitled "Transmission of Electromagnetic Waves in Hollow Tubes of Metal."[16] At the same meeting Dr. George C. Southworth of the Bell Telephone Laboratories announced that he and his colleagues had been working along similar lines and that two papers on hyperfrequency wave guides—the first presenting experimental results; the second, mathematical theory—had appeared in the April 1936 issue of the *Bell System Technical Journal.*[17] Barrow was promoted to the rank of assistant professor in 1936 and continued his intensive development of microwave theory and applications. He supervised the doctoral studies of Lan Jen Chu (1938), who made important contributions to the theory of wave transmission in hollow tubes of elliptical and rectangular cross section and in the Barrow electromagnetic horn. Barrow also supervised seven master's theses in 1938, including that of Donald E. Kerr; in 1939, seven master's theses, including that of J. J. Jansen; and in 1940, six master's theses, including Frank M. Lewis's study of instrument landing. In 1941 he transferred his activities to the MIT Radar School (described in chapter 13), becoming its director in 1943.

Bowles's comprehensive program of research involving navigation in fog and the theory and application of microwaves furnished the foundation for a special

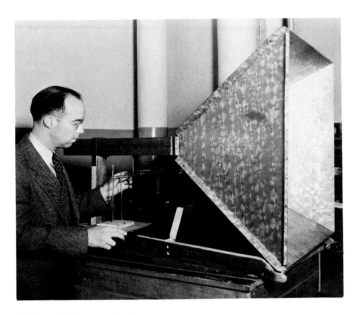

Wilmer L. Barrow with horn antenna.

study of instrument landing for airplanes. This work, involving the cooperation of Professor C. Stark Draper of the Department of Aeronautical Engineering, was summarized by Hazen in the president's report for October 1939:

The Instrument Landing Research for the Civil Aeronautics Authority has been carried to a successful conclusion. In it the applicability of the Barrow horn radiators for producing radio beams and the first application of centimeter waves to the problem have been demonstrated. Ingenious adaptation of the aeroplane gyro by Professor Hall simplified the instrument aspect of the problem. The first application of the Stanford new ultrahigh-frequency generator, the Klystron, was made here in this research through the courtesy of Stanford University, the Sperry Gyroscope Company, and the United States Army. This project has, in fact, shown the benefit derivable from cooperation between departments within the Institute, and between the Institute and industry, and the Army.

As a result of its advantageous position in the ultrahigh-frequency field and the outstanding importance of the subject, the Department, through a grant from the Executive Committee, is now to develop an ultrahigh-frequency laboratory. The Department is also fortunate to be enabled to collaborate with the Loomis Laboratory at Tuxedo Park and with Professor Stratton of the Department of Physics in a research on the propagation of ultrahigh-frequency waves, and to undertake two other researches in the ultrahigh-frequency field, one sponsored by the International Telephone and Telegraph Company and the other by the Sperry Gyroscope Company. Both this and the related Instrument Landing Research are directed by Professor Bowles.

This research led to the presentation of a paper, "The CAA-MIT Microwave Instrument Landing System," by Bowles, Barrow, and others at the winter convention of AIEE in January 1940.[18] A final contract with the Civil Aeronautics Authority leading to a complete experimental microwave instrument landing installation at the Boston Municipal Airport had to be broken off prematurely in 1941 because of the emergency requirements of the Second World War.

In addition to the projects of the Electrical Engineering Department, the Physics Department carried on at Round Hill a nuclear research program under Dr. Robert J. Van de Graaff and his colleagues L. C. and C. M. Van Atta.[19] The first large (seven-million-volt) Van de Graaff generator was housed in the hangar built for the airship *Mayflower*. This generator was later moved to the MIT campus and then to the Museum of Science in Boston.

The successful program in radio, fog, and aerial navigation research under the leadership of Bowles represents one of the outstanding segments of the department's history. It was the groundwork that enabled MIT to establish and operate rapidly and efficiently the MIT Radar School and the MIT Radiation Laboratory, important in the conduct of the Second World War.

Chapter 7

The Industrial Cooperative Program: William H. Timbie and Course VI-A

For many years the so-called cooperative course in electrical engineering, Course VI-A—in which students took alternate terms out in industry learning what engineering was "really about"—was associated with a distinguished man who sported a fine white goatee, William H. Timbie. He put a personal stamp on the program that was unlike anything else at MIT. In an educational environment that many students called a factory, grinding out specialists in diverse fields, Timbie stood out as a man who was warm, genial, and, most especially, a humanist scholar who had not lost his bearings in the swift current of science and technology. He could urge the merits of working in real factories to young students just as cogently as he could converse with them on the marvels of Homer. He got to be a great friend of Van Bush, and they sailed together, enjoyed telling stories, and together wrote a textbook, *Principles of Electrical Engineering,* that became famous in engineering circles as "Timbie and Bush." Timbie had been trained as a classicist and then had learned the practical end of things electrical while teaching before he came to MIT.

The cooperative course in electrical engineering was significant to the development of the department in many ways. Though MIT was not the first school to try out this experiment in education, it became an outstanding proponent of it, seeing in it an integral aspect of alternative education. The course represented, like graduate research, a link to the problems of industrial and business development, and tied theory to practice in the most direct way possible. Undergraduates learned at first hand what their elder brothers out there were doing by spending alternate terms working with them, somewhat as interns; back at school, they could see by contrast what their non–co-op contemporaries were missing.

In theory, the system was perfect; in practice, it took a lot of care and attention—in matching students to industrial situations and in finding the right people in the cooperating companies to hold up their end of things. It was also a somewhat fragile system, for as the tides of economic growth or economic recession affected industry and business, the numbers of students that companies could accept either grew or declined. When a company was having to lay off engineers, it could not rationalize to its employees the hiring of untrained counterparts. This, too, became part of the young co-op students' education.

Social issues of different kinds came in as well, and changed over time. In an earlier epoch, for instance, there was still manifest in American industry a reluctance to hire Jews. Young Jewish students embarking on the cooperative program discovered early the painful possibilities of their chosen career path. Should they drop out, try to "pass," or confront? In coping with such issues—with Jews, with blacks, with women, and so on—students, professors, company managers, and executives grappled, for better or worse, with the intersection of science, engineering, and human ethics in the most direct way.

Thus, electrical engineers in particular at MIT had to take the so-called technology-and-society issues very seriously at an early stage of their education. This is not to say they achieved any grand and obvious successes straightaway, but it was a tempering and a learning experience. And each generation of co-op students learned different aspects of the industrial and economic reality—earlier students learned what it was like to work in real factory environments; others learned about industrial research, development engineering, and corporate management; others learned the arcane arts of winning government contracts by assessing RFPs (Requests for Proposal); and still oth-

William H. Timbie, who took charge of the MIT Cooperative Program, Course VI-A, in 1919 and administered it until 1947, exerted a genial influence on generations of co-op students. As a young teacher, he looked so young that he decided to sprout a goatee, which distinguished him throughout his life.

 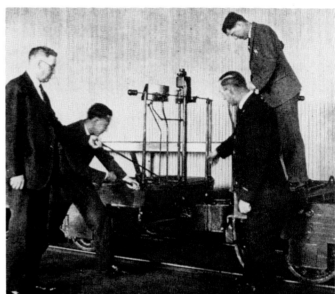

Over the years of the Cooperative Program its students
worked at a variety of assignments in the electrical industries—research, production,
engineering, business—imbib-
ing early other insights into the
makings of an industrialized society than technical alone.

Chapter 8

Frozen Motion: Harold Edgerton and the Stroboscope

The career of Harold Edgerton at MIT is another case of student involvement in research opening up new horizons in electrical engineering. In Edgerton's case it was initially his own research on a technical problem in electrical machines, and then his continuing research along with his students, that would lead to vital wartime applications, to the formation of a new company, and to a new vision in the photographic arts.[1] Even the large entertainment industry built around the use of stroboscopic lights in discos can be traced to his early research and development work with stroboscopic phenomena and intense light sources. His long-time collaboration with photographer Gjon Mili and his more recent work on deep-sea cameras with Jacques Cousteau show how his influence in the pictorial arts has continued to spread. Moreover, as a creative educator, Harold Edgerton has been deeply influential in the careers of many young engineers and scientists.

Tackling Tech

At the time Harold Eugene Edgerton applied for admission to the master of science program at MIT, in April 1926, he was on the "test course" of the General Electric Company in Schenectady, New York. He had graduated from the University of Nebraska the previous June and had spent summer vacations, starting in his high-school years, working for power companies in Nebraska and Iowa. He brought to the GE test course such interest and competence that he was given charge, for a six-month period, of the night shift in the building where the big motors were tested. Arriving at MIT in September 1926, he lost no time in making the acquaintance of Professors Waldo V. Lyon and Ralph R. Lawrence, who headed the research and instruction in alternating-current machinery. The Bush-Hazen integraph was just becoming operational and Edgerton was excited by its possibilities.

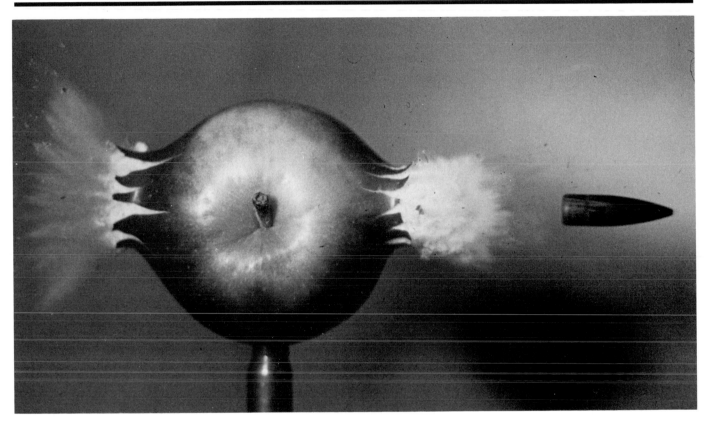

Science and Art: Photographs made with stroboscopic light, a technique developed by Harold E. Edgerton, allowed scientists, engineers, and artists to see and study the sequential structure of a great range of dynamic phenomena.

Apple and bullet.

Acrobat.

Tennis serve by Gussie Moran
(multiflash).

Hummingbirds.

Fan with smoke.

MIT from across the Charles
River.

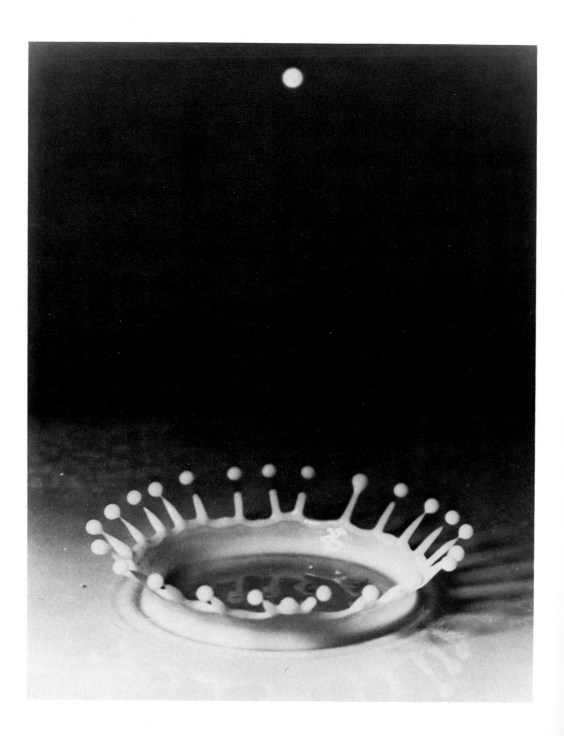

Milk drop splash.

In the department's Vail Library he read avidly of researches on synchronous motors, and his attention was arrested by the 1924 and 1925 articles of H. Cotton in the British *Journal of the Institution of Electrical Engineers*, giving the nonlinear differential equations of transients in synchronous machines—equations that could be solved only by laborious point-by-point calculations. Here, he saw, was a problem for the new integraph. On November 19 he submitted an outline for his master's thesis, "Abrupt Change of Load on a Synchronous Machine," signed by Lyon as his supervisor and by Bush for the Graduate Committee.

Calculations were made on the Bush-Hazen integraph and these were compared to tests made on a 5-kW, 4-pole, 60-Hz synchronous, cylindrical-rotor, sine-wave generator, used as a motor. On the same shaft with the test motor were a second similar 5-kW machine and a DC motor, normally used for driving the sine-wave generators. In the Edgerton tests the DC machine was used as a generator to supply abrupt loads. Both calculations and tests showed that, when load was applied abruptly, the motor would break out of synchronism with only about 70 percent of the pull-out load it would carry if added slowly. If, for a specified load, abruptly applied, the "swing" curve reached a maximum and then decreased, stability was maintained. If the curve continued upward, stability was lost; the motor pulled out of synchronism. (The picture of the integraph that appears in the introduction to Part II was posed while Edgerton was making his thesis runs.)

In May 1927 Lyon and Bush encouraged Edgerton to publish his results;[2] the Edgerton solutions led to wide usage in studies of transient stability.[3] When, as an instructor in 1929, Edgerton supervised the master's thesis of Frederick J. Zak, he urged that the results be published in the *Journal of the Institution of Electrical Engineers,* where he had first found the suggestions for his own thesis.[4] The problem, "Synchronizing an Induction Machine," used integraph solutions rather than the laborious point-by-point methods of the earlier British writers.

In 1929–30 Edgerton suggested to student Kenneth D. Beardsley that, as a master's thesis, he develop a stroboscope for use in the dynamo laboratory. In this thesis Beardsley reviewed the various instruments in use, mentioning especially the neon stroboscope that had been developed by the General Electric Company. This was a bulky, complex device, and its pink flashes were used to observe rapidly moving objects as though they were standing still or moving slowly. The pink neon light gave visible images but was not the right color for photography. Beardsley used a Cooper-Hewitt mercury lamp as his light source, and its blue and ultraviolet illumination enabled him to make some fairly good photographs.

In the fall of 1930 Edgerton used a standard 20-ampere thyratron as his light source. This stroboscope gave a mercury-arc flash of about ten microseconds' duration 60 times per second when the AC terminals were connected to a 60-Hz source. Edgerton arranged to observe stroboscopically the motion of the salient-pole rotor of a 160-horsepower, 60-Hz, 10-pole synchronous motor in the substation of the MIT Dynamo Laboratory. White cardboard bands were fastened to the poles and were marked alternately *N* and *S*. Although normally rotating with a peripheral speed of about 95 miles an hour, the poles seemed to stand still as successive flashes illuminated momentarily each of the five pairs of poles. Paul Fourmarier, a graduate student from France, had studied this salient-pole synchronous motor, showing with integraph calculations and oscillograms its performance as it attained almost synchronous speed by induction motor action and finally pulled into synchronism after the DC field was energized. When the paper based on the

Edgerton and stroboscope,
1931.

Fourmarier thesis was presented at the January 1931 meeting of AIEE, the audience was able to see in motion pictures the rotor action described in the paper.[5] For a pulling-into-step sequence, the poles first seemed to be rotating backward more and more slowly until the DC field was switched on. The pole structure then jumped ahead, oscillating about a final apparent stand-still position. The discussion of the paper, most of which dealt with calculations of motor action, reveals how Vannevar Bush remained alert to new research directions. Regarding the motion pictures, he observed, "It is interesting to note that the light intensities produced are very large indeed. Mr. Edgerton used a stroboscope which had a period of illumination of only ten microseconds occurring once each cycle and yet was able to obtain sufficient light for photography. The instantaneous candle power of the source is hence enormous, as may readily be computed. This appears to have interesting possibilities for further development."

When Alfred M. McClure and Lawrence F. Stauder, two engineers in the GE Lynn Engineers program (see chapter 7), were looking for a thesis to complete their master's degree course, Edgerton suggested that they develop a portable mercury-arc stroboscope. The 500-V DC source was obtained by connecting a booster DC generator in series with the 230-V laboratory DC supply. A major modification in the McClure-Stauder thesis was a rectifier circuit to provide the DC charging source. In their thesis of 1931, "The Mercury-Arc Stroboscope," the authors state, "the first all-alternating-current apparatus was set up by Mr. Edgerton and Mr. McClure for demonstration purposes in connection with lectures given by President Compton and Dr. Killian."

In the spring of 1931 Edgerton completed his own doctoral research, "Benefits of Angular-Controlled Field Switching on the Pulling-into-Step Ability of Salient-Pole Synchronous Motors." The document included the motion-picture studies made with the mercury-arc stroboscope. During that same spring he decided to remain with the EE Department as a teacher, despite an attractive offer from GE, and he supervised the bachelor's thesis of Gordon S. Brown and Kenneth J. Germeshausen on a pulling-into-step problem. While working on his thesis, Germeshausen introduced several improvements in the Edgerton stroboscope.

Kenneth Germeshausen made a short visit to Europe following his graduation. He found upon his return that the Depression had largely eliminated employment opportunities. No funds were available to employ additional research assistants at MIT; but Edgerton persuaded Professor Dugald Jackson to take Germeshausen on as an assistant without stipend, hoping that industrial projects could be found to provide him an acceptable income. Not only did Germeshausen find consulting work in local manufacturing plants, but he also helped the students in their laboratory projects and theses. Edgerton tells of an investigation that Germeshausen made in the old Lever Brothers soap factory (where Technology Square now stands). A claim of patent infringement made necessary an understanding of the process of making soap powder by blowing liquid soap into a chamber containing hot air. Pictures showed the lightweight soap particles in the process of formation.

Edgerton's 1931 notebook records that he first noticed the stroboscopic effect in the fall of 1927 while taking voltage-angle curves on a 20-ampere pool-type thyratron. The light of the tube showed the blades of an electric fan that was keeping the tube cool. The stroboscopic effect was dormant in Edgerton's mind until the following spring, when he developed it into an "open-house" demonstration. Several round cardboard discs were prepared with different numbers of equally divided sections and arranged to be rotated on the shaft of a motor, which could be run at such speeds as would display interesting stroboscopic patterns.

Patents and Products

On November 27, 1931, Bush wrote Edgerton: "The Institute policy with regard to patent matters of this sort is not yet established. It is not felt, however, that it is at all desirable to ask you to postpone any actions which you might like to take, until the policy is adopted, for this will take some time. Accordingly, it is considered satisfactory for you to proceed at the present time, if you wish, and file patent applications in your own name in this matter, and the commercial aspects of this development can be considered to be your own property."

Germeshausen describes how one of their difficulties was resolved: "The question was how to proceed, since good patent attorneys are expensive and we were not rich. Professor Bowles came to our rescue. He introduced us to David Rines, an excellent patent attorney noted for assisting inventors with insufficient funds. He said he would file and prosecute the patents if we would pay the out-of-pocket costs, such as filing fees. Then when and if we began to collect income, we would pay him what we thought the effort had been worth. There was just a handshake on this, no written agreement."

On June 11, 1932, Edgerton wrote Bush: "I have an agreement with General Radio Company for the manufacture and sale of stroboscopes of the mercury-arc type which I have developed. I have filed an application for patent and have proceeded with this agreement as confirmed by your letter of 11/27/31." A copy of the General Radio agreement signed by J. Warren Horton, chief engineer, was included with Edgerton's letter to Bush. (Horton was a professor in the EE Department before and after World War II.)

Because Edgerton and his wife were planning to drive back to Nebraska in the summer of 1932, he suggested to General Radio that if they could have a model of the new stroboscope ready for this trip, he would make demonstrations along the way. After a couple of design alterations a model of suitcase size was produced, and the Edgertons set out with the strobe on the back seat of their Model A Ford. "These summer demonstrations," Edgerton testifies, "built up a demand for the stroboscopes before they were available on the market."

The General Radio Company—whose manufacturing plant was in the MIT neighborhood—announced in December 1932 its Model 548-A Edgerton strobo-scope with separate mercury-vapor lamp. In the *General Radio Experimenter* of August 1935, Herman H. Scott, VI-A, S.B. '30, S.M. '31, described the company's Type 631-A Strobotac (stroboscopic tachometer), with its self-contained cold-cathode neon flash lamp, developed largely by Germeshausen with the cooperation of Edgerton and Harold S. Wilkins of the General Radio Company.

Consulting Partnership

Herbert E. Grier, a Course VI graduate in 1933, developed the high-speed drum camera in his master's thesis research under Edgerton's supervision. It was at this time that Edgerton arranged for Grier to be appointed as a research assistant without salary, and thus began the consulting partnership of Edgerton, Germeshausen, and Grier. For many years this was an informal partnership, with Edgerton on the MIT faculty and Germeshausen and Grier serving as research associates without MIT compensation. As the Strobe Lab grew, the partners expanded their consulting services and were able to support financially a larger lab staff and provide Institute-wide applications of stroboscopy, flash photography, and advanced

Edgerton training a stroboscopic light on a machine tool in order to photograph its ''chatter'' during operation, 1937.

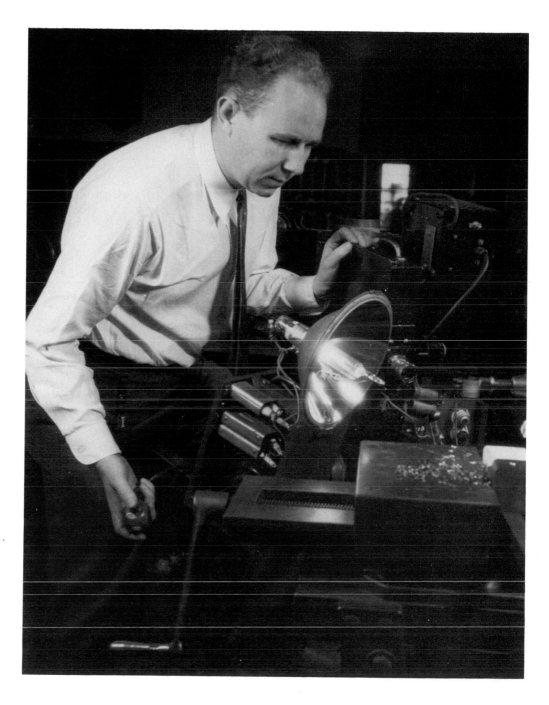

motion-picture techniques. Jackson immediately saw exciting new revelations to come out of these researches and wrote supportive notes to President Compton and professors in several departments. Edgerton was encouraged to apply for MIT funds to support the strobe programs; on one occasion in 1934 he wrote directly to President Compton requesting $1,750 for summer research on larger and more powerful sources of stroboscopic light and on high-speed camera studies, such as: (a) the determination of the final velocity and shape of a drop of liquid falling through still air; (b) the nature of the formation of tracks in the Wilson Cloud Expansion Chamber apparatus; (c) the flight of insects and birds; (d) stress-strain curves of materials during impact; and (e) the flow of air through fans and propellers.

At the end of each calendar year, Edgerton gave Professor Jackson a financial statement showing operating expenses of the Strobe Lab and the income of each of the partners. Each partner had his own particular abilities and interests. Germeshausen designed and constructed many kinds and sizes of flash tubes and experimented with discharges in the rare gases neon, argon, krypton, and xenon. Grier's emphasis was on the engineering application of equipment, much of which he built in the department machine shops.

The master's thesis (June 1938) of Joseph F. Keithley, VI-A '37, is a good example of an Edgerton-supervised research. With the assistance of the three partners, Keithley built three U-shaped tubes, one to contain argon, another a mixture of krypton and xenon, and the third krypton. Applied voltages and other parameters were varied and the resulting light spectra observed. Argon had been used for black and white photography since 1935, but xenon-filled flash lamps with filters to absorb some of their blue light gave a good approximation to daylight and became useful for color photography.

J. Ralph Jackman of the MIT Photographic Service called on Edgerton for help when his attempt to photograph active children gave blurred negatives. Argon-filled flash lamps solved the problem. Eastman Kodak Company announced their "Kodatron" to the public in 1940. This studio flash unit, designed by the three partners at MIT, was manufactured by the Raytheon Company. After World War II a few more powerful color Kodatrons were similarly produced.

The famous *Life* magazine photographer Gjon Mili, VI '27, began his use of the flash lamps in 1937, after Edgerton had shown him what they could do for his art.[6] Germeshausen made the argon flash lamps and Grier designed and built the equipment used by Mili in his series of *Life* pictures (1939–1943). Joseph Keithley, employed at that time at Bell Telephone Laboratories, helped Mili on weekends to set up and operate the strobe lights.

The way Edgerton attracted projects to the Strobe Lab and got them financed is exemplified in the research on glass fracture. In the fall of 1938 Dr. John C. Hostetter of the Hartford-Empire Company, Hartford, Connecticut, looked at some pictures showing panes of glass in the process of forming cracks. These pictures had been made by Frederick E. Barstow, a graduate student in physics whose master's thesis had been supervised by Edgerton. Libbey-Owens-Ford and Corning had supplied glass panes for the experiments. Barstow had had the assistance of all three partners and the use of Professor Moon's Illuminating Engineering Laboratory.

Hostetter took the lead in procuring funds from his company and five others, Pittsburgh Plate Glass, Libbey-Owens-Ford, Hazel-Atlas, Corning, and Owens Illinois, to set up a research account in February 1939 to extend Barstow's work into a study of the stresses in the glass while the cracks were developing. Hazen, the department head at that time, stated that the results of this program would be published as a scientific investigation; in 1941 Edgerton and Barstow published a report describing the apparatus and showing photographs of stress or sound waves in the glass.[7] It was found that the stress waves traveled more than three times as fast as the circular cracking front and so preceded the cracking phenomena.

The War Period

As World War II loomed, Edgerton began his development of night aerial reconnaissance photography. The method in use up to 1939 was to drop an explosive chemical flash bomb from the plane and try to get a picture of the ground during the short flash of intense light. In the summer of 1939 Major George W. Goddard of the Air Force visited the Strobe Lab and asked whether a strobe light could be used for night photography from a plane 2,000 feet above ground. Extrapolating from past experience, Edgerton replied that there were no "in-house" tubes that would do the job, but that he believed such a tube could be built. Soon a contract was signed and the experimental equipment was delivered at Wright Field. Thus began the development of six distinct types of flash tubes used by the Army and Navy during World War II.

Of course, a 2,000-foot height was not adequate for night reconnaissance over the battle areas of France and Germany. The design of larger flash tubes required intensive study and experiment. The war years brought Edgerton and his team into risky and sometimes frustrating situations. For example, the first flash installation in the Liberator B-24 bomber had three tons of capacitors, generators, and associated circuitry, making take-off almost impossible. How Edgerton overcame the obstacles with the cooperation of Nela Park, Wright Field, and persons in the European Theater of Operations is detailed in John Ely Burchard's *MIT in World War II*.[8]

During the war Germeshausen joined the MIT Radiation Laboratory and became chief of modulator switch development; Grier joined Stark Draper's group on gun sight development. Both partners continued to give Edgerton part-time assistance. Through the close working relationship of the partners with Raytheon, they (in particular Grier) became aware of some of the development problems with what was called the "firing set" of the atomic bomb, which Raytheon was manufacturing for Los Alamos. It appeared that much of the partners' experience in the generation and precise control of high-energy short pulses could be applied to firing-set design. As a result, Los Alamos requested the partners to aid them in the development of new and improved firing sets. Since the partnership had no administrative structure, it was decided to contract the work to MIT, under the supervision of Grier.

Incorporation

At the end of the war MIT decided against having such highly classified work on campus. Since the Atomic Energy Commission wanted the work continued, the three partners formed a corporation and became a prime contractor to the AEC. This was done in October 1947, with Grier as president, Germeshausen as vice-president and treasurer, and Edgerton as chairman of the board of directors—he held this position until 1965. At the time of incorporation the Strobe Lab was in Building 20-D, formerly occupied by the Radiation Laboratory, and the corporation headquarters were set up in the Hood Building. In 1949 Edgerton, Germeshausen, and Grier, Inc. (EG&G), moved off the MIT campus and established an office at 160 Brookline Avenue in Boston. Grier took the leadership in setting up an office and laboratory in Las Vegas, close to the Nevada test site of the AEC. The early years of the corporation were dedicated to research and testing for the AEC and other government agencies.

"Doc" Edgerton could be a humorous showman. Here, in 1960, he delivers a lecture to visiting parents at MIT.

The trio who built EG&G: Harold E. Edgerton, Herbert E. Grier, and Kenneth J. Germeshausen.

With the preoccupation of getting EG&G going, one might think that the Strobe Lab at MIT would suffer neglect. Not so. In the fall of 1949 Edgerton announced his program of studies in theory and applications. A list of apparatus on loan, with or without operators, was included. The prospectus included the following statement: "The research work on stroboscopic light at MIT has paid its way from the start, which incidentally was during the Depression of 1930. In fact, those strenuous times may have given an incentive to the project, since men and time were available." Edgerton continued his classroom teaching in electronics and by January 1949 had supervised theses for 50 students.

The words and pictures of this chapter can only show samples of Edgerton's extensive interests and accomplishments. His role as a creative teacher and experimenter who could become involved personally in all kinds of problems, technical and otherwise, contributed to the building of the EE Department in many ways. Every year there are seniors who "just can't get started on the thesis." When such a disheartened senior pays a visit to "Doc," he comes away not only with a thesis problem but with its solution already under way. "Let's get going!" "What's holding you?" These are familiar Edgerton prods. They often work.

After the Electrical Engineering and Computer Science Department (EECS) brought together in the Fairchild Building in 1973 many of its scattered activities, it still felt the need for more space and planned the construction of a new building. When Professor Gerald L. Wilson, department head, announced the ground breaking in 1981, he said, "The EG&G Education Center is the gift of Doc Edgerton, Ken Germeshausen, Herb Grier, and EG&G, Inc. With the ground breaking, EECS is taking the first official step toward establishing the facility and, equally important, toward honoring our great friends who have made it possible."

In 1983 Harold Edgerton was still active at MIT as Institute Professor Emeritus, his long career marked by many honors from national and international organizations. He was the recipient of the National Medal of Science and the Founders Medal of the National Academy of Engineering for 1983. And, as of his 80th birthday in 1983, students and admirers named the corridor outside his laboratory in Building 4 "Strobe Alley."

Chapter 9

Network Analysis and Synthesis: Ernst A. Guillemin

Of the many people who came to MIT to study electrical engineering in the Jackson era and were to remain as teachers, one of those who would make a great contribution to the field through his scholarly approach to it was Ernst Adolph Guillemin. Unlike Edgerton, who was a born experimenter, or Bush, who was a leader of research, or Bowles, who was an extraordinary engineer, Guillemin exercised his creativity with the concepts and theories that supported and illuminated the practice. He brought his vision of network theory into a form graspable by students, and thereby influenced the direction and thinking of a generation of bright young engineers.

His textbooks, and thereafter textbooks by his former students at different schools, would make him, as one later tribute acknowledged, ''the embodiment of the inspired ideal teacher and scientist of integrity.'' A summary of Guillemin's philosophy of education and his attitude toward theory appears in his preface to his textbook *Introductory Circuit Theory*:

I have always held that, where the teaching of basic concepts and procedures is concerned, no distinction should be made between the so-called ''elementary'' and ''advanced'' methods. We refer to things as being ''advanced'' only so long as we understand them insufficiently well ourselves to be able to make them clear in simple terms. Once we understand a subject fully and clearly, it is no longer difficult to make it understandable to a beginner. And, if we do not warn the beginner beforehand, he will not be able to distinguish when we are teaching him the ''elementary'' methods and when the ''advanced.'' Such a distinction will reside only in the teacher's mind; to the student both will be equally novel and equally clear.

Network Analysis and Synthesis

Guillemin came to MIT in the fall of 1922 after receiving a bachelor's degree in electrical engineering at the University of Wisconsin. He embarked on a two-year master's degree course as an assistant, with most of his teaching time assigned to the Dynamo Laboratory, an experience that reportedly impelled him more strongly in a theoretical direction. His master's thesis in 1924, "A Three-Phase Thermionic Tube," was supervised by Bush. The tube was an extension of the Hull magnetron and was controlled by a three-phase rotating magnetic field.

Because of his scholastic achievements, MIT awarded Guillemin a two-year Saltonstall traveling fellowship, which he used at the University of Munich, Germany, where in 1926 he was awarded a doctorate for his thesis entitled "Theorie der Frequensvervielfachung durch Eisenkern Koppelung," completed under the supervision of physics professor Arnold Sommerfeld. Returning to MIT in 1926, he was made an instructor and taught chiefly under the supervision of Gustav Dahl in the field of electric power networks and power system stability. At midyear 1927–28 he was assigned to aid Bush in the supervision of graduate students. In the spring of 1929 Bowles invited Guillemin (promoted to assistant professor in 1928) to teach subject 6.312, which at that time included communication transmission lines, repeaters, balancing networks, and filter theory. This he did, in addition to teaching power system stability for the Lynn Engineers at the General Electric Company plant.

Communication engineering, with its consideration of the frequency domain and its consequent expanded opportunity for mathematical and physical analysis, attracted him. He brought together the scattered fragments of classical network theory and in November 1930 completed the manuscript for *Communication*

Ernst A. Guillemin.

Networks, volume 1, his first book, which made available to undergraduate students the formulation and solution of network equations.[1] He stressed the use of exponential functions in both steady-state and transient solutions. He presented what he called the "direct method" of introducing initial conditions into transient solutions by supplying charges in capacitors and currents in inductors, a procedure that recurred in the "state variable" approach of the 1960s. He was insistent that students understand the physical phenomena of networks before using the Fourier or Laplace transformations.

Guillemin's leadership in the development of filter theory and related topics gave rise to an evolving set of class notes that became the second volume of his *Communication Networks*, published in 1935.[2] This textbook, which begins with a history of electrical conduction by means of wires, points out that "Heaviside's work forms the backbone for all subsequent investigations of an engineering nature." Guillemin's clear and student-oriented writing style is in fact reminiscent of Heaviside's work.

A thorough discussion of "long lines" leads into the concept of a two-terminal network with its driving-point impedance. The realization of a physical network corresponding to a driving-point reactance function is introduced, first through Foster's Reactance Theorem[3] and then through the Cauer Extension.[4] Two-terminal-pair networks lead into the development of filter theory. Guillemin also includes the work of Otto J. Zobel of the Bell System, especially with respect to filters.

The importance of the wave filter in the design of communication circuits suggests the need for a perspective on the work of George A. Campbell. The research and development activities of the Bell System were early established in Boston under the direction of Dr. Hammond V. Hayes. In his quest for exceptional research talent in 1897 Dr. Hayes employed George

A. Campbell, who had received a bachelor's degree from MIT in 1891 and during the next six years had studied at Harvard, Göttingen, Vienna, and Paris. Campbell was convinced that the best way to increase the quality of long-distance cable telephone transmission was to load the lines with inductances, and he worked out the theory of the loaded cable, completing a confirmatory experiment in the Boston laboratory in September 1899.[5] Loaded cable lines connecting Jamaica Plain and West Newton by way of Boston showed the expected gain in transmission efficiency and were put into public service on May 18, 1900. Campbell's first laboratory loaded cable had an upper cutoff frequency of 11,000 Hz. In experimenting with a possible reduction in cutoff frequency, he suddenly realized that the telephone line was essentially a low-pass filter, and by 1915 he had worked out a comprehensive theory of the electric wave filter, a patent for its invention being issued to Campbell in 1917.

Professor Bowles has postulated that the pioneering work of Campbell, Hayes, and John Stone predated the Pupin invention of the loading coil. Legal interference proceedings between the Campbell and Pupin patent applications began soon after the issuance of the Pupin patents on June 19, 1900, and were in the courts until April 6, 1904, when the Pupin patents were sustained. The American Telephone and Telegraph Company purchased the rights to the Pupin patents in order to be legally covered during the litigation. The facts are presented and discussed in a recent history of Bell System science and engineering, showing that the Bell System applications were based on Campbell's work.[6]

Up to 1930, communication engineers concerned with filter design had spoken of the realization of a frequency characteristic. For example, the 1926 doctoral thesis of Wilhelm Cauer at the Technische

Hochschule of Berlin under the supervision of Professors G. Hamel and K. W. Wagner was entitled "Die Verwirklichung von Wechselstromwiderständen vorgeschriebener Frequenzabhängigkeit." The MIT doctoral thesis of Yuk Wing Lee in 1930, supervised by Professor Norbert Wiener, was entitled "Synthesis of Electrical Networks by Means of the Fourier Transforms of Laguerre's Functions." The term *synthesis* resulted from a conversation in which Lee proposed to Bush the idea that since the determination of a network characteristic for a given network is called analysis, the reverse process might be called synthesis. Bush agreed that this was logical and encouraged Lee to use the new term, even though it had not appeared previously. Guillemin believed Lee to be the first to use the word in this sense. Another important result of Lee's thesis was the demonstration that the real and imaginary parts $P(\omega)$ and $Q(\omega)$ of an admittance function $Y(\omega) = P(\omega) + jQ(\omega)$ are not mutually independent but are Hilbert transforms of each other. In the following year, 1931, the doctoral thesis of Otto Brune, under the supervision of Guillemin, was entitled "Synthesis of a Finite Two-Terminal Network Whose Driving-Point Impedance Is a Prescribed Function of Frequency." In the dedication of his book *Synthesis of Passive Networks* (1957) Guillemin ascribes to Dr. Brune the laying of "the mathematical foundation for modern realization theory."[7] In 1932 Charles E. M. Gewertz completed his doctoral thesis, "Synthesis of a Finite Four-Terminal Network Whose Driving-Point and Transfer Functions Are Prescribed."

Gathering together the work of Brune, Cauer, Foster, Bartlett, and others, Guillemin embarked on his long career in the development of modern network synthesis, a field in which he inspired a host of graduate students, who have become leaders throughout the world. One of his students, Professor David F. Tuttle, Jr., of Stanford University, published his 1,163-page synthesis textbook in 1958.[8] Volume 1 of a projected two-volume sequence, it treated one-terminal-pair networks. The second volume was to treat two-terminal-pair networks, as included in Tuttle's Stanford course. Volume 2 was not completed, but during a sabbatical year, 1961–62, in Spain, Tuttle published in Spanish a book that included two-terminal-pair networks. This was translated into English and published under the title *Electric Networks* in 1965.[9]

Professor Mac E. Van Valkenburg of the University of Illinois published his *Introduction to Modern Network Synthesis* in 1960.[10] According to his preface, "Authors of books on network synthesis owe a special indebtedness to Professor Ernst A. Guillemin of Massachusetts Institute of Technology. . . . One of Professor Guillemin's contributions to this field has been to devise explanations for the subjects of synthesis in a form especially suited for student understanding and textbook presentation. He has also endowed a generation of teachers and engineers with an enthusiasm for his subject." Another student and close friend of Guillemin was Dr. Louis Weinberg, who published his book on networks in 1962.[11] Although observing that the contributors to the growth of his ideas are "too numerous to list," he asserts, "Two important influences should be singled out, however: Professor E. A. Guillemin and Professor R. M. Foster. . . . [The author's] steadfast enthusiasm for network theory is due in no small measure to their friendship and to the examples set by their careers. Professor Guillemin was the author's inspiring teacher at MIT and Professor Foster's correspondence and conversations with the author have been continually stimulating and suggestive."

Guillemin characterized his brand of network studies as limited to linear, passive, lumped, finite, and bilateral qualities, although he supervised theses dealing with active, nonlinear, and nonbilateral networks. After

Guillemin with Dr. M. V. Cerrillo
in the Research Laboratory of
Electronics.

his retirement in 1963, two of his former students, Professors Harry B. Lee and Richard D. Thornton, established new subjects to include active and nonlinear aspects of network theory.

Guillemin also brought advanced concepts into network analysis and popularized the sophomore introductory subject through his lectures. He was in charge of this subject in the spring term of 1949, teaching one section along with other instructors (Davenport, Van Rennes, Scott, Weinberg, Löf, Lucal, Jensen, and Whipple). Students were complaining that they were finding the textbook *Electric Circuits*, compiled by the EE staff, "hard going"; they requested further exposition. The following spring Guillemin instituted his two lectures per week for the whole group of sophomores, supplemented by two recitation sections of about 20 students per section. Because the sophomore curriculum did not require differential equations until the second term, an orientation subject, 6.05, requiring only algebraic equations for the solution of resistance networks, had been developed over the years by Professors Frazier and Fano.

By 1952 Guillemin had organized this material on resistance networks in such a way as to include a thorough treatment of network geometry (trees, cut sets, tie sets, choice of variables as loop currents or node voltages), the writing of equilibrium equations, the use of matrices and determinants in solving network equations, and some of the network theorems, such as the Thevenin and reciprocity theorems. At this point Guillemin's next textbook, *Introductory Circuit Theory*, was published,[12] and a full two-term course in network theory with accompanying instructional laboratory was inaugurated. The second term was devoted to steady-state and transient solutions of resistance-inductance-capacitance (RLC) networks.

Although authors were not named in the "Blue Series" of textbooks published by the department in the 1940s, Guillemin was the principal author of the vol-

Guillemin lecturing.

ume called *Electric Circuits*. As a sequel to that series, he published in 1949 *The Mathematics of Circuit Theory* under his own authorship.[13] A postwar publication, following the developments of the Radiation Laboratory, this book supplied the mathematics for transmission lines, wave guides, and antennas, as well as for general multibranch networks. It was used largely to strengthen the mathematical background of first-year graduate students. His last book, *Theory of Linear Physical Systems, from the Viewpoint of Classical Dynamics, Including Fourier Methods,* was published in 1963.[14] In the later years this material was taught, chiefly by Professor Thomas G. Stockham, as graduate subject 6.55, Linear System Analysis. A memorial book, *Aspects of Network and System Theory*, was written and published by his former students in 1970.[15]

During his teaching years Guillemin served as consultant for several manufacturing companies; in the war years he spent about half his time as general consultant and lecturer in the Radiation Laboratory. Perhaps the most significant of his contributions to radar circuitry was his design of a pulse-generating network. Among the recognitions of his achievements were the IRE Medal of Honor award in 1961 and the AIEE Education Medal in 1962.

Although Guillemin set up a small company in Cambridge, which later moved under EG&G management (see chapter 8), and though he received many recognitions of his influence in network theory, it was teaching that remained closest to his heart. It was said that from his teaching ''the productivity of his students has in turn multiplied his own contributions many times.'' Later in life he was to say that the teaching of network theory to young people and the experience of their companionship in that endeavor were ''utterly fascinating.''

Chapter 10

The High-Voltage Research Laboratory: John G. Trump

In the first decades of this century, as electrical transmission voltages steadily increased, there was growing work in both industry and universities on various high-voltage phenomena. Notable in the industrial sphere, for instance, was Steinmetz's artificial lightning laboratory at the General Electric Company. The most famous early high-voltage laboratory in the university setting was that developed by Harris J. Ryan at Stanford, where Ryan had headed the Electrical Engineering Department from 1905 to 1931. In Ryan's laboratory high DC voltages were produced by switching charged capacitors into a series connection, thereby causing their individual voltages to add together (Marx Generator). High AC voltages were produced by transformer action. The Ryan type of high-voltage laboratory was expensive, but many EE departments across the country thought they must have one. At MIT the matter was constantly under discussion until the move to Cambridge in 1916. The problem of high-voltage testing was solved when the neighboring Simplex Wire and Cable Company offered MIT the use of its high-voltage equipment.

The really significant high-voltage work at MIT, however, dates from 1931 and is associated with two people who arrived that year, John George Trump, who came to the EE Department to pursue doctoral studies, and Robert J. Van de Graaff, who came to join the Physics Department as a research associate. Trump had received his undergraduate EE degree in June 1929 from the Polytechnic Institute of Brooklyn, where he held an instructorship in electrical engineering for two years while he studied at Columbia University for a master's degree in physics, awarded in 1931. During that same period a new type of high-voltage generator had been invented by Van de Graaff, an Oxford Ph.D. (1918) who worked as National Research Fellow at Princeton University from

1929 to 1931. President Compton, who had been at Princeton until 1930, arranged to have Van de Graaff join the MIT Physics Department in 1931 to develop the new generator. A small model of the machine was soon in operation at MIT, and plans were under way for a giant model to be constructed at Round Hill (see chapter 6). Van de Graaff and his physics colleagues, L. C. and C. M. Van Atta, set up a program to study atomic and nuclear structure.

While the Van de Graaff generator was under development at MIT, Ernest O. Lawrence and M. Stanley Livingston were building their cyclotron at Berkeley, reaching one million volts in February 1932. Physicists were curious to know whether the atomic nucleus contained protons and electrons, and if so, how they were arranged. Ernest Rutherford believed a new particle having no electric charge was involved; this particle, the neutron, was discovered by James Chadwick at the Cavendish Laboratory in Cambridge, England. Nuclear structure unfolded rapidly with the announcement of the hydrogen atom of mass two, deuterium, by Harold Urey of Columbia University; the fission of lithium by John D. Cockcroft and Ernest T. S. Walton at the Cavendish Laboratory; and the discovery of the positron by Carl D. Anderson at the California Institute of Technology. All of these advances made the year 1932 a "miraculous year" for physics.[1] M. Stanley Livingston was to join the MIT physics faculty in 1938, continuing his development of particle accelerators.

Trump's career was set in motion when Vannevar Bush suggested that he become acquainted with Van de Graaff. A close and lasting friendship ensued and they worked together for many years on high-voltage

John G. Trump at the beginning of his career in high-voltage research at MIT.

phenomena. Vacuum and high-pressure gas insulation were later developed by Trump in connection with the design of compact machines for use in numerous applications of high-energy particles, including the treatment of malignant tumors.

The Doctoral Thesis

Van de Graaff, in studying the physics of his air-insulated electrostatic generator, did some calculations on the force between two parallel plates with a high-voltage gradient between them. He had constructed in early 1932 a vacuum cylinder about 18 inches in diameter and was able to maintain a vacuum of about 10 millimeters. He proposed to Trump the design of a vacuum-insulated electrostatic energy converter that would hopefully be of a size comparable with that of existing electromagnetic converters.

Trump found Van de Graaff's ideas exciting and on October 6, 1932, submitted his proposal for the design and construction of a 60-Hz synchronous motor to fit into Van de Graaff's vacuum cylinder. First would come the design calculations, and then an actual motor would be built in the department's machine shop. Van de Graaff would supervise the thesis; entitled "Vacuum Electrostatic Engineering," it was submitted on December 18, 1933. Several types of vacuum machine were analyzed, including a constant-voltage DC generator. The AC synchronous motor, believed to be the first of its kind, was constructed and mounted on the end plate of Van de Graaff's vacuum tank. First operated on August 9, 1933, as a synchronous motor at a peak AC voltage of 73 kV and a maximum power of 55 W, it was also run as a generator, delivering useful power to the line.

Part of the thesis was a further investigation of Van de Graaff's original ideas for a DC vacuum transmission line. Computations showed that such a line could transmit a million kilowatts a thousand miles at a million volts with an energy loss of 2.5 percent. A vacuum line would comprise conductors buried underground; would have no corona, no lightning faults, and no instabilities; and would be less expensive than a corresponding AC line.

A patent on the electrostatic generator was filed on December 16, 1931, by Van de Graaff and issued on February 12, 1935. His electrical transmission system patent was filed on July 5, 1932, and issued on December 17, 1935. These novel ideas stirred up a lot of interest among power engineers, and discussions about "a thousand-mile vacuum" led to the idea that future developments of these ideas should probably be covered by MIT patents. (Later, in fact, after MIT established a patent policy, inventions such as magnetic storage for digital computers became important sources of income for MIT.)

After the completion of his thesis, Trump, as a research associate in the EE Department, continued his vacuum insulation studies as part of the department's general research program. In the spring of 1935 an air-insulated, 750-kV generator was completed to work into a metal vacuum system. The research was basic to the understanding of breakdown phenomena in vacuum insulation.

Cancer Treatments

In the early summer of 1935 Trump's vacuum system was also used to produce high-voltage X rays. A water-cooled platinum target was maintained at ground potential, and an electron current of about one-half milliampere at slightly over half a million volts was recorded. The steady and easily controlled voltage and current were observed by physicists and medical men, who were convinced that here was an effective source for very high voltage X rays. As a result of this demonstration, Dr. Richard Dresser of Har-

Vacuum tank rolled away to
show Trump's 60-Hz (3,600-
rpm) vacuum-insulated synchronous motor, with AC
power supplied by transformer
at left.

The 700-kV vacuum system, powered by a 750-kV multi-belt electrostatic generator (*right*), completed in 1935.

vard Medical School's Collis P. Huntington Memorial Hospital approved the construction of a million-volt, three-milliampere, air-insulated X-ray generator to be used there for the treatment of cancer and other diseases. Completed in 1937, the machine produced more homogeneous and penetrating X rays than had been available up to that time. During the first three years of operation, over a thousand patients were treated with definitely favorable results, especially in cases of deep-seated malignancies.

Encouraged by the success of this first hospital unit, MIT built a second unit, pressure insulated and therefore smaller, for the Massachusetts General Hospital. Young physicians from the hospital worked with the engineers and physicists to become acquainted with the clinical use of the 1.25-MV machine before it was delivered to the newly constructed George Robert White Memorial Building in the spring of 1940. It gave continuous service for 16 years.

Trump's high-voltage activities were interrupted by his participation in the microwave radar developments of the MIT Radiation Laboratory (see chapter 13). He was director of the British Branch Radiation Laboratory from February 1944 to May 1945. When he returned from his war duties, his High-Voltage Research Laboratory (HVRL) became an acknowledged graduate research facility of the Electrical Engineering Department. HVRL has continued its association with the medical profession and with many of the high-technology companies established around Cambridge and Boston in the postwar years.

Trump's program of high-voltage applications has become worldwide.[2] On the MIT campus he has cooperated with the Department of Food Technology in the sterilization of foods, serums, and drugs. Working with hospitals and commercial drug laboratories, he has aided physicians in the sterilization of human tissues and bones for transplant operations, and in the inactivation of viruses and enzymes. He has maintained a close relationship with physicists, working with them in such fields as nuclear structure, solid-state phenomena, solid and gaseous insulation, fission and fusion phenomena, and inertial guidance. The impact of his machines and researches has even reached the People's Republic of China.

In 1980 Trump was succeeded as director of HVRL by Professor James R. Melcher.

Chapter 11

The Laboratory for Insulation Research: Arthur R. von Hippel

When Arthur R. von Hippel came to MIT in the fall of 1936 at the invitation of President Compton, he brought with him a rich background for the contributions he would make in what he was to call molecular science and molecular engineering. As an undergraduate and graduate student at the University of Göttingen, where his father was a professor of criminal law, he witnessed the new discoveries of Wolfgang Pauli, Werner Heisenberg, and Erwin Schrödinger as these young physicists brought quantum-mechanics corrections to Niels Bohr's famous model of the atom. But von Hippel, being of a practical turn of mind, joined Göttingen's Institute for Applied Electricity, where in 1924 he was awarded the Ph.D. degree, summa cum laude, for a thesis entitled "The Theory and Investigation of the Thermomicrophone." He developed measuring instrumentation for the design and testing of this device.

As assistant to Professor Max Wien in Jena (1924–1927) he was asked to investigate the magnetic properties of sputtered metallic films, which assignment led him to look deeply into the sputtering process itself and to publish four papers on the subject in *Annalen der Physik*. After a year as Rockefeller Fellow in Physics at the University of California, he lectured at Jena (1928–1929) and at Göttingen, where he investigated the mechanism of electrical breakdown in gases, liquids, and solids in Professor James Franck's Physikalische Institut (1929–1933). An initial paper with Professor Franck, entitled "Electrical Breakdown and Townsend's Theory," was followed by his own series of three papers on breakdown in solid insulators in *Zeitschrift für Physik*. During this productive period he saw that the breakdown mechanism could best be revealed through the study of material having ordered structure, such as single crystals. Concentrating on the alkali-halide crystals, which could easily be grown in his laboratory, he discovered "avalanche breakdown" in solids, observing the destruction paths proceeding backward from anode to cathode in specific crystallographic directions. He also discovered that sparks building up in gases from positive electrodes had a quite different pattern from those at negative electrodes.

Political events in Germany were to push his investigations elsewhere. Unwilling to swear allegiance to Hitler, he and his wife, Dagmar, Franck's daughter, left Germany for a stay of 18 months (1933–34) in Istanbul, Turkey, where he became professor of electrophysics. A decisive period in his career followed (January 1935 to September 1936), spent as guest professor at the Niels Bohr Institute of the University of Copenhagen. Professor Bohr became one of von Hippel's heroes, as Franck had been at Göttingen.

Then President Compton personally invited von Hippel to come to MIT; a physicist with such a background, Compton believed, could likely make important contributions to the Department of Electrical Engineering.

Initiation into the EE Department

Von Hippel's arrival at MIT in September 1936 was hardly auspicious. The department was not at all aware of his earlier work or of his plans, and no preparations had been made to accommodate his research. He was assigned a small windowless room in the basement of Building 10 while instructor Gordon Brown, in charge of the research laboratories, rearranged the space in Room 10-395, inside which an enlarged Room 10-371 (the former cinema integraph room) was made available in November. The instruments and other equipment that von Hippel had brought from Europe were set up in this laboratory. Other equipment was soon added: a vacuum system,

a 20,000-volt DC source, and special devices that von Hippel designed and built locally. He was immediately helpful in the final phases of the doctoral theses of H. Y. Fan "(Transition from Glow Discharge to Arc") and Dean A. Lyon "(Electric Strength of Extremely Thin Insulating Films"), presented for June 1937 degrees. Two significant researches during von Hippel's first year at MIT were on the breakdown of glass under high-voltage stress and on the emission of electrons from metals into solid insulators.

By May 1937 the MIT community was convinced that von Hippel's revolutionary ideas could indeed make a significant contribution to electrical engineering; Professor Moreland decided to have him take over, starting in the fall of 1937, undergraduate subject 6.26, Insulation, thereby relieving Professor Moon to devote most of his attention to the new Course VI-B, Illuminating Engineering. Here in 6.26 von Hippel introduced his new point of view, showing that the dielectric properties of materials were determined by the behavior of electrons, ions, atoms, and molecules. A new graduate subject, 6.64, Electric Insulation, emphasizing major advances in insulation, was inaugurated by von Hippel in the spring of 1938. So attractive were his research and teaching that already in 1937–38 he supervised three master's theses.

Research Reports for November 1939 reported, "Recognizing the situation and the promising start that had been made, the Institute some time ago inaugurated a *laboratory for insulation research* in the Department of Electrical Engineering. In development since November, 1937, this laboratory bridges between physics and electrical engineering by attacking the problems of insulation from the atomistic standpoint." As von Hippel's research expanded, it embraced staff and students in physics and chemistry as well as in electrical engineering. Not only did this Laboratory for Insulation Research (LIR) thrive and expand

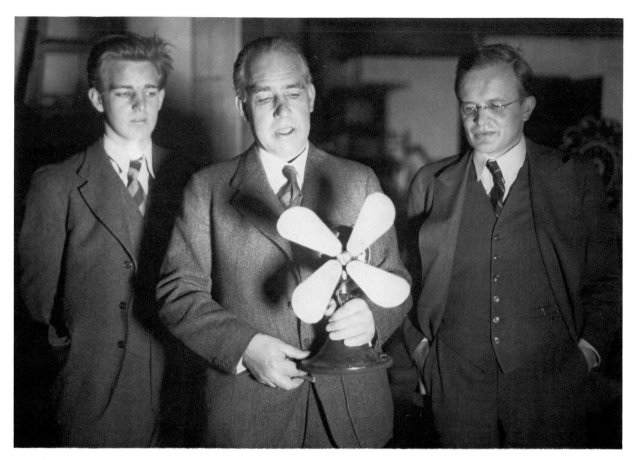

Arthur R. von Hippel (*right*) with
Niels Bohr and Bohr's son.

for more than three decades, but von Hippel's book *Molecular Science and Molecular Engineering* brought profound and significant changes to the materials technology of those years. Young men of unusual ability came to work with von Hippel. For example, Julius P. Molnar, who later became executive vice-president of Bell Telephone Laboratories and subsequently president of Sandia Laboratories, did his physics thesis, "The Absorption Spectra of Trapped Electrons in Alkali-Halide Crystals," with the materials, apparatus, and guidance in LIR. Others investigated the initial stages of electric breakdown in gases through a study of Lichtenberg figures; examined the propagation of electrons in single crystals as a wave phenomenon; and studied with von Hippel the importance of order versus disorder by comparing crystals with glasses, extending this work to study the effect of temperature, the action of frequency, and the transition from insulator to semiconductor and metal.

Wartime Activities

Von Hippel had an opportunity to demonstrate his anti-Hitler position when, in the early days of World War II, the president of the International Telephone and Telegraph Company sought aid in producing selenium rectifiers. The company's facilities in Germany had been lost, and its attempt to reproduce them in New York had been catastrophic—about 90 percent of the production had to be thrown away. By studying the phase transitions of selenium from the red insulating to the gray metallic form, and the rectifying action of the boundary, von Hippel and his coworkers designed a method to produce good rectifiers rapidly by electroplating. The production time was shortened from a thermal anneal of two days to an electrolytic formation in twenty minutes. They also studied the making of selenium photocells and their spectral response, and retained for MIT the patent right to convert solar into electric energy by this method.

Positive Lichtenberg figure and its growth pattern.

In September 1941 von Hippel became a citizen of the United States—just three months before the Pearl Harbor attack pushed America into World War II. During the war LIR assumed responsibility for the development and measurement of radar dielectrics, as well as for the initiation of their proper manufacture and application. The laboratory had to create measurement techniques and equipment to determine the dielectric constant and loss in all kinds of materials as a function of temperature and of frequency, from DC through the microwave region. The decimeter and centimeter ranges were practically unexplored territory. New types of standing-wave equipment, such as the MIT Coax instrument, had to be developed and their manufacture arranged, as well as their distribution to American and Allied government and industrial laboratories. After the war the Coax instrument was manufactured commercially; it remains today a standard tool for the measurement of dielectric properties at centimeter wavelengths. The classified reports called ''Tables of Dielectric Materials'' are still a major reference in the field. They were prepared under the leadership of William B. Westphal, who joined LIR after receiving his bachelor's degree in 1942. As leader of the Dielectric Measurements Group, he established an international reputation in dielectric measurement and continued his work in this area until 1980.

Polymers like polystyrene and polyethylene were quite new at that time, having been used as filler materials for rubber tires and for some household items; they had to be upgraded to low-loss radar dielectrics and their useful temperature range extended by additive agents, since the Navy persisted in running radar cables through the boiler rooms of battleships. Many other materials—plastics and rubber, ceramics and glasses, single crystals and polycrystalline materials—were needed and had to be made in the laboratory or by industry.

In developing ''high dielectric-constant ceramics,'' von Hippel discovered the ferroelectricity of barium titanate ($BaTiO_3$) and used this material to produce high-voltage capacitors and ceramic delay lines. Studies on these dielectrics led to the application of dielectric heating for rapid wood curing. The making of selenium photocells pulled LIR into war research on infrared photocells of the thallous sulfide type. Classified LIR wartime reports and subsequent publications summarize these developments.

In the course of the war it became clear that a close liaison was required between LIR and the government agencies responsible for the procurement and proper application of dielectric materials. Thus the laboratory joined with the Army, the Navy, and the War Production Board to form the War Committee on Dielectrics. The committee met in Washington once a week during the later war years, and after mutual trust had been established, it successfully handled a number of emergency situations. This relationship led to an Army–Navy–Air Force three-service contract with LIR to start peacetime materials research.

The New Peacetime Program
During the war von Hippel was a member of the Radiation Laboratory (see chapter 13); but he kept his LIR organization intact, so that when the Radiation Laboratory disbanded in late 1945 LIR was in a position to continue its research. It was at this juncture that von Hippel perceived clearly the challenge of transforming the whole field of materials research, using as a basic approach his book *The Molecular Designing of Materials and Devices*. The idea was to create new materials to order by synthesizing them out of the basic building blocks at the molecular level. This is molecular engineering. The understanding of the building blocks and their interrelationships is molecular science. In his postwar program von Hippel was more and more insistent on understanding the behavior of materials as they occurred in nature, on

understanding how the observed phenomena happen and how they can be influenced by molecular means.

Von Hippel wrote, "The molecular engineer starts his synthesis of materials from *atoms* as the fundamental building stones." Bohr had originally modeled the atom as a positively charged nucleus with electrons rotating around it as the planets move around the sun. Bohr was quick to understand the quantum mechanics of Heisenberg, Schrödinger, and Pauli, and his "orbits" became atomic "orbitals." Orbitals were conceived as shells of probability distributions of electron positions, solutions to the Schrödinger wave equation, concentric about the nucleus. The outer electrons determine the valence or combining properties of the atom when the outer shells are not filled. Many physical properties of various elements have been learned from their atomic structure.

The next, more complicated, building block beyond the atom is the molecule. When two hydrogen atoms unite to form a hydrogen molecule, new forces come into play. Each atomic electron is no longer confined to its atomic orbital, but the two electrons of the molecule share an orbital relationship with the two nuclei. As von Hippel says, "Intelligent construction of the next set of building blocks, the molecules, requires insight into the rules of interaction between atoms: why certain molecules prove stable and others unstable, why certain combinations form voluntarily and others only under coercion, why certain partnerships between atoms lead to small, saturated molecules of defined structure and others grow into macrostructures by continuous addition." These relationships were thoroughly understood and taught to von Hippel's students and colleagues.

The study of atomic and molecular structure led into the investigation of other kinds of structure. For example, in the early 1950s manufacturers of transistors

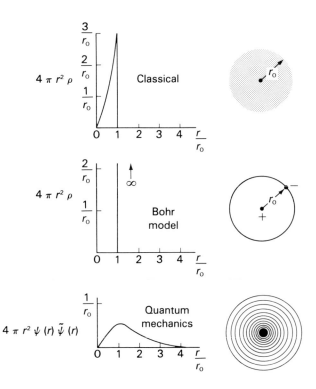

Three models for electron charge distribution in the hydrogen atom: the classical model, an electron cloud of uniform density; the Bohr model; the quantum-mechanics model (normalized to unit electron charge).

were having problems in producing pure crystals of germanium and silicon. It was at this time (1952) that von Hippel brought Alexander Smakula (Göttingen Ph.D. 1927) into LIR. Smakula, a specialist in growing single crystals, contributed to solving the crystal-growing problem. (Incidentally, Smakula had been the inventor of the antireflection coating for optical lenses so widely used today in camera lenses.)

The molecular approach brought great insight into the electromagnetic behavior of all kinds of materials. Electrical conduction in single (nearly pure) crystals, in semiconducting (doped) crystals, and in metals was studied in depth. Electrical conduction and breakdown in gases, including natural lightning phenomena, were investigated in the laboratory and in the field. Lichtenberg figures gave testimony to the sequence of events in such phenomena. These studies led to an understanding, not only of the disruptive and destructive qualities, but also of the useful qualities of gaseous breakdown—as in neon, mercury, and sodium lamps, as well as in the thyratron and ignitron tubes used as control devices and switches.

Not only did von Hippel build an interdisciplinary staff of chemists, physicists, metallurgists, and electrical engineers; he also built an international one by inviting scientists from abroad to spend extensive periods in LIR. Studies of ferroelectric materials were strengthened and expanded with the coming of Berndt Matthias, a postdoctoral fellow who had done research on ferroelectricity at the Swiss Federal Institute in Zurich. The electrical, electromechanical, optical, and domain properties of barium titanate single crystals provided topics for a number of doctoral theses and led to important publications, among them a paper by von Hippel in the *Reviews of Modern Physics* in 1950. Walter Merz, also from the Swiss Federal Institute, contributed to the ferroelectric research. Matthias and Merz, after leaving the laboratory, went on to establish worldwide reputations as leaders in the field of ferroelectricity and superconductivity.

LIR group discussing the operation of a magnetometer. *Left to right:* von Hippel, Stanley Kingsbury, David J. Epstein, Perry Miles, Robert Hunt, Archie MacMillan, Alexander Smakula.

David J. Epstein, who was to develop the undergraduate core curriculum subject 6.08, Molecular Engineering, during the Gordon Brown era (chapter 19), entered MIT as a graduate student in 1947 with a research assistantship in LIR. He received the master's degree in 1949 and the Sc.D. in 1956, with a thesis entitled "Magnetic Lag in Ferrites," both the S.M. and Sc.D. theses being supervised by von Hippel. Although Epstein became von Hippel's leader in magnetic materials, he had a comprehensive grasp of the whole field of molecular engineering, as evidenced in his classroom and laboratory teaching of materials engineering to undergraduates. The educational laboratory dealt with the role of materials in modern devices: junctions in semiconductors, transistors, dielectric and magnetic amplifiers. Forrester's coincident-current magnetic-core memory for digital computers used ferrites as core materials (see chapter 17).

Richard B. Adler, a member of the staff in the Research Laboratory of Electronics (RLE) since 1946 and of the department's faculty since 1949, was strongly influenced in his subsequent transistor research and teaching by von Hippel's molecular point of view. Adler became the first leader of the new Lincoln Laboratory's solid-state and transistor group, which he headed as long as it remained housed on campus in RLE (1951–1953). Leading up to the extensive use of transistors in digital computers, Adler supervised, for example, several master's theses on transistor magnetic-core drivers as these were being built in the Digital Computer Laboratory, and he pioneered (with S. J. Mason, C. R. Hurtig, and W. E. Morrow) in the development and use of a new nonlinear circuit model for point-contact transistors.

Adler's leadership in transistor theory and applications gave him an excellent background for helping Gordon Brown in his program to broaden the field of energy conversion. To revitalize the electric power field, teachers had to make it attractive to bright young people. Given the success of science during World War II, and the strong component of science in the new semiconductor materials and devices field, the bright young people were certainly drawn in those directions. Thus, it was argued, if electric power could be upgraded into a more fundamental and broader discipline than merely the study of transformers, transmission lines, and rotating machines, and if this more general discipline could involve the new semiconductor materials and the rising molecular engineering point of view, then surely it would be irresistible to this new generation.

In addition to Brown's encouragement, Professor Pierre Aigrain of the University of Paris pointed out that the efficiency of thermoelectric converters had been drastically underrated by Lord Kelvin. Moreover, the semiconductor technology brought about by advances in the transistor field offered real promise of making significant improvements in the physical properties of semiconductors by using the strategies of molecular engineering to make them more effective thermoelectric converters.

In the late 1950s a small group studying thermoelectric processes and materials was set up in RLE, cooperating with a French group at the Laboratoire Centrale des Industries Electriques in Paris. Aigrain participated in the RLE group and led the one in Paris. Somewhat later, the Energy Conversion Laboratory was set up in Building 10 by Professors David C. White and R. B. Adler to develop thermoelectric conversion as a practical means of electric power generation and refrigeration. Another member of the team was Paul E. Gray, whose doctoral thesis on the dynamics of thermoelectric machines (1960) was supervised by White. Aigrain continued to conduct related research in France and visited the MIT project regularly, giving inspiration and guidance. Major American participants from outside MIT were the Navy and corporations like Westinghouse and General Electric. Very

Richard B. Adler (*left*), leader in solid-state studies, with student.

Herbert H. Woodson, Thomas A. Stockham, Jr., unidentified student, and David C. White in a Building 10 laboratory.

intensive work in this field was also carried on in the Soviet Union during these years.

For eight years, beginning in 1960, Adler served as technical director of the Semiconductor Electronics Education Committee (SEEC). An international university- and industry-sponsored educational development effort, headed by Professor C. L. Searle with other MIT participation by Professors P. E. Gray, A. C. Smith, and R. D. Thornton, the SEEC developed seven volumes of textbooks, a number of pedagogical laboratory experiments, and four educational films, in an integrated package. The SEEC was designed to bring the solid-state electronics revolution, which had started with the invention of the transistor in 1948 at Bell Telephone Laboratories, firmly into the electrical curricula of the universities, worldwide—a task in which it was largely successful.

The period 1958–1966 saw the parallel development of the Center for Materials Science and Engineering from the seeds sown by LIR, RLE, the Energy Conversion Laboratory, and Lincoln Laboratory (see chapter 17); it saw, too, the involvement of more young people in issues concerning the generation of electric energy than had been involved in the conventional field of electric power for many previous years. Ultimately, however, it became clear that "energy conversion" was not a particularly helpful unifying concept, because the underlying principles of electromechanics, magnetohydrodynamics, and conversion from heat to any other form of energy are quite different in character. Revitalization of the electric power field for the long term had to await the emergence of solid-state power electronics, cheap computation, and alternative fuels, so that radical changes in the whole system— as a system—could begin to take place in an economic environment of expensive imported petroleum.

After about five years of von Hippel's postwar program, he believed the time had come to share the new molecular approach with workers outside MIT. In his book *Dielectrics and Waves* he set forth the "macroscopic" phenomena involved in dielectric measurements and then presented the new "molecular" point of view.[1] His "conviction that scientists, engineers, manufacturers, and users of dielectrics should learn to speak each other's language and appreciate mutual problems, failures, and advances was the driving impulse which led to the Summer Session Course of LIR at MIT in September 1952." The lectures, given by specialists from government, industry, and LIR, were reviewed and edited into a book, *Dielectric Materials and Applications*,[2] published in 1954. Problems were "brought into the limelight which could stimulate much fruitful new activity." The book closed with about 150 pages of a first public issue of the "Tables of Dielectric Materials." Other MIT investigators— John G. Trump, Dudley Buck, and William N. Papian—contributed chapters (see chapters 10 and 17 of this volume).

Again in 1956 von Hippel organized a summer session group to share experiences. The lectures were again edited into a book, *Molecular Science and Molecular Engineering*, and published in 1959.[3] Von Hippel says, "My colleagues, each one in his own field, have added the depth of understanding and experience demanded by our universal theme." Out of 22 contributors besides von Hippel, we mention a few who are involved in our present historical review: Epstein contributed a chapter entitled "Ferromagnetic Materials and Molecular Engineering"; Adler wrote a survey of semiconductor diodes and transistors; Smakula's chapter was "Growth and Perfection of Single Crystals."

Now the bandwagon was really rolling. But to get an honest appraisal of insights and capabilities achieved, LIR sponsored in 1963 a summer course entitled "The Molecular Designing of Materials and Devices." There were 82 registered participants and 37 lecturers, covering subjects from atoms and molecules to the design patterns of the most complex materials and devices. From abroad came Helen D. Megaw of the Cavendish Laboratories in Cambridge; Carl Wagner, director of the Max Planck Institut für Physicalische Chemie in Göttingen; Hendrik A. Klasens and Gert W. Rathenau of Philips Eindhoven; Heinz Raether of Hamburg University; and Alfred von Engel of Oxford University; a scintillating array of experts from American government, American industries, and MIT were present as well. The resulting book, also entitled *The Molecular Designing of Materials and Devices*, was published by the MIT Press in 1965.[4]

The field of materials research was examined critically during the 1958–59 academic year by Professors von Hippel, Adler, Nicholas J. Grant of Metallurgy, and John C. Slater of Physics, leading to the development of a program of financial support for a consolidated MIT effort. In the preamble to von Hippel's laboratory report for January 1961, he mentions a federation of four EE laboratories as an intermediate step toward a truly interdepartmental Center for Materials Science and Engineering. These were his own LIR; the Computer Components and Systems Group in Building 10, directed by Ewan W. Fletcher; the Energy Conversion Laboratory already mentioned; and the Materials Applications Section of the Electronics Systems Laboratory (ESL) under James G. Gottling.

The 1961 report of President Julius A. Stratton announced that in June the U.S. Advanced Research Projects Agency (ARPA) had negotiated a contract to supply a large sum that would be incorporated into the larger program of the Center for Materials Science and Engineering, already operating in the laboratories of several departments in anticipation of the new building that was to house the center.

The building, the largest constructed since the original MIT complex, was finished in 1965; in 1966 it was named the Vannevar Bush Building. Although the larg-

Vannevar Bush Building, home of the Center for Materials Science and Engineering. The concept of research "centers" in which the interests of different specialists could be federated has been growing in importance at MIT during the past two decades. The building for the Center for Materials Science and Engineering, completed in 1965, has brought together scientists and engineers from many departments and research groups at MIT.

est departments in the center were Metallurgy and Materials Science, Electrical Engineering, and Physics, the new facilities were made available to the Institute-wide community. The research groups (crystal physics under Smakula, magnetics under Epstein, magnetic spectroscopy under Perry Miles, ceramics under George Economos) and three other LIR groups moved to the new building with von Hippel's blessing. Von Hippel continued to carry on considerable research in his Laboratory for Insulation Research until 1980, when his laboratory (then in Building 38) was completely terminated. The Arthur R. von Hippel Reading Room in the Bush Building was "dedicated to Arthur von Hippel, Institute Professor, pioneer in interdisciplinary research and materials science and engineering," on November 19, 1973.

Many materials problems that relate to electrical, electronic, and magnetic devices have been solved in the Center for Materials Science and Engineering. The materials thus considered by members of the department have ranged widely, including, for example, the lead salts $PbTe$, $PbSe$, and PbS, used for thermoelectric energy conversion; more recently, a variety of oxides selectively doped with rare-earth and transition-metal ions that function as sensitizers and activators for photo- and cathodo-luminescence; and amorphous semiconductors such as are used in xerographic copying, vidicon tubes, and electronically alterable memories.

The Center for Materials Science and Engineering enables about 40 professors and their research groups from several academic departments to work together interactively, sharing central facilities. The directors of CMSE have been Professors Robert A. Smith (Physics, 1960–1968), Nicholas J. Grant (Materials Science and Engineering, 1968–1977), Mildred S. Dresselhaus (Electrical Engineering and Computer Science, 1977–1983), and J. David Litster (Physics, 1983 to the present).

Mildred S. Dresselhaus, director of the Center for Materials Science and Engineering from 1977 to 1983.

During the war years and thereafter, the radomes on the roof of Building 6 came to symbolize for a generation of students MIT's involvement in war research efforts.

Part III

The War Years

Introduction
Threshold of
Major Change

It has often been remarked that World War II was the most technological war in all history. Yet many wars have brought forward new technologies, which were transformed into peaceful enterprises thereafter. In the Renaissance Leonardo da Vinci conceived a giant machine of swirling blades that would mow down enemy soldiers; it was never built, but it became the prototype for the hand lawnmower of a later day. Technologies designed for war have found peacetime utilization, and technologies designed for domestic purposes have become the implements of warfare. It has worked both ways. The telegraph, out of which electrical engineering emerged, was used primarily for railroad and business communications, but became an important weapon in the American Civil War, principally in the North. The atomic bomb, which was built and used in World War II, led to a new source of electric energy at a time when demand for electricity was continually increasing in the postwar era. The Planned Position Indicator (PPI), radar, LORAN, servomechanisms, and early digital computers, whose origins appear in the chapters in this section, have become in the postwar era vital tools for safe air and sea navigation and in the machine tool industry. The digital computer has become a mainspring of business and industry, appearing in new office systems and in the home. These are but a few examples of the vital technologies in which MIT was involved during the war years.

As an educational institution MIT was profoundly affected by its commitment to the war effort. Before the war there were already strong and growing external commitments: the Bush differential analyzer, the Rockefeller differential analyzer, the

Hazen Cape Cod Canal model, Bowles's involvement in patent cases, Bush's involvement with the Raytheon company. But the big push into the external arena came during the war, when MIT took on the responsibility of the Radiation Laboratory and became involved in government and industrial research and development projects, in military activities off campus and on. In the accelerated Radar School MIT trained many thousands of radar officers, who would take command of radar equipment that was still in the laboratory while they were being trained. All these activities ultimately changed engineering education in general and electrical engineering in particular. Though the story of the Radiation Laboratory (chapter 13) has been well documented elsewhere and represents a commitment by MIT as a whole, the brief account in this volume examines especially how members of the Electrical Engineering Department were involved in its activities. The development of the Servomechanisms Laboratory (chapter 14) and of Whirlwind (chapter 15) represented new commitments specifically by the EE Department.

Chapter 12

Organizing for War Research

NDRC and OSRD

As the war clouds gathered over Europe in 1938, 1939, and 1940, an extraordinary group of scientists and engineers met frequently in Washington to conduct the business of certain high-level national committees. Dr. Frank B. Jewett, who had been an instructor in physics and electrical engineering at MIT from 1903 to 1905 and who was in 1939 president of both the Bell Telephone Laboratories and the National Academy of Sciences, was one of these men. Another was Dr. Richard C. Tolman, Course X S.B. '03, Ph.D. '10, who had been an instructor in theoretical chemistry at MIT while working for the doctorate. He had written an excellent expository book on relativity in the early 1920s, when the first courses in Einstein's Theory were being given in America. In 1939 he was professor of physical chemistry and mathematical physics and dean of the Graduate School at the California Institute of Technology. He became so concerned about the Hitler threat that he moved to Washington in early June 1940 to be ready for national service. Vannevar Bush, president of the Carnegie Institution in Washington and chairman of the National Advisory Committee for Aeronautics, President James B. Conant of Harvard University, President Isaiah Bowman of Johns Hopkins University, and President Karl T. Compton of MIT were also in this group; they saw clearly that the United States was not prepared, either as an arsenal of defense for the attacked countries or as a participant in the war against Hitler. They shared the conviction that the encounter would be intense and that in order to win it the allied nations would have to produce superior and innovative weapons as well as excel in military strategy. They set up a plan for a new organization that would lead in the development of weapons while the armed services and industrial establishments were taken up with fighting the war and manufacturing the necessary equipment.

The EE Department on the eve of the war. Among the youthful electrical engineers who would contribute to war research are, in the front row, M. F. Gardner (*second from left*), Gordon S. Brown (*seventh from left*), and Harold L. Hazen (*ninth from left, kneeling*), who had just been appointed department head in 1938; in the back row, standing, Karl L. Wildes (*second from left*), Jay W. Forrester, (*fourth from left*), Edward Moreland (*tenth from left*), and L. F. Woodruff (*eleventh from left*). At the extreme right, standing, is Truman S. Gray. Standing above in the center background is Ralph D. Bennett.

In early June 1940 Bush presented to President Roosevelt a proposal for a civilian organization, to be recruited largely from the institutions of higher learning, "to conduct research for the creation and improvement of instrumentalities, methods, and materials of warfare." The president was quick to recognize the potency of this group that had offered its services, and he followed Bush's proposal very closely in setting up a special committee. On June 15 Roosevelt signed letters of appointment for members of the National Defense Research Committee (NDRC), established by order of the Council on National Defense (made up of cabinet members) on June 27, 1940. Bush was named chairman, the other members being Tolman, Conant, and Compton. Ex officio members were Jewett, as president of the National Academy of Sciences, and Conway P. Coe, as Commissioner of Patents. Also appointed to NDRC were Brigadier General George W. Strong and Rear Admiral Harold G. Bowen.

At its first official meeting in July 1940 the committee elected Tolman as vice-chairman and chose as its executive secretary Irvin Stewart, a former member of the Federal Communications Commission and currently director of the Committee on Scientific Aids to Learning. Carroll L. Wilson, who had served for five years as assistant to President Compton and who was known to have a special knack for untangling knotty problems and making things run smoothly, became Bush's deputy.

The committee lost no time in getting to work. Divisions and sections were set up: Tolman, as chairman of Division A, was to deal with ordnance and armor; Conant, of Division B, with bombs, gases, and chemistry problems in general; Jewett, of Division C, with transportation and communications; Compton, of Division D, with fire control, radar, detection, and coun-termeasures; and Coe, of Division E, with patents and inventions. Section 2 of Division D is of special interest in the history of MIT's EE Department; its concern was gun fire control and it was headed by Dr. Warren Weaver, director of the Rockefeller Foundation's Division of Natural Sciences. As soon as he heard about NDRC, Weaver wrote Bush that he was "anxious to be of service and prepared to take on a full-time job." Weaver selected initially as his key associates Professor Samuel H. Caldwell, who was in charge of the differential analyzer development at MIT; Dr. Thornton C. Fry, mathematical research director at Bell Telephone Laboratories, who had presented in 1926–27 a course on applications of probability theory as lecturer in the MIT Electrical Engineering Department; and Edward J. Poitras, VI-A S.B. '28, S.M. '29, an expert in automatic control and designer of the controls for the 200-inch telescope on Mt. Palomar. He soon added Duncan J. Stewart, chief engineer of the Barber-Colman Company of Rockford, Illinois.

The first project to be tackled by Section D-2 was the design and development of an electrical director to replace the mechanical director for 90-mm antiaircraft guns. With the cooperation of the Coast Artillery Board, Section D-2 within a few weeks had worked out with the Army and the Bell Telephone Laboratories the specifications for this electrical director. The experimental model was known as the T-10 and was tested by the Coast Artillery Board in November 1941. In February 1942 the Army standardized the director as the M-9 and immediately put it into production, subsequently achieving great success in the field. In 1944 the M-9 director—in conjunction with the later NDRC developments, radar SCR-584 and the proximity fuse—was very successful in shooting down German buzz bombs over England. It was Section D-2 that contracted with Gordon Brown's MIT Servomechanisms Laboratory for the design of a speed gear servo for the 37- and 40-mm gun mounts.

In the spring of 1941 President Roosevelt was having serious problems in organizing medical research for wartime applications. By this time Bush had done such a masterful job with NDRC that the president decided he wanted Bush to take under his wing the organization of medical research. Bush at first demurred, but had received such strong backing from the president that he finally relented and took on the task of organizing the Committee on Medical Research (CMR) as a principal subdivision of the Office of Scientific Research and Development (OSRD), established by executive order in June 1941. (NDRC was the other principal subdivision of the newly created OSRD.) Bush was appointed the director of OSRD. There was little change in NDRC at this point, except that Dr. Conant was moved up to its chairmanship, with Professor Roger Adams, head of the Chemistry Department at the University of Illinois, filling the vacancy as chairman of Division B.

Bush's OSRD was an ideal organization for the tasks facing it. Its structure allowed it to achieve its objectives most efficiently: the research groups were flexible and adaptable to the changing needs of warfare; a top-level administrative office handled fiscal and contractual matters. Dr. Irvin Stewart was made executive secretary of OSRD, as well as of each of the constituent organizations, NDRC and CMR. Bush's deputy, Carroll Wilson, was moved up to OSRD, and Chairman Conant of NDRC was aided, first by George W. Bailey as special assistant, and then by an increasing office staff as the need arose; Edward L. Moreland, dean of engineering at MIT, became consultant in June 1942, and from August 1942 throughout the war he was the executive officer in the Chairman's Office, recognized by both Bush and Conant as the "active executive head of NDRC." As his load grew with the increasing research and production responsibilities, he took on deputy executive officers,

technical aides, and executive assistants to share his administrative tasks. After the atomic bomb attacks on Japan, a course of action recommended by Bush to end the war, and after the Japanese surrender, Moreland headed a scientific intelligence survey in Japan, analyzing Japanese military research and development.

The most striking feature of Bush's OSRD was his personnel structure; he engaged competent, loyal, and devoted persons to staff his pyramidal organization from the apex downward. First of all, President Roosevelt had complete confidence in Bush, supporting him generously and protecting him from interference. In response, Bush "took the attitude that loyalty to the Chief must be absolute and untarnished."[1] Bush's close acquaintance with the men at the top and the mutual confidence among them proved to be major factors in the high achievement level, sense of urgency, and complete security of this gigantic undertaking.

The Japanese attack on Pearl Harbor on December 7, 1941, and the subsequent declaration of war on the United States by Germany and Italy brought such an increase in NDRC activity that by the fall of 1942 an expansion and reorganization became necessary. Eighteen divisions were formed in December 1942, Division 7 being designated as the Fire Control Division to continue and expand the work of the former Section D-2. Harold L. Hazen, head of MIT's Electrical Engineering Department, was appointed chief of Division 7. Warren Weaver, former chairman of Section D-2, felt strongly that he ought not to "hang on" and be guilty of "looking over the new chief's shoulder," but finally did accept membership in Division 7 after considerable urging by Hazen. Weaver's principal leadership, however, was given to the setting up and administration of a new group called the Applied Mathematics Panel (AMP), with Dr. Mina Rees as his chief technical aide. The panel supplied mathematical services, not only to the Fire Control Division, but to the whole NDRC organization and the armed services.

Hazen's Division 7 office was at his regular MIT location, with a secured room across the corridor housing government-cleared secretaries and safe filing cabinets. Professor Karl L. Wildes, technical aide to the chief and secretary to Division 7, was in administrative charge of the secured area. The office of Edward Poitras at NDRC headquarters under Section D-2 became the Washington office of Division 7. Soon after the initial organization, Hazen added as division members Ivan A. Getting, representative of the MIT Radiation Laboratory; Preston R. Bassett of the Sperry Gyroscope Company; and Albert L. Ruiz of the General Electric Company. Bassett and Ruiz did not represent the manufacturers but acted as personal experts in the fire-control field.

Hazen's analytic ability and his experience in automatic control and instrumentation enabled him to master rapidly the fire-control problem. He was initiated into the former activities of Section D-2 through visits arranged by Stewart, Caldwell, Poitras, Fry, and Weaver to contractors already deeply involved in research projects, including the Bell Telephone Laboratories, General Electric, Bausch and Lomb, and Columbia University. An account of the activities of Section D-2 and Division 7 appears in chapters 3 through 9 of *New Weapons for Air Warfare,* part of the official seven-volume OSRD history series, *Science in World War II.*

For MIT an important long-term outgrowth of the work of Section D-2 came about through the involvement of mathematician Norbert Wiener in the M-9 electrical director project for antiaircraft ground fire. One of the early problems with this director was to measure how well it performed its functions Errors in performance were of two distinct kinds, those that arose from the

Harold L. Hazen (*right*) with Carlton Tucker, his academic executive officer.

dynamics of the director itself and those caused by human inaccuracies in tracking the target. Under a Section D-2 contract with the Barber-Colman Company of Rockford, Illinois, a dynamic tester was developed that, with ideal input data, measured the dynamic accuracy of the director and led to improvements in the design of its components. The second problem was to average the jumpy data coming from the human tracker so as to provide smoothed target-course data for the director. Professor Wiener of the MIT Mathematics Department worked on this problem of data smoothing under an NDRC contract negotiated by Section D-2 in parallel with staff members of the Bell Telephone Laboratories. Even with smooth tracking, another important aspect of antiaircraft fire is the prediction of the target position between the time of firing and the expected hit. In the early fall of 1940 Wiener gave serious attention to data-smoothing and prediction problems.

Wiener first attacked the prediction problem in terms of servo theory; but he found that a network that gave good prediction was unstable and that a stable network was not a good predictor. He then turned his attention to statistical methods and produced a monograph entitled ''The Extrapolation, Interpolation, and Smoothing of Stationary Time Series with Engineering Applications,'' which Weaver immediately saw as a brilliant and significant analysis of the prediction and smoothing problems. This monograph was bound in a yellow jacket when issued on February 1, 1942, as Report to the Services No. 19, and thereafter was known as the ''Yellow Peril.'' This report would prove to have a significant influence on the education of electrical engineering students. Dr. Y. W. Lee, a former thesis student of Wiener, returned to MIT in 1946 and by the fall of 1947 offered a graduate subject based on the ideas of the Wiener report. A landmark in the history of communication and control engineering, Lee's graduate instruction and research

Karl Wildes, Hazen's NDRC executive officer.

in the statistical theory of communication attracted a whole generation of the department's keenest graduate students to the exploration of this new field.

Also working with Wiener on the prediction problem was engineer Julian Bigelow, who in 1943 coauthored with Wiener and Arturo Rosenblueth the paper "Behavior, Purpose, and Teleology,"[2] which would become one of three famous papers that laid the foundations of the new field of cybernetics (see chapter 16). Bigelow went on to Princeton in 1946 to become chief engineer on the Institute for Advanced Study computer.

Professor Caldwell, one of the original members of Section D-2, devoted himself to problems of national defense from the summer of 1940 to the end of the war. At first he was involved in the broad field of fire control; but as time went on, he saw the increasing importance of airborne defensive and offensive fire-control equipment. This was a much more difficult problem than ground-based or even ship-based fire control because the base of operations was rapidly moving all over the "wild blue yonder." Because of the plane's own speed the gun had often to be aimed behind the target in order to hit it. Several contracts with colleges and industrial laboratories were initiated to achieve an understanding of the airborne fire-control problem and to break it down into component problems that could be separately attacked. In February 1942 Caldwell set up a contract with the Franklin Institute in Philadelphia "to provide personnel and laboratory facilities for experimental work" in the general field of airborne fire control. Caldwell's new technical aide, George A. Philbrick, left the Foxboro Company to assume the technical responsibility for this contract. The competence and speed with which the Franklin Institute group got into action is illustrated by the development of an aerial torpedo director, which was carried all the way from conception to flight testing during two months—July and August 1942—a development feat that would take years under normal circumstances. The Navy standardized this torpedo director as the Mark 32 by the end of 1942. Although it went into production, it was not actually used in combat during the war.

Many members of the Electrical Engineering Department were involved in other war activities. Gordon Brown's work in the Servomechanisms Laboratory (see chapter 14) brought him and his staff into contact with the armed forces and with the Radiation Laboratory; he was also a consultant to Hazen's Fire Control Division of NDRC. Another outstanding contribution was made by Professor Ralph D. Bennett, who left MIT in July 1940 to build up the staff of the Navy's Mine Laboratory (later the Naval Ordnance Laboratory) in order to solve the "degaussing" problem of American ships and keep them from setting off magnetic mines. Bennett became director of the Naval Ordnance Laboratory, which designed the mines used in the Pacific theater. Professor Truman S. Gray spent the summer of 1941 working on instrumentation aspects in Bennett's Washington laboratory. Professor William H. Radford was consultant to the new Missiles and Optics Division of NDRC. Professor L. F. Woodruff became full-time consultant to the Army Ordnance Department. Other department personnel worked with the Radiation Laboratory and the Radar School (see chapter 13).

The Radiation Laboratory

The great wartime laboratory with which MIT was most associated in the public mind was the Radiation Laboratory. From it emanated the radars that made the difference in aerial and sea warfare, and from it, after the war, emanated a great series of books—a cornucopia of knowledge generated during the war research—that became a potent influence in postwar engineering and science. For MIT itself the Rad Lab experience was decisive—it marked a real change from the long, slow growth of the past in science and engineering and brought about an enormous change of scale, especially in the way the people involved thought about the potentialities of research and development. It had an especially profound impact in the field of electrical engineering.

How the Radiation Laboratory came to be situated at MIT, and the projects it carried out during the war, are sketched in this chapter. What is not always clear is that the Rad Lab was not an MIT laboratory as such, but a laboratory *at* MIT, which was largely under the jurisdiction of a division of the NDRC. In fact, President Karl T. Compton of MIT had to be persuaded that the lab should be at MIT. Moreover, the Rad Lab was headed mostly by "outsiders," the director being Dr. Lee A. DuBridge, who came from the University of Rochester to take on the task. Scientific personnel were recruited from organizations across the country. After the war some of the Rad Lab members joined the MIT faculty, including a future MIT president, Jerome B. Wiesner. Valuable Rad Lab equipment was acquired for use in the postwar Research Laboratory of Electronics (see chapter 16).

Although the Institute was not directly involved in the technical management of the work of the laboratory, it took on the responsibility for its administration. (This kind of management of research and development laboratories for the government by other organizational

MIT Radiation Laboratory director Lee DuBridge (*center*) and assistant director I. I. Rabi (*right*) with an American copy of the British-invented cavity magnetron, crucial to the development of radar. E. G. Bowen (*seated, foreground*), who headed British radar development, was a member of the Tizard mission, which secretly brought the magnetron to the United States in 1940.

James R. Killian, Jr., who was involved in the administration of the MIT Radiation Laboratory during the war years. Killian describes Rad Lab as a "breathtaking" moment in MIT's history. It was, with other wartime laboratories, to introduce the MIT community to a new scale of research and development and a deepening involvement with the federal government.

entities has become a fairly common structure in the postwar environment.) This turned out to be a monumental management task, since the laboratory grew quite quickly and ultimately involved 4,000 people. MIT, which handled the Rad Lab finances, salary reviews, and so forth, had in effect to set up a dual management structure, which became progressively complex as the lab grew. Because Compton was often away from the Institute on war-related business, the Rad Lab administration fell largely on the shoulders of Dr. James R. Killian, then executive assistant to Compton; Horace Ford, MIT treasurer; and Nathaniel Sage, head of the Division of Industrial Cooperation.

Killian, subsequently president of MIT (1949–1959) and President Eisenhower's science advisor in the Sputnik-crisis era, reflects that the Rad Lab was a "breathtaking development" for MIT at that time. He points out, however, that the "overwhelming scale and size and drama associated with the Rad Lab . . . tended to obscure two other enormously important activities"—namely, the work of the Servomechanisms Laboratory and of Stark Draper's Instrumentation Laboratory, both of which were indigenous MIT activities and drew on MIT personnel. They were important, he asserts, not only for their war work but also for their contributions to education. Draper, he notes, was a "superb teacher as well as being one of the great engineers of our time"; and Gordon Brown, through his Servo Lab experience, became "a dramatically successful leader in the development of the Electrical Engineering Department."

Although the Radiation Laboratory was abruptly "demobilized" at the close of the war, it perpetuated itself in another way. By the time of the victory in Japan it had become clear to everybody what a germinal role a really large laboratory could exercise in an educational environment. It was difficult to imagine MIT without Rad Lab. Thus, when Professor John Slater, then head of the Department of Physics, proposed that MIT should set up a successor laboratory as a joint project of the Departments of Electrical Engineering and Physics, there was almost spontaneous agreement. (This is not to say that there were not serious issues involved in setting up and operating such big laboratories in an educational situation; we take up those issues later.)

Given the profound polarization of opinion in our times about nuclear energy and nuclear radiation, there is a fine irony and a reflection of things past in the name "Radiation Laboratory." It was chosen as a sort of decoy, because in that period atomic and nuclear physics were regarded by most scientists as harmless activities! In fact, the MIT Rad Lab, established in the fall of 1940, was set up to explore and develop microwave radio techniques and to design microwave equipment for use in warfare. The principal frequencies used were S-band in the vicinity of 3,000 MHz (10 cm), X-band around 10,000 MHz (3 cm), and K-band around 30,000 MHz (l cm). Frequencies somewhat lower than these (longer wavelengths) are called ultrahigh frequencies (UHF) and were used in most radio detection sets prior to 1940.

The Radiation Laboratory story is told briefly in James Phinney Baxter's *Scientists against Time* (1946) and is reported more thoroughly in the *Summary Technical Report* of Division 14, NDRC, in three volumes of the 70-volume *NDRC Summary Report,* published in 1946. Professor Henry E. Guerlac of Cornell University was the laboratory's official historian and wrote a comprehensive account covering the early history of microwaves as well as the wartime developments and field applications. This work, "History of Radar in World War II," was not published but is available in a somewhat abbreviated microfilm form in the Library of Congress. Our account is indebted in part to Guerlac.

Radar in the Battle of Britain

The military situation of 1939 and 1940 made the microwave development assumed by the Radiation Laboratory an urgent matter. Germany had invaded France, as well as Poland, the Netherlands, Belgium, and Norway, and by June 1940 had concluded an armistice with France. Great Britain's south coast was thus threatened by the German Luftwaffe, operating from new bases in the north of France. England's only hope of survival was to mount a successful attack against the German air force.

Despite the seemingly overwhelming odds, the British scientific and technical establishment had been mobilized in the right direction. Already by the middle 1930s, British politicians and scientists were anxiously aware of the threat to democracy in Europe occasioned by the rising power and audacity of Hitler and Mussolini. The Royal Air Force was greatly strengthened, and means for the detection of invading airplanes were studied in earnest. A committee of distinguished scientists was set up to intensify the research in air defense under the chairmanship of Sir Henry T. Tizard, Fellow of the Royal Society and rector of the Imperial College of Science and Technology. By the time this committee held its first meeting in January 1935, Robert Watson-Watt, superintendent of the Radio Department of the National Physical Laboratory, had made calculations indicating that an airplane would reflect enough energy from a transmitted radio signal to make its presence detectable. By the end of March Watson-Watt had organized a radio research station at Orford Ness and recommended the use of short pulses of 50-meter waves, rather than continuous waves, in the detection problem. One of the three experimenters at this early laboratory was Dr. Edward G. Bowen, formerly engaged in cosmic-ray research at the University of London and later to be intimately connected with MIT's Radiation Laboratory.

Work on the British radio direction-finding technique (RDF) progressively expanded as young physicists were recruited, until several hundred were involved at several centers and testing stations. The crude devices of 1935 gave way to sophisticated instruments that could detect aircraft and surface ships at increasing distances. Transmitter frequencies went from 6 MHz (50 meters) in 1935 to 200 MHz (1.5 meters) by the beginning of 1940. By the time German daytime attacks began in August 1940, followed by night attacks in September, they had come under the surveillance of a chain of RDF stations. Though Royal Air Force planes did not yet have RDF equipment, their pilots fought heroically against the German aerial onslaught. By spring 1941, however, aircraft interception Mark IV sets had been mounted in about two-thirds of the British night fighters, and these fighters destroyed more than half the German planes they encountered. Similar RDF equipment—aircraft-to-surface-vessel (ASV) systems—was found to be effective in the detection from the air of ocean vessels and surfaced submarines.

Before the Battle of Britain a device known as the Plan Position Indicator (PPI) had been developed and demonstrated with reasonable success at a frequency of 200 MHz (1.5 meters). This instrument displayed on the face of an oscilloscope a map of the land area or objects being investigated, and became one of the most valuable components of the RDF system. It was evident that a sharper PPI image could be realized if the transmitted radio beams could be made narrower, but this required an increase in frequency above 200 MHz.

It was Denis M. Robinson (S.M. '31, University of London Ph.D. '29) who by February 1940 had developed an experimental setup at 600 MHz (50 cm), but with much less power than in the 200-MHz sets. Robinson, now given charge of the new "Centimeter

Group" of the Air Ministry Research Establishment (AMRE), continued to work on the 50-cm system during the spring months, even though he was aware of promising work at the University of Birmingham on a klystron tube that could produce about 400 watts of 10-cm continuous-wave power. In anticipation of a 10-cm generator Robinson's group developed a 10-cm antenna with a parabolic reflector. By May a klystron tube became available, and Robinson's group was strengthened through the addition of several nuclear physicists from the Cavendish Laboratory at Cambridge. By June a resonant-cavity magnetron was furnished from Birmingham, and the essential features of an aircraft-interception program at 3,000 MHz (10 cm) were worked out with Bowen in consultation.

Although the Luftwaffe had been repulsed with the aid of the 200-MHz systems by this time, the microwave development was to play an important role in the later World War II activities of both the British and the Americans. The name RADAR (Radio Detection and Ranging) was suggested by Commander S. M. Tucker of the United States Navy. It was adopted by the British to replace the term RDF in 1943, following its adoption by the Americans in 1940.

Workers in radio detection and direction finding, aware of the advantages of narrow radio beams and small equipment, were continually in quest of a microwave generator tube that would give an adequate amount of power. One of the most promising early microwave generators was the magnetron, invented by Dr. Albert W. Hull of the General Electric Company in 1920. Experimenters in several countries made modifications of this tube. By 1929 Kinjiro Okabe in Japan had produced, with a split-anode magnetron, waves under 6 cm in length but with insignificant power. By 1939 various American industrial laboratories working with the Army and Navy had pushed magnetron development to the microwave range with 10 to 20 watts of power.

Another promising tube was the klystron, in which a pair of resonant cavities (the ''Rhumbatron,'' developed by Professor William W. Hansen of Stanford University) were built into a vacuum tube as suggested by Russell M. Varian, a Stanford research associate. The development of the klystron was carried out at Stanford by Russell Varian and his brother Sigurd. Late in 1937 they had a 10-cm klystron operating with about one watt of continuous-wave power.

Hansen and the Varian brothers used the new tube in airplane detection and blind-landing experiments. This was the same klystron used in Edward Bowles's blind-landing experiments at MIT (see chapter 6). (Hansen had been a National Research Fellow at MIT from 1933 to 1935; as Sperry Gyroscope Special Research Fellow at MIT during the academic year 1940–41, he presented graduate subjects on klystron oscillators in the spring term.)

The breakthrough in the quest for a powerful microwave generator came at the University of Birmingham in England, where in 1939 Professor M. L. Oliphant and his group of physicists set about improving the klystron and the magnetron with a goal of I kW of 10-cm power. After a considerable improvement in the klystron at 10 cm, no way seemed possible to produce a kilowatt of power. At this point the group hit upon the ingenious idea of introducing the resonant-cavity principle of the klystron into the magnetron, which was able to handle large anode currents. Physicists J. T. Randall and A. H. Boot of Oliphant's group are credited with the development of this idea and with the initial operation of the resonant-cavity magnetron in February 1940. The British General Electric Company did the engineering and produced a manufacturable tube by June 1940. Other 10-cm designs followed, with increased peak powers from 5 to 50 kilowatts.

The Tizard Mission

A working model of the resonant-cavity magnetron was demonstrated to the Americans by the top-secret Tizard Mission, a group of British scientists who arrived in Washington early in September 1940 to seek American aid in their radar program. Among the visitors were representatives from the British armed services and three civilian scientists, Sir Henry Tizard, Professor J. O. Cockcroft of Cambridge University, and Dr. Edward G. Bowen, who had headed the British radar developments. The mission first met with representatives of the American military services, and the radar activities of both nations were mutually disclosed. The first demonstration of the resonant-cavity magnetron in America, which took place on October 6 at the Bell Telephone Laboratories (BTL) in Whippany, New Jersey, produced about 6 kW of 10-cm power. (The American Arthur Samuel played an important role in the success of this first demonstration.) The electronics experts of BTL began the next day to plan for the American manufacture of the new tube. On October 12 and 13, following a session with Bush and Compton in Washington on the 11th, Bowen and Cockcroft of the British group, together with Dr. Alfred L. Loomis, Edward Bowles, and Professor Ernest O. Lawrence of the University of California—three experts on microwaves—and Carroll Wilson representing NDRC, outlined a program for American microwave development, including the immediate setting up of a civilian laboratory patterned after the British laboratories. Bowen, who had been so prominent in the British developments, remained to work with the new American laboratory.

As described earlier, Karl T. Compton's Section D-2 of the NDRC dealt with gun fire control; his Section D-1 was concerned with radio detection and countermeasures. During the summer of 1940 Bush, Compton, and Loomis investigated the status of radio detection research in Great Britain and in the military laboratories, universities, and industries of America. In consultation with Army and Navy investigators, they reached an early decision that NDRC should take on the problem of pushing radio detection into the microwave region while the armed services continued to develop their fairly successful longer-wave sets and get them into production. In setting up Section D-1, Compton selected Loomis as its chairman; a lawyer-scientist, he had by this time established in his own laboratories at Tuxedo Park, New York, an energetic research program in microwaves. Section D-1 was soon called the Microwave Committee, with Edward Bowles as its executive secretary. The other members were Dr. Ralph Bown of the Bell Telephone Laboratories, H. Hugh Willis of the Sperry Gyroscope Company, R. R. Beal of the Radio Corporation of America, George F. Metcalf of the General Electric Company, John A. Hutcheson of the Westinghouse Electric and Manufacturing Company, and Ernest Lawrence.

The Microwave Committee and the armed services were intensely interested in Bowles's hyperfrequency radio program. The Stratton-Houghton-Radford studies on signal transmission through fog seemed promising, not only for the blind landing of airplanes but also for locating land and ocean objects from the air—and therefore for bombing through cloud cover.

The Bowles program had several other facets. Dr. Wilmer L. Barrow had "primary responsibility for research and expenditure of funds on investigations of transmitting and receiving antennas [for aircraft detection] sponsored by the Sperry Gyroscope Company." Frank D. Lewis, S.B. '37, S.M. '40, was full-time assistant on this study. Walter W. Mieher was Barrow's full-time assistant on the study of hollow pipes, horns, and resonators; in September Daniel S. Pensyl and Henry J. Zimmermann, both S.M. '42, were added as aides in this project. Professor von Hippel and Shepard Roberts, S.B. '38, S.M. '39, Sc.D. '46, were assigned to research on measurements of properties of dielectric materials at microwave frequencies, supported by a grant obtained by Bowles from the International Telephone and Telegraph Company. Bowles also supervised Jackson H. Cook, VI-C '36, and Arnold P. G. Peterson, S.M. '37, Sc.D. '41, who was working under General Radio Company sponsorship.

Alfred Loomis had become interested in ultrahigh-frequency research in 1939 after being assured by Compton and Bowles that this field was promising and important. He decided to sponsor an MIT program under Professor J. A. Stratton and Donald E. Kerr, VI-C S.B. '37, S.M. '38, VIII Ph.D. '50, on the use of microwave radiation, to begin in the fall of 1939. By April and May 1940 plans were under way for a summer research program at the Loomis Laboratories at Tuxedo Park. About half of the Tuxedo Park group were MIT men: William H. Ratliff, Jr., S.M. '39, Frank D. Lewis, Donald E. Kerr, and William G. Tuller, VI-A S.B. and S.M. '42, Sc.D. '48, with participation by Bowles. Most of this work used an 8.6-cm klystron and involved the detection of moving targets by beating the reflected wave with the transmitted wave—the Doppler effect.

Ernest O. Lawrence, who had received the Nobel Prize in physics in 1939, was so well known and highly esteemed by the national community of physicists that he was singularly qualified to lead the Microwave Committee in recruitment. It was he who suggested that the new microwave laboratory take the name Radiation Laboratory (ironically, because atomic and nuclear physics were regarded as harmless at that time).

The first scientist to be enlisted by Lawrence was Professor Kenneth T. Bainbridge (VI-A S.B. '25, S.M. '26; Princeton Ph.D. in physics, '29) of Harvard's Physics Department. By the middle of October the Microwave Committee, in conference with Compton and Bainbridge, decided to offer the directorship of the proposed microwave laboratory to Professor Lee A. DuBridge, chairman of the University of Rochester's Physics Department and dean of its Faculty of Arts and Sciences. Lawrence and Loomis very quickly enlisted DuBridge to accept this responsibility, even though the Radiation Laboratory had as yet no form or substance. No time was lost in seeking out, through personal contacts, leading physicists to fill the key positions. The last week in October about 600 nuclear physicists assembled for a conference at MIT. Professor Barrow took advantage of this occasion to present, with his fellow workers, afternoon seminars on ultrahigh-frequency phenomena. Some of the most prominent men of the conference were enlisted to join the new laboratory.

The MIT Rad Lab Is Established

The choice of MIT as the site for the Radiation Laboratory was a logical consequence of Bowles's comprehensive program of ultrahigh-frequency and microwave research and teaching. It would be neutral ground with respect to the military services and industrial firms and was conveniently located with respect to coastal target areas. On October 16 Bush, Bowles, Loomis, and Jewett agreed to propose this to Compton, who, after considerable persuasion, agreed to the MIT location and telephoned Killian to see whether space could be made available in the MIT buildings. On the 18th a full meeting of the Microwave Committee in Washington, attended by officers of the Army and Navy, confirmed the plans to establish the MIT Radiation Laboratory. On October 25 NDRC approved the Microwave Committee's program and allocated $455,000 for the first year of operations. The first space occupied by the laboratory was Room 4-133, where Bowles had set up his ultrahigh-frequency laboratory under the immediate supervision of Professor William M. Hall in consultation with Barrow. A group of Bowles's research assistants had their desks in this room. Here Dr. DuBridge established his office.

Organizational meetings were held in November 1940. Six technical working groups were set up: Section 1 was to work on pulse modulators under K. T. Bainbridge (Harvard); Section 2 on transmitter tubes under I. I. Rabi (Columbia); Section 3 on antennas under A. J. Allen (University of Pittsburgh); Section 4 on receivers under L. A. Turner (Princeton); Section 6 on cathode-ray tubes under W. M. Hall (MIT); Section 7 on klystrons under F. D. Lewis (MIT). Section 5 was to be concerned with microwave theory, but no such group was organized at the beginning, although several of the Rad Lab staff actually functioned as a theory group.

By the summer of 1941 J. A. Stratton and other members of the Microwave Committee saw the need for an organized program in theory to assure the sound development of equipment. While working with Melville Eastham's Navigation Group on the long-range navigation (later called LORAN) program, Stratton pointed out that ground-wave transmission could reach greater distances if lower frequencies, say 150 meters (2 MHz), were used rather than the ultrahigh frequency (30 MHz) of the early experiments. Experimental and theoretical studies at the lower frequencies were carried out in the summer of 1941 by Stratton's assistants Donald Kerr and Richard B. Lawrence (VI-C '40, VIII Ph.D. '42), John A. Pierce of Harvard, and others, laying the groundwork for the development of a navigation grid that by 1943 and 1944 guided the North Atlantic convoys and covered central Europe.

In early 1942 Kerr organized a group to study radio propagation. Beginning in March 1942, J. A. Stratton and R. A. Hutner issued a series of reports dealing with radio transmission at various frequencies, chiefly over seawater. It was the lower-frequency (2-MHz) LORAN that provided skywave synchronization (SS) to cover airplane navigation over land at night; some 22,000 sorties of blind bombing were carried out over Europe by the British and American air forces beginning in September 1944. This SS LORAN system was also used by the U.S. Army Air Force in the China-Burma-India theater.

Don Kerr was very active in the development of 3-cm radar and in August 1943 showed that it could detect the periscopes of submarines at a distance of three or four miles. His propagation group pioneered in "seeing" distant storm activities with radar, and this blossomed into the new science of radar meteorology. Kerr, with his prior experience in the Bowles blind-landing program, studied all existing blind-landing systems with Luis Alvarez and, following the 3-cm radar development, worked on the development of a system for landing airplanes known as "ground control of approach" (GCA), in which most of the equipment was on the ground so that the plane was "talked down" as the pilot received instructions on his radio receiver. The "first PPI operator" (on the ground) picked up the incoming plane at 15–20 miles with a 10-cm search set; the "second PPI operator" brought the plane into the landing course; and finally the "approach controller" guided the pilot down the glide path through use of a high-precision 3-cm system.

British radar development suffered from a lack of engineering personnel, the young physicists having to learn from experience many of the practical design techniques that would have been rules of thumb for engineers; the MIT Radiation Laboratory had a favorable proportion of young engineers (many with their highest degrees from MIT Course VI), and these men were used effectively to bring the equipment to practical operational quality.

Albert M. Grass, S.B. '34, was with Rad Lab during most of its existence and worked on the design of radar scanning methods, range and test circuits, and servomechanisms. He developed a photoelectric auto-

matic range device and in 1942–1943 was involved in the tests of the famous SCR-584 gunlaying radar, working chiefly with its preproduction model, the XT-1.

George Hite, S.B. '41, having been a student technician with Kerr, Kenyon, Lewis, and others in the Bowles blind-landing project, joined Rad Lab upon graduation; he worked on a delayed ranging circuit for the PPI, on an automatic tracking circuit for the Mark 35 radar, and with Dr. Britten Chance on precision ranging devices. Hite's classmate, Earl H. Krohn, quipped that the ranging unit was "not designed by chance."

Earl H. Krohn and H. A. Kirkpatrick were project engineers on a range radar for the Mark 51 Naval Gun Director and its successor, the Mark 56 director, for which Krohn was involved with the design, development, and finalizing of the various radar "black boxes."

John E. Meade, S.B. '32, was project engineer for an experimental model designed to demonstrate to the Navy and its contractors that a small microwave radar was feasible for shipborne fire-control systems. He accompanied the set in its tests at the Naval Proving Ground, Dahlgren, Virginia, in the early fall of 1941. In 1942 Meade joined the Navy and continued his interest in shipboard applications of radar.

Robert M. Alexander, S.B. '39, was Rad Lab's consultant and project engineer working with industry and the Navy in the construction of the ASH and AIA-1 3-cm radar sets, used by the American and British navies. John H. Tinlot, B.S. '43, designed a new type of receiver for the AC version of an experimental radar beacon system.

Donald G. Fink, S.B. '33, was associated with Melville Eastham, beginning in July 1941, in the early years of the LORAN navigation system. In March 1943 Fink became head of Division 11 (Navigation), replacing Eastham. During the early years Fink was active in setting up siting stations for standard LORAN in the Aleutian Islands. With the development of skywave synchronized (SS) LORAN in early 1943, he took this new navigation system to the ETO in July, August, and September. In October 1943 he gave up his position in Rad Lab to accept a full-time appointment in Bowles's Washington office. In November he left on a mission to the Far East, where he laid out SS LORAN sites. Installations of SS LORAN equipment were promptly made by the Navy. The final months of Fink's service were spent on Dr. Stratton's Committee on Air Navigation and Traffic Control.

Getting U.S. Radar into Combat Use

By early 1942 Secretary of War Henry I. Stimson was convinced that microwave radar could become a powerful new weapon of warfare if it could be manufactured and introduced into the active theaters without delay. He asked Bush to recommend a man who could act as his personal adviser in recognizing and correcting faults in the Army radar program. On April 6, 1942, Edward L. Bowles was appointed expert consultant to the secretary of war. Stimson soon recognized that he had in Bowles a man of extraordinary abilities. Not only had Bowles conducted at MIT an intensive research program in ultrahigh-frequency communications and the blind landing of airplanes, but he also had an intimate acquaintance with key personnel at the Bell Telephone Laboratories, the Western Electric Company, and other industrial research and manufacturing organizations. He had been instrumental in setting up in the early days of Rad Lab an arrangement whereby the Raytheon Manufacturing Company in Newton provided model shop facilities for the engineering and production of the earliest 3-cm magnetron tubes. This had led to an order in the spring of 1941 for Raytheon to produce one hundred 3.2-cm magnetrons and some of the first aircraft interception systems. It had also led to the setting up of

Rad Lab's model shop in the fall of 1941 at 230 Albany Street, Cambridge, under the name of the Research Construction Company, Inc., a subsidiary of the Research Corporation of New York. Here limited numbers of radar components and systems could be produced quickly and competently under the direction of Rad Lab personnel. To ensure the propriety of its engineering and manufacturing activities, the company operated under a committee that included high officials of major industrial organizations, such as Bell Telephone Laboratories, Sperry, General Electric, RCA, Westinghouse, and the Research Corporation, with Melville Eastham of General Radio as chairman. During the course of the war over $11 million worth of equipment was delivered by the model shop to the services, Rad Lab, and other OSRD contractors.

With his Washington appointment Bowles was aware that he had accepted heavy responsibilities. Over the summer and fall of 1942 he built up an expert staff: as his personal assistant and executive officer he chose a former student, Allen V. Hazeltine, VI-A S.B. '37, S.M. '38; appointed as full-time expert consultants were Julius A. Stratton, Harold H. Beverage (RCA's vice-president in charge of research and development), and David T. Griggs (who had played an active part in Rad Lab's development of airborne fire-control radar); Dr. Louis N. Ridenour, assistant director of Rad Lab, and Dr. Ivan A. Getting, head of Rad Lab's fire-control group, were made part-time consultants. Other Rad Lab and industrial leaders were placed in active theaters of Army Air Force operation. Frank D. Lewis returned from the London OSRD mission to serve as consultant on radio recognition, identification, and countermeasures. Bowles was considered by Rad Lab its "permanent ambassador for microwaves."

In August 1943 Bowles was asked to direct all the communications in the Army Air Force. Although he considered it important to maintain his activities as expert consultant in Secretary Stimson's office, he was also made special consultant to AAF Commanding General H. H. Arnold and worked closely with him in policy matters and in the planning and initiation of a wide range of technical projects.

Julius A. Stratton, who had moved from the Electrical Engineering Department to the Physics Department in 1930, carried heavy responsibilities as expert consultant to the secretary of war. Stratton's first mission, with H. H. Beverage as partner, was to check the communications for the North Atlantic air lanes that carried supplies from America to Great Britain. In September 1942 these communications experts visited air bases in Maine, Labrador, Iceland, and Greenland, and brought about improvements in the communication facilities.

By November 1942 enemy bombing of American cities had become unlikely, and the Army Air Force was giving most of its attention to the offensive program in the European theater of operations. To step up the effectiveness of air defense radar, Stratton assembled a group from Rad Lab and the principal industrial radar laboratories. Two important achievements of this group were, first, an increase in cooperation between personnel in the research and development centers and those in the testing laboratories at Orlando, Florida; and second, the creation of a program for supplying the practical radar needs of the AAF—a program that was adopted without change by the AAF board at Orlando. The work of the air warning group was so useful that Bowles was asked to establish two other committees under Stratton's chairmanship, one on radar aids to bombing and the other on airborne radar fire control.

A conference in the spring of 1944 between Bowles and General Carl Spaatz, commander-in-chief of the U.S. Strategic and Tactical Air Force (USSTAF), made it clear that, although radar was being used for bombardment through cloud cover, much more could be done to maximize the strategic and tactical advantages of new weapons and new developments in aircraft. In order to achieve the ability both to fly successful combat missions and to transport supplies and personnel safely in all kinds of weather, a coordinated, systematic program had to be set up that would include the research and development program. When Bowles returned from Europe late in May, he had assumed the responsibility to organize a massive attack on this problem. After consultation with Stratton and General Arnold, Bowles set up the Committee on Air Navigation and Traffic Control under the chairmanship of Stratton. Original members of this committee were Dr. W. L. Barrow (Sperry Gyroscope Company), H. H. Buttner (Federal Telephone and Radio Laboratories), W. Littlewood (American Airlines), A. E. Raymond (Douglas Aircraft Company), and D. G. Fink and N. H. Frank from Bowles's office. W. H. Ratliff joined as technical aide in January 1945. The deliberations and recommendations of this committee were intensive and far reaching, extending into civilian transport after the close of the war. While Bowles was away on a trip to the Far East in late 1944, Stratton acted as his deputy on all radar matters.

In addition to the Bowles-Stratton-Barrow communications group, others of the electrical engineering faculty were involved in Radiation Laboratory activities. Professor John G. Trump had begun as technical aide to Karl T. Compton when NDRC was organized in the summer of 1940. When Compton's Section D-1 turned its attention exclusively to microwave developments and became known as the Microwave Committee under the leadership of Alfred L. Loomis, Trump became involved with the setting up of the Radiation Laboratory. After Rad Lab got under way,

Trump became the NDRC representative at the laboratory; when in April 1942 Bowles was appointed expert consultant to the secretary of war, Trump became secretary of the Microwave Committee. He was the only MIT member of Rad Lab's steering committee, the laboratory's policy-formulating body. To quote from Guerlac's history, "His office, which by the time of the reorganization of OSRD in December 1942 had grown to consist of five staff members and five secretaries, and had come to be called the Division 14 or Radar Division Office, became a very important administrative unit in the radar program."

Trump's term as director of the British Branch of the Radiation Laboratory (BBRL) was preceded by the following events. In March 1941 Conant and Carroll Wilson had established the London mission of NDRC (later OSRD), leaving Dr. Frederic L. Hovde, of the University of Rochester, in charge. The following month a reciprocal British group was established in Washington, D.C., as the British Control Scientific Office under the direction of Dr. Charles Darwin, director of Britain's National Physical Laboratory. Through these offices exchanges of scientific information and personnel took place, including early visits by Bainbridge and Rabi to England. In May Frank D. Lewis of Rad Lab joined the London mission as radar liaison representative, continuing there until he was called to Bowles's office in Washington in January 1943 to take charge of radio-related problems. The increasing cooperation in radar development between Rad Lab and the Telecommunications Research Establishment (TRE) in England led to NDRC approval in October 1943 of the British Branch of the Radiation Laboratory. The first director of BBRL was Dr. Lauriston C. Marshall of Rad Lab, who got activities under way in December 1943. By this time David Griggs, formerly active at Rad Lab in the development of airborne radar, but now on the Bowles Washington staff, had

accompanied the first twelve B-17 planes equipped with H_2X radar to England and, through his technical skill with the radar sets and the training of crews, had demonstrated convincingly that this new three-centimeter, self-contained radar H_2X would be a powerful instrument for bombing through overcast.

In December 1943 Bowles and Stratton, with others from their Washington office, flew to North Africa, South Italy, and England to engage in the formulation of plans for Air Force communications in the European theater, especially for the bombing program of the Eighth and Fifteenth Air Forces. It was on this mission that Bowles and General Spaatz worked out a plan whereby a group of specialists in certain technical fields would be assigned by Bowles and attached directly to USSTAF headquarters, where they would aid in getting new devices and techniques into military operations. While responsible to General Spaatz, this Advisory Specialist Group (ASG) would have access through Bowles's office to the latest technical information from the MIT Radiation Laboratory, the Harvard Radio Research Laboratory (countermeasures) under the direction of Frederick Terman, and the American industrial laboratories.

Plans for the impending invasion of France and for the formation of the Advisory Specialist Group led to a decision in February 1944 by Bowles, DuBridge, and the Rad Lab Steering Committee to expand BBRL and to make Trump its new director. Bowles, Trump, DuBridge, Ridenour, and others flew to England and set about the reorganization of the American groups based there, in view of their relation to the new Advisory Specialist Group. These American groups were the OSRD London mission under Bennett Archambault, BBRL under Trump, and ABL-15 (the British branch of the Harvard Radio Research Laboratory) under Victor H. Fraenckel. In March 1944 General Spaatz issued a directive to the commanding general of the Eighth Air Force officially establishing the Advisory Specialist Group. This directive had been drawn up with the advice of Bowles and Trump, naming the initial group as DuBridge, Ridenour, Griggs, Fraenckel, and H. P. Robertson (a bombing specialist), and describing the technical backgrounds of these specialists. D. K. Martin (Bell Telephone Laboratories) and H. W. Hitchcock (Pacific Telephone and Telegraph Company) were soon added as communications specialists. Trump soon had the enlarged BBRL running in a businesslike way after the pattern of the Cambridge organization. To quote Guerlac, "Trump, by virtue of his position as head of the operating organization, and by personal ability, reliable judgment, quiet charm, and great patience, gradually became one of the most respected and influential scientific figures in the entire theater."

After the Allied invasion of Normandy, which began on June 6, 1944, Trump and Ridenour were disheartened to learn that Rad Lab in Cambridge felt that BBRL support was important only until D-Day. Trump visited the Ninth Air Force in France, arriving the day Paris fell to the French, August 23, 1944. He saw clearly that with the defeat of German military power still ahead, the bombing commands would need a continued high level of civilian aid in radar matters, and upon his return to England set about working out plans for a service base to be established near Paris with personnel from both BBRL (radar) and ABL-15 (countermeasures). At a luncheon on September 1 with General Spaatz and other high-ranking military officers, Trump expressed his ideas about the continued importance of radar and civilian field service. After lunch General Spaatz signed a letter to General Arnold prepared by the Advisory Specialist Group saying, in part, "As the intensity and scope of our operations increase, our need for the specialized help BBRL is supplying us grows. I should like to see the organization augmented, not decreased." During a two-month respite in the United States, Trump urged Rad Lab to support BBRL strongly, not only for its importance in

finishing the conquest of Germany but for its field experience in such recently developed systems as Microwave Early Warning (MEW), navigation systems (SS LORAN), and self-contained bombing systems (Eagle), which were in prospect for the Pacific theater.

When Trump returned to BBRL on November 12, he acquainted himself with activities in the ETO during his absence. On November 14 General Spaatz received a cable from Bowles, saying in part,

At my instance the following proposal has been approved by Dr. L. A. DuBridge, Director of NDRC Radiation Laboratory: The facilities and personnel of BBRL are hereby placed at the disposal of Military and Naval Forces operating in ETO to aid in the application of existing radar and radar equipment and in the modification of equipment to meet new requirements. To insure that this service shall be effective and that the activities of BBRL shall be coordinated closely with current and impending military operations, Dr. John G. Trump, Director of BBRL, has been appointed an Expert Consultant in my office, and will be assigned as a member of your Advisory Specialist Group. As such he will study requirements and plans in ETO and on the basis of information obtained therefrom will establish programs and priorities of BBRL, and in his capacity as Director of the Laboratory will supervise the implementation of these programs.

As the Allied armies advanced, Trump, Griggs, DuBridge, and other Americans eventually conferred with German radar and radio scientists as they compared the German and American radar activities. In anticipation of the impending V-E Day (May 8, 1945) Trump was back in the United States on April 27. His ETO experience gave him the background to redeploy personnel from Europe and to supply specialists from Rad Lab for action in the Pacific theater. For these tasks he was made chief of a new Rad Lab field service division and also became assistant director of Rad Lab.

One of the most sophisticated radars to be developed by the MIT Radiation Laboratory toward the end of World War II was the V-beam, designed to perform early-warning tasks as well as provide ground control for aircraft.

With the new focus of American power in the Far East, Bush asked Compton to leave his position as OSRD chief of field service and to establish a Pacific branch of OSRD in Manila. Bowles, as expert consultant to the secretary of war, arranged for Moreland to become "technical, industrial, and scientific advisor to General of the Army Douglas MacArthur" to assist in making American resources available in the support of military operations. He would, of course, cooperate with Compton's Pacific branch of OSRD. Compton and Moreland flew on the same plane to Manila to assume their new posts only a few days before V-J Day (August 14, 1945) and so were not much involved in war activities. They remained to head a scientific intelligence survey of Japan, however, contributing "notably to the preservation of valuable records and information and to the study of Japanese research and developments."

As final Allied victory came within sight, Bowles had given much thought to a peacetime program that would recognize the lessons of the war years. In his view the nation should be kept so thoroughly prepared that adjustment to a war footing could be brought about quickly. Such planning would also enhance peaceful pursuits, as educational institutions and industrial organizations would coordinate their activities more effectively than in the past. The military forces would be kept strong and abreast of scientific and engineering advances by attracting into their ranks top-level graduates of science and engineering schools and by sending promising young officers into schools for postgraduate study. "We must systematically and deliberately couple these elements together," wrote Bowles, "so as to form a continuing, working partnership, and thereby lay the foundation for maintaining our national integrity." Bowles received the Distinguished Service Medal on November 14, 1945. He remained in Washington until 1947, when he returned

Edward L. Bowles receives the Distinguished Service Medal from General Arnold, November 14, 1945.

to MIT as consulting professor of electrical engineering; in 1952 he became consulting professor of industrial management. He continued to have heavy responsibilities in the national government and was active in the affairs of the Raytheon Company and other industrial firms.

Other EE Staff

Professor Arthur R. von Hippel devoted his personal energies and his Laboratory for Insulation Research (LIR) during the war years to dielectric problems of Rad Lab and the armed services. As early as April 1940 a preliminary report by von Hippel and Shepard Roberts had been presented at the Washington meeting of the American Physical Society on a new method of measuring dielectric constant and loss in the range of centimeter waves. This research was originally sponsored by the International Telephone and Telegraph Company, and the authors acknowledged advice in the planning stages from J. A. Stratton and L. J. Chu. As part of von Hippel's ultrahigh-frequency program, Chu completed a paper, "Wave Guides with Dielectric Sections," in March 1941. Subsequent researches were carried out under von Hippel's own NDRC contract. Measurements at frequencies up to 3×10^{10} Hz were devised by LIR's measurements group under the leadership of William B. Westphal, and many new instruments were developed. Dr. von Hippel and the LIR staff furnished consultation to industrial makers of microwave equipment. Twenty-four papers were eventually published covering the wartime work of this laboratory.

Professor Harold E. Edgerton was well known for his applications of stroboscope and flash photography. His advice was sought on defense problems, and he was appointed a consultant to the Radiation Laboratory. His Medal of Freedom citation says, in part, "His untiring effort, resourcefulness, and competence made aerial night reconnaissance of enemy-held territory a reality under adverse weather conditions. . . .

Professor Harold Edgerton's stroboscopic and flash photography techniques performed a vital role in nighttime surveillance during World War II. This aerial photograph shows an important road and rail junction in Normandy on June 6, 1944.

Professor Lan Jen Chu, a pioneer in sophisticated radar systems at Rad Lab, who also made fundamental contributions to electromagnetic theory and formulated the small-signal power theorem, a basic principle in plasma physics.

The results of his endeavor have provided the U.S. Army Air Forces and Ground Forces with vital intelligence information that previous equipment of this nature was not capable of obtaining." In February 1944 Edgerton was appointed expert consultant in Bowles's office, and he was in the ETO for six months, developing flash photography techniques for nighttime surveillance. Edgerton's research associate, Kenneth J. Germeshausen, supervised Rad Lab's development of hydrogen thyratrons from the summer of 1941 to the end of the war. Acting as a switch to unload the energy stored in a pulse-forming network into a magnetron tube, the hydrogen thyratron was switching 150 kW of peak power by the end of 1942 and, by late 1944, 1,000 kW. Professor Ernst A. Guillemin was Rad Lab consultant on the design of pulse-forming networks.

Professor Lan Jen Chu was involved in Rad Lab's design of antennas, including the special-purpose type for scanning in airborne and shipborne radar systems. He was active in Rad Lab's studies of magnetrons and wave propagation and was consultant to Frederick E. Terman's Radio Research Laboratory at Harvard, which dealt mainly with radio countermeasures. As expert consultant to the secretary of war, Chu was made a member of Bowles's Advisory Specialist Group and sent to China on the staff of General A. C. Wedemeyer, commander of U.S. forces there. A native of China and a pioneer in microwave research, Chu was eminently qualified to assist the Chinese war establishment in the introduction of new radar equipment.

Other members of the electrical engineering staff were associated with Rad Lab for short periods. For example, Professor L. F. Woodruff worked during 1940–41 on servo design for the XT-1 radar; he left Rad Lab to work with Barrow in the MIT Radar School. He was later appointed civilian consultant to the U.S. Army Ordnance Department. Robert M. Fano, S.B. '41, Sc.D. '47, joined the laboratory in June 1944 after teaching measurements and communications at MIT; he worked on radio-frequency modulators and microwave filters.

Rad Lab Disbands

In 1944 and 1945 Rad Lab was at the height of its achievement, having become one of the world's outstanding research and development centers. Within months of the war's end, however, it would be no more. At the V-J convocation of August 14, 1945, in MIT's Great Court (now Killian Court) it was announced that Rad Lab's mission had been accomplished and that the laboratory would be brought to an "orderly termination" on December 31, 1945. The last months were spent in completing or transferring to the armed services or other contractors the studies and equipment that were under way. A publications office had been organized in June 1945 to prepare the documentation of the scientific and engineering achievements of Rad Lab, making them available to the educational and industrial communities. This material was eventually published by the McGraw-Hill Company in 27 volumes, with an index of the Radiation Laboratory Series—reports written during the war. Many papers were now submitted to scientific and technical journals. For a list of Rad Lab's research divisions and their constituent groups, with the names of division and group leaders as of 1945, see table 13.1.

Of great importance to MIT was the Basic Research Division organized on January 1, 1946, to carry on Rad Lab's research in physical electronics, microwave physics, and the behavior of materials at microwave frequencies. Engineering applications were made to transportation and communications. This new division was headed by Dr. Julius A. Stratton, with Dr. Albert G. Hill as associate director. It was operated under OSRD until July 1, 1946, when its activities and property were taken over completely by the Research Laboratory of Electronics (see chapter 16).

The MIT Radar School

Another aspect of MIT's role in World War II is seen in the specialized training programs it ran, including the influential MIT Radar School. In response to a demand for aeronautical engineers in 1939, following Hitler's thrust into the Low Countries and France, President Compton and the staff in aeronautical engineering made available in the summer of 1940, without charge, a ten-week course to train junior aeronautical engineers. Fifty-one graduates of various other engineering courses were given special training in this course and before the conclusion of their studies were signed up for employment in the manufacturing industries. The increased need for such specialized training brought about, in early 1941, a national program of courses sponsored by the United States Office of Education, known as Engineering Defense Training (EDT). Dean Moreland was appointed adviser to the northeastern district (Region I), and headquarters were established at MIT to administer the specialized free courses given in the Boston area. The institutions cooperating in the instruction were Harvard University, Northeastern University, Tufts College, and MIT. Congress appropriated $9 million to cover the "out-of-pocket" costs of these specialized courses. In the summer of 1941 Congress appropriated $17.5 million more and named the new program Engineering Science and Management Defense Training (ESMDT). Boston College and Boston University were added to the Region I group of institutions. By July 1942 the United States was at war and the annual congressional appropriation for the Engineering Science and Management War Training (ESMWT) program was raised to $30 million. More than a half-million students across the country received training in these specialized programs.

Table 13.1
Radiation Laboratory Research Divisions, 1945

Division 4—Research
I. I. Rabi
41—Fundamental Developments, E. M. Purcell
42—Propagation, D. E. Kerr
43—Theory, G. E. Uhlenbeck
44—Experimental Systems, J. L. Lawson
45—Special Dielectrics, O. Halpern

Division 5—Transmitter Components
J. R. Zacharias
51—Modulators, H. D. Doolittle
52—Transmitters, G. B. Collins
53—RF, A. G. Hill
54—Antennas, L. C. Van Atta
55—Test Equipment, F. J. Gaffney
56—Component Engineering, M. M. Hubbard
57—Special Problems, J. C. Slater

Division 6—Receiver Components
L. J. Haworth
61—Receivers, S. N. Van Voorhis
62—Indicators, C. Sherwin and J. Soller
63—Precision, B. Chance
64—Trainers, R. L. Garman
65—MTI, R. A. McConnell

Division 7—Beacons
L. A. Turner
71—RACONS, A. Roberts
72—Identification, M. D. O'Day

Division 8—Fire Control and Army Ground Forces
I. A. Getting
81—Systems, W. L. Davenport
82—Systems, R. P. Scott
83—Servos, N. B. Nichols
84—Theory, R. S. Phillips
85—Design, J. S. White

Division 9—Airborne Systems
M. G. White
91 T. W. Bonner
92 M. G. White
93 W. M. Cady

Division 10—Ground and Ship
J. C. Street
101—Mechanical Engineering, M. B. Karelitz
102—Ship Applications, J. S. Hall and R. E. Meagher
103—Special Applications, R. M. Emberson
104—Ground Applications, E. G. Schneider

Division 11—Navigation
J. A. Pierce
111—Laboratory, A. J. Pote
112—LORAN Operational Research, J. A. Pierce
113—Field Engineering and Procurement, W. L. Tierney

Division 12—Field Service
J. G. Trump

British Branch
J. G. Trump

The MIT Radar School, largest of the specialized training programs at the Institute, began in June 1941 as a project of the Electrical Engineering Department under the national EDT program; from July 1943 the operating expenses were financed through Navy and Army contracts with MIT. Here a total of more than 8,800 officers, enlisted men, and civilians were trained in radar techniques.

Many members of the EE Department contributed to the development of the Radar School. The directors of the school, first Professor Wilmer L. Barrow and later Professor Carlton E. Tucker, were both from the EE Department, as were several associate directors, Professors William H. Radford, Henry J. Zimmermann, and Malcolm S. McIlroy. Both McIlroy and Radford wrote excellent chronicles. McIlroy's "History of the MIT Radar School in Relation to Army Training" focused on Army programs, whereas Radford's "Technology's Radar School" emphasized Navy programs.

In terms of engineering education the MIT Radar School was far more influential at MIT, and at other schools, than its highly specialized and short training periods might suggest. Out of the school came a widely acclaimed treatise, "Principles of Radar," by Professors Reintjes and Coate of the EE Department. Also, a course that Professor Zimmermann was to introduce as a graduate subject in 1946, Pulse Circuits, was a direct outgrowth of the Radar School teaching material. This subject, in turn, had a major impact on all EE curricula in the United States in the immediate postwar years. It led to a realization that circuit analysis should be taught from the viewpoint of generalized waveform inputs and that circuit analysis should be invariant, whether the input to the circuit is sinusoidal or nonsinusoidal.

Chapter 14

Servomechanisms: The Bridge to a New Period

Of the many laboratories established within the Electrical Engineering Department over its long history, one of the most fruitful and influential was the Servomechanisms Laboratory. Founded on the eve of World War II, the Servo Lab performed a vital wartime role in the development of gun-positioning instruments; among its postwar activities were the development of automatic controls for one of the nuclear reactors at the Brookhaven National Laboratory, systematic studies aimed at process control, and the creation of the field called numerical control, which would help bring about a major revolution in the machine tool industry by the late 1950s. Thereafter, with the growing availability of large and fast digital computers in the MIT Digital Computer Laboratory (an offspring of the Servo Lab), automatic programming systems for numerical control were devised. These, in turn, would spawn the more generalized Computer-Aided Design Project, which by 1960 was able to rely on large general-purpose interactive computers. By then the computer revolution was under way, penetrating every field of scientific and engineering endeavor, a development we will pick up in later chapters. Though the Servomechanisms Laboratory changed its name to the Electronic Systems Laboratory in 1959 (the year Gordon Brown became dean of engineering), and though it widened its activities in many areas, such as library automation, it retained a central identity in automatic control and a close coupling with the teaching program. Two important laboratories, the Digital Computer Laboratory and the Dynamic Analysis and Control Laboratory, grew out of the old Servo Lab. The lab not only proved an important link to the future; it also ran counter to the prevailing spirit of academic research, owing in large part to the pragmatic character of the man who first headed it, Gordon S. Brown.

With the departure of Vannevar Bush for Washington, the EE Department had, in fact, reached a significant turning point, for Bush had been the key instigator of change in graduate research programs. The Servo Lab could be seen as providing something of a surrogate for Bush's influence in the person of Brown, a member of the new generation of engineers who had worked closely with Bush and Hazen and who had imbibed the outlook of Dugald Jackson.

In the Servo Lab, says Brown, the presiding spirit was a drive to get things done on time. Unlike many other academic laboratories (except for that of Draper, who was pursuing similar lines), this one undertook real engineering development projects on contract, a novel experience even for a school of engineering at that time. Thus, asserts Brown with pride, the faculty and students worked "in an interdisciplinary environment on relatively new and authentic problems that had not been tackled before, in ways whereby they carried responsibility to get them done." In weapons development, he observes, "we'd never seen a hydraulic control system in our lives until we actually had a 40-mm gun control in the lab." The gun control Brown's lab developed went to the Army and then out to actual combat operation. Eventually, 40,000 of them were manufactured by American industry. "This was contrary," Brown continues, "to the academic doctrine that you should never do anything in a rush, you should never have deadlines, you should be allowed to live in an ivory tower world of dabbling." A fundamental shift of outlook was taking place in the wartime period—a true practicum was biting deeper into engineering training.

In retrospect, the story of the Servomechanisms Laboratory and its successor, the Electronic Systems Laboratory, seems to exhibit an almost ideal interaction of research and education. Real problems, whether of war or peace, attracted bright students and staff to work together in an atmosphere of professional engineering attainment. While engaged in these pursuits, the staff and the students prepared themselves to take on increasingly responsible positions in industry and in society generally.

As Brown rose in the department he was able to translate the unique experience of the Servo Lab into a thoroughgoing electrical engineering educational reform—some called it a revolution. "As an undergraduate," says Brown, "I spent my machinery days measuring the steady-state performance of motors and generators, just as hundreds of my predecessors had, all working on machines that had been designed and built decades before. Now, in the Servo Lab, we had an opportunity to come to grips with the dynamics of machines and to build machines of our own. As a result of our work in the whole area of feedback control systems, which Hazen and Bush had pioneered, we were able to bring about a radical change." The study of feedback control was to have vast ramifications in the postwar years in areas as diverse as the numerical control of machine tools and neurophysiology; the field was largely uncharted when Brown seized the opportunity to get into it.

Brown, a native of Australia, had come to the United States in 1929 at the age of 22. He received the bachelor of science degree from MIT in 1931, having been a member of the Course-VI honors group. In his graduate program he tackled some of the tough mathematical problems of that period, and for his Sc.D. thesis in 1938 he developed the cinema integraph, a computational machine that used motion pictures rather than cardboard or metal "function masks." When he became an assistant professor of electrical engineering in 1939, he had been in charge of the department's Research Laboratory for several years and so was intimately involved in many of the

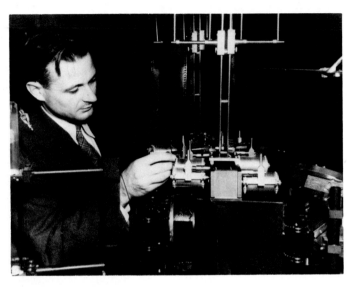

The young Gordon Brown working on the cinema integraph.

ongoing research projects. At this time Hazen proposed that Brown enter the field of servomechanisms, resulting in Brown's pioneering instruction in this field and the founding of the Servomechanisms Laboratory. One of MIT's early contracts with the National Defense Research Committee in October 1940 resulted from Brown's proposal for investigations in servomechanisms. By the war's end in 1945 Brown was doing over a million dollars of servo business per year.

Servomechanisms Development

Although servomechanisms had been in use for at least a century—Watt's flyball governor on one of his steam engines is perhaps the earliest example—the theory of their functioning was not much advanced when Bush and Hazen began their researches in the 1920s and 1930s. It was hardly suspected then just how important they would become: concepts of feedback control in conjunction with information theory (which was being developed during this same period) were to be crucial architectural elements in the field of cybernetics, which began to take form in the postwar period.

At MIT the first work in servomechanisms was done in connection with the early integraphs (see chapter 4). These devices utilized servomechanisms of the on-off variety, while the automatic curve follower and the cinema integraph contained servos actuated by variable but continuous signals. In the 1930s Bush had encouraged Hazen to write a treatise on the theory and design of servomechanisms based on Hazen's research and knowledge in the field. Bush thought such a treatise might well become a classic that would lay the foundation for further servo development; and indeed subsequent writers have considered the Hazen papers the basis for the development of today's rigorous servomechanism, or feedback control, theory.

The word *servo* was not new with Bush or Hazen; it was a familiar term in the late 1920s and early 1930s when autopilots for airplanes were under development. The term can be traced back to Joseph Farcot, a graduate of the Ecole Centrale in Paris, who invented the servo-moteur in 1868 and received a prize for it in 1872. J. C. Maxwell in 1868 was probably the first to analyze mathematically Watt's flyball governor and other "moderating" and "governing" devices; E. J. Routh provided another early study in 1877.[1] Hazen remarks that N. Minorsky, in a paper on continuous control published in 1922, "gives an excellent analysis of the rudder-hull dynamic system in the ship-steering problem." This paper points out the inadequacy of mere positional control and stresses the importance of including rates of change in the automatic steering process.[2] In his two papers published in 1934, Hazen gives a systematic quantitative treatment of each of the categories—relay-type servos, definite-correction servos, and continuous-control servos—and shows how the theory is applied in the design of a specific servo that was needed to balance against each other the light fluxes of two photoelectric cells in the cinema integraph.[3] The design comprised a DC amplifier (the subject of part of Gordon Brown's S.M. thesis in 1934) and a servomotor with a low-inertia rotor and a mechanical (Nieman) torque amplifier. This servo was also used in the automatic curve follower for the differential analyzer. The curve follower was described in the S.M. thesis of Harold A. Traver (VI-A S.B. '32, S.M. '33) and in a paper by Hazen, Brown, and Jaeger in the December 1936 *Review of Scientific Instruments*. It was exhibited at the Chicago World's Fair of 1932–33 (as was Professor Trump's Van de Graaff generator) and at the Harvard-MIT meeting of the American Association for the Advancement of Science in December 1933. The Franklin Institute awarded Hazen its Levy Gold Medal in 1936 for his publications on these servo developments.

A more ominous occasion for further servomechanism work was suggested by political events in Europe. In the late 1930s it became evident to the U.S. Navy that there was a need for accurate and rapid gun-positioning equipment on ships. Although the Navy had the world's best fire-control equipment, the increasing speeds of airplanes and the complexity of ship maneuvers demanded a further advance in fire-control techniques. Hazen's papers showed that MIT could make a significant contribution in the field of servomechanisms. Accordingly the Navy requested in the spring of 1939 that a group of four naval lieutenants be accepted for a special course in servomechanisms and fire control, presumably to be taught by Hazen. However, Hazen's new responsibilities as department head necessitated that someone else take on the servo teaching. Gordon Brown, by then an assistant professor, had worked closely with Hazen on the servos in the differential analyzer, the cinema integraph, and the automatic curve follower, and was therefore next to Hazen in servo experience. Hazen asked Brown to take over the course, which would deal with such subjects as gyroscopic stabilization, follow-up mechanisms, and positioning machinery for heavy 16-inch batteries and deck-mounted 5-inch/38 guns. By the time Brown started the course in the fall of 1939, the Germans had invaded Poland.

The officers assigned to MIT in September 1939 were Lieutenants Edwin B. Hooper, Lloyd M. Mustin, Horacio Rivero, and Alfred G. Ward. Four graduate students, J. L. C. Löf and E. C. Mitchell of the Electrical Engineering Department and Y. J. Liu and A. R. Maxwell of the Aeronautical Engineering Department also registered for the course. John W. Anderson, a graduate student and research assistant, was assigned to the servo work half-time. George C. Newton, Jr., a third-year undergraduate student, who would later become associate director of the Electronic Systems Lab, assisted as a part-time employee. The Hazen and Minorsky papers were studied in depth and laid the foundation for the analysis of follow-up systems. Dr. Stark Draper took an active interest in the new servo class; it was one of his doctoral students, Y. J. Liu, who contributed the charts for the solution of cubic and quartic algebraic equations that appear in the analyses of certain dynamic systems. Dr. Hugh Willis of the Sperry Gyroscope Company, at that time interested in Draper's gyroscopic lead-computing sight, furnished some of the first experimental equipment. Brown recalls,

During the second term we began to pull together a laboratory. I somehow fell heir to a five-horsepower hydraulic piston-pump piston-motor assembly. Harry Lawrence [of the EE shop] found a large bed-plate for me, and we borrowed or begged a five-horsepower, 600-RPM motor to drive it. This was an enormously bulky piece of equipment, and it was all assembled on the floor of my office, Room 4-232. The second-term subject, 6.606, had two hours of class and three hours of laboratory each week and the student group was enlarged through the addition of S. C. Russell [Course II], C. Wang [Course VI], and C. C. Lawry, Jr. [Course VI-A]. That winter, Selden B. Crary of the General Electric Company delivered an AIEE paper on the voltage regulation of large AC generators.

During the rest of the academic year Brown and his associates worked out a unified methodology for the analysis of the broader category of feedback systems, embracing not only the follow-up variety but also the regulator variety used in governing the constant speed of steam turbines and in the voltage regulation of AC generators. The title of the course was changed to "Dynamic Analysis of Automatic Control Systems" for the year 1940–41.

The four naval fire-control officers did their thesis research in pairs—Hooper and Ward, Mustin and Rivero—under Brown's supervision. Brown recalls,

Mustin was a pistol-shot addict and had the conviction that a gyroscope could be used to make a gun sight for the 20-mm Oerlikon deck mount widely used for the air defense of naval vessels. Knowing that Draper had invented and built a working model of a two-gyro gun sight which, by this time, had become a 'little black box,' I suggested to Mustin and Rivero that they talk their problem over with Draper. Draper was unusually well grounded in mathematics and physics; he also had an uncanny ability to come up with a simple and practical solution to a seemingly complex problem. Gun directors of 1939 were indeed complex devices, with ball-and-disc integrators and other slow-acting components. Draper saw that these were too sluggish to cope with fast-flying torpedo planes. His solution was to place the little black box directly on the gun, so that the computing mechanism could displace the reticle of the gun sight. When the gunner, in tracking the target, held the reticle of his gun sight on the moving target, the motion of the gun and gunsight would quickly and continuously cause the gun barrel to point in the right direction for a hit. In the winter of 1940–41 Jay Forrester, Al Hall, Jack Silvey, and I witnessed a demonstration of this sight at Watertown Arsenal. The little black box was mounted on the barrel of a 22-caliber rifle which, in turn, was mounted on a plumber's pipe-threading stand. As the marksman kept his reticle aimed at a moving piece of cloth, the gyros computed the lead angle for the gun barrel and the bullets hit the target. However, in 1939–40 the Armed Services were so impressed with their sophisticated and costly devices that they were not interested in the Draper sight, but after Pearl Harbor and the sinking of the British ships Repulse *and* Prince of Wales, *the Navy was awakened to the value of the Draper invention. Out of this came the Navy's famous Mark 14 sight, which in its production model aboard the USS* South Dakota *brought down so many Japanese planes on October 24, 1942, in the Battle of Santa Cruz.*

Lieutenants Mustin and Rivero submitted their thesis, "A Servomechanism for a Rate Follow-up System," on May 16, 1940. In the thesis proposal, dated March 7, 1940, was the following statement: "Although the servomechanisms involved will not represent confidential material, the specific field of their application, their particular requirements, and the background treatment involved will necessitate the special classification." The thesis was indeed classified "confidential" and was not available at MIT until July 1972. To set the work in historical perspective, Mustin prepared, in January 1971, a memorandum to accompany it in the Navy files. Here he pointed out that the title was deliberately uninformative; he explained that "central to the thesis is its concept of an antiaircraft fire-control device of which a primary component of the solution would be relative target motion as measured by the precession of constrained gyroscopes. . . . This approach would permit development of a device having very fast solution times, accuracy adequate for the close-in problem, simple and compact in size (on the order of less than one cubic foot for the complete computational package), and capable of substantially better control accuracy than any other devices then in use or known to be under development for similar purposes." "Materials in several of Dr. Draper's gyro and instrumentation courses taken at MIT by Lts. Rivero and Mustin were substantial factors in the evolution of the concepts seen in this thesis," he acknowledged; however, "Dr. Draper did not suggest any gun-control applications at the time."

The Hooper-Ward unclassified thesis was entitled "Control of an Electro-Hydraulic Servo Unit." In it Hooper presented a concept of criteria for minimum errors that turned out to be of fundamental value.

In the summer of 1940 Brown spent considerable time at the Sperry Gyroscope Company and was introduced to people at the Ford Instrument Company and the Waterbury Tool Company. He came back to MIT with his car full of equipment, including some one- and two-horsepower hydraulic piston-type pump motors and hydraulic transmissions. Sperry became much interested in the growing servo activity at MIT, and in the early fall of 1940 Dr. Hugh Willis of Sperry asked Brown to undertake some "fresh fundamental research" in servo design, giving special attention to an all-hydraulic system. The Sperry agreement was officially signed with MIT's Division of Industrial Cooperation (DIC) in December, although the research had started in September.

A strong research team was already established. In the first term, 1940–41, J. W. Anderson was continued half-time; A. C. Hall was assigned half-time to the Servo Lab while continuing to carry administrative responsibilities in the Measurements Laboratory under Professor T. S. Gray; D. P. Campbell, a new graduate student from Union College, was assigned full-time. G. C. Newton continued his part-time employment and in the second term wrote his bachelor's thesis under Brown's supervision. By December Brown had recruited J. O. Silvey, S.M. '39, from the Lombard Governor Company in Ashland, Massachusetts, a skillful man with mechanical devices; and Sperry had detailed to Servo Lab one of their men, Ed Dawson, who had had previous experience in the development of hydraulic valves used to actuate small power pistons.

Early studies were on the small hydraulic transmissions; but soon, Brown recalls,

It became desirable, in Sperry's judgment, to introduce us to the more modern development in hydraulic transmissions, and we were supplied with some of the hand-built models of what was to become the hydraulic transmission used in the 50-caliber twin-turret mounts on B-24 and later on B-17 planes. These were driven by 3,600-RPM, 24-volt DC motors, and the hydraulic pump-and-motor assembly was all contained in a seven-inch cubic aluminum box which was relatively light in weight. I remember well my surprise when I found that, operating at a thousand pounds per square inch, these devices could develop about five horsepower, all in that tiny space, whereas a year and a half earlier I had begun to put together a five-horsepower hydraulic-drive system where the power assembly weighed several hundred pounds.

The 37-mm antiaircraft gun had been standardized in 1927 by the Army Ordnance Department and had been put into production in 1938. Development was continuing at the Watertown and Frankford Arsenals, and the Sperry Gyroscope Company was under contract to aid in this project. By the fall of 1940 the MIT servo research activities were so attractive to Sperry that an arrangement was made for the Watertown Arsenal to furnish a 37-mm mount to Brown for laboratory experimentation. The mount was assembled in Room 10-066, next to the elevators in the basement of Building 10. The space was so limited that when the gun was horizontal it could not be rotated the full 360 degrees in azimuth. (This limitation ultimately produced a useful design innovation.)

By the beginning of the second term, 1940–41, Brown had recruited J. W. Forrester, who had spent a year and a half as research assistant in Trump's High-Voltage Laboratory. Forrester, Anderson, and Campbell were assigned to full-time duty in the Servo Lab, and A. C. Hall's time was increased to three-fourths. By early March 1941, two drives, one in azimuth and one in elevation, had been mounted on the 37-mm gun and were working well. At a conference one afternoon in Room 10-066 various ways were sketched by which the tracking error might be reduced to zero, and it was Forrester who conceived the beginnings of the error-correcting mechanism for which a joint patent was later issued to Forrester and Brown. It proved a widely usable device and was applied to various servomechanism configurations.

The rapidly expanding servo research activities soon outgrew their two small rooms. A new building at the back of the campus, intended for use by the Military Science Department as storage space, was nearing completion. Brown moved his lab into the west end of the building in April 1941.

Wartime Servo Lab Projects
The beginning of American participation in the war in late 1941 brought home the importance of the research going on in the Servo Lab. Wartime projects included improvements in a British oil-gear servo, design of a speed-gear servo, construction of prototypes, construction of azimuth and elevation control units, design and construction of a servo for a fuse-setter rammer, design and construction of mount power drives for 40-mm guns, and many other gun-related developments.

Evolution of Servo Theory
Under the exigencies of wartime research work, the documentation of new findings in servo theory and their incorporation in the educational program went forward under Gordon Brown's supervision in an ad hoc fashion. Things were evolving so fast that text material, worked out in the form of notes, was continually under revision.

Campbell assisted Brown in the classroom instruction in 1941–42 and advanced rapidly as an effective expositor of servo theory and its applications, teaching a section of 6.606 in the spring of 1943 and a section of 6.605 that summer. The text material received contributions from those who taught sections of subjects 6.605 and 6.606, notably Professor Harold I. Tarpley of Pennsylvania State University, H. Tyler Marcy, George C. Newton, Jr., William M. Pease, Ernest W. Therkelsen, and Robert B. Wilcox. Finally in 1948 Brown and Campbell's *Principles of Servomech-*

anisms was published as a textbook for senior and graduate students. The central theme of the book was "dynamic synthesis." It began with the transient (differential equation) approach in the time domain and then showed how to set the time functions over into the frequency domain through the Laplace transformation. The frequency-domain analysis and synthesis were included as developed in the doctoral thesis of A. C. Hall, "The Analysis and Synthesis of Linear Servomechanisms," first classified in 1943 as a "restricted" document and declassified in 1946. Chapter 11 of the Brown-Campbell book, entitled "A Method for Approximating the Transient Response from the Frequency Response," was part of the thesis research of George F. Floyd (VI-A S.B. '43, XVIII S.M. '47, VI Sc.D. '49), supervised by Brown and Campbell. As early as the fall term, 1945–46, Campbell offered a new undergraduate subject, 6.213, Automatic Control Principles and Applications, and included a section on the process control system.

The Postwar Period: Process Control and Nuclear Reactor Instrumentation

Despite the advances made in servomechanisms during the war years, industrial applications of automatic control to instrumentation and manufacturing processes were frequently still mere appendages, added to regulate this or that aspect of a complex process. There was no true integration of functions. Campbell became concerned about this problem and set out to develop a systematic approach growing out of the Servo Lab experience with military fire-control system design. He regarded the complete process of a manufacturing plant as a dynamic system to be treated in a scientific and mathematical way.[4] Campbell made process control his specialty and developed a graduate course by that name. His book *Process Dynamics* was nearly ready to go to press at the time of his death in January 1957; his friends Page S. Buckley of the Du Pont Company, Leonard A. Gould

Drawn into the teaching of servomechanism theory, Brown supervised its practical wartime application in gun-related and other projects.

and Herbert M. Teager of MIT, and H. Sandvold of the Norwegian Hydroelectric Company took up the task of carrying the book through to publication in 1958.

Leonard A. Gould (S.B. '48, S.M. '50, Sc.D. '53) was closely associated with Campbell and as a professor continued to provide strong leadership in this field. His experience with the chemical industries brought him the conviction that the process-control engineer could greatly increase his effectiveness if he would first study the process itself, setting up its mathematical model and introducing suitable simplifications, and then look to control theory for an applicable control technique. This order of approach was in contrast to the prevailing method of first considering a general inclusive control theory and then looking for applications among the chemical processes.

Out of Gould's approach came a new body of control theory, more pertinent and therefore more powerful in the solution of chemical control problems. Gould's book *Chemical Process Control*, published in 1969, set forth this closer link between control theory and its applications. Staff and students of the Department of Chemical Engineering joined Gould and his Electrical Engineering Group in the Electronic Systems Laboratory (ESL); this cooperation resulted in a high level of research and educational activity. Gould also pursued an interest in advanced feedback-control theory and coauthored with George C. Newton, Jr., and James H. Kaiser the 1957 book *Analytic Design of Linear Feedback Controls*, which provides an excellent history of the control art. Gould was appointed an associate director of ESL in 1970.

The war's end did not stop the flow of officers detailed to the department for special graduate programs. In the fall of 1949 eighteen officers were registered, including the usual four in naval fire control. Military research in the Servo Lab continued at a significant level, but peacetime applications of automatic control required increasing attention. Of that period Gordon Brown wrote,

Brookhaven National Laboratory [established by the Atomic Energy Commission] was coming into existence in the summer of 1947, and there was need for someone who would undertake the development of the instrumentation and rod-drive equipment for the graphite nuclear reactor that was being installed there. This was a fairly large program, and it ran into about a million dollars. Bill Pease, John Fiore, Joe Ivaska, Bob Bittenbender, Jim McDonough, Bill Grim, and other MIT men, with a large group of people loaned to the laboratory by Jackson and Moreland, actually designed, developed, and manufactured 185,000 pounds of equipment, including hydraulic motors, hydraulic power drives, and other devices to insert and withdraw the rods, and rather unique "Scram" equipment for rapid shutdown in case of an emergency. Professor Truman Gray, who had stood by the educational system during the war, had been promised a year off after the war and elected to take this leave in 1947–48 as an employee of Servo Lab where he took charge of designing and developing the reactor's instrumentation. He continued as a consultant to Brookhaven until 1951. When the reactor was brought up to critical state, the equipment worked extremely well.

Numerical Control of Machine Tools

Another significant activity of the Servo Lab began, Gordon Brown recollects,

just when Bill Linvill, VI-A S.B. '45, S.M. '46, Sc.D. '49, had finished his doctoral thesis, "Control Systems Operating on Sampled Data." It so happened that John Parsons telephoned me one afternoon from Worcester to ask if I could furnish him with a power drive that would take intermittent data. After a little discussion of what he meant by power drive, it seems that he had in mind something of a couple of horse-

Professor Truman S. Gray (*seated on lab bench*) with members of the Electronic Instrumentation Group in the MIT Servomechanisms Laboratory. *Left to right:* David R. Whitehouse, Professor Thomas F. Jones, Professor Gray, Hardy M. Bourland, Richard H. Spencer, and Professor Albert B. Van Rennes.

power, and what he meant by intermittent data turned out later to be pulses that would come from some punched cards, or the like, that would carry the instructions for the control of a machine tool. I told him that about $25,000 was a round figure for a program of that kind, and that was the beginning of the program which became Numerical Control. This activity has brought great distinction to the Institute and at the same time was one factor in a major revolution of the whole machine tool industry.

The first, or basic study, phase of this project was sponsored by the Parsons Corporation of Traverse City, Michigan, and ran from July 1949 through June 1950. Parsons was under contract to the Air Materiel Command of the U.S. Air Force to design and construct a milling machine directed by data punched on IBM cards, as originally conceived by John T. Parsons. The Servo Lab was to study the general problem of controlling a machine tool from numerical data and to assist Parsons in the design of certain of the system components. Assistant Professor William M. Pease, VI-A S.B. '42, S.M. '43, was made project engineer and James O. McDonough, VI-A S.B. '43, S.M. '47, assistant project engineer for this undertaking. In their search for an original, simple descriptive term for their new development, Pease and McDonough coined the name *numerical control*. Among the many feasible machine-control systems considered by the Servo Lab group, one seemed most promising and was studied in depth. In June 1950 Servo Lab issued a report to Parsons containing design and performance specifications for a numerically controlled milling machine that would receive its operating data from a punched paper tape.

Parsons then authorized Servo Lab to enter upon the second phase, which called for the design and construction of an experimental milling machine according to the specifications of phase one. A standard Cincinnati Hydro-Tel milling machine with a working space of 60 by 30 by 15 inches was furnished to MIT by the

Air Force. Its three conventional controls for positioning the cutting tool were removed and replaced by three hydraulic power servos receiving simultaneous but individual commands from their respective command synchros in the *director*. With this arrangement the cutting tool could be guided over a prescribed space path.

The director, electrically connected to the machine and therefore offering remote operation, was constructed on panels in six standard relay racks to afford ready accessibility in the experimental version. The director performed several functions. It read the binary numbers that had been recorded on a 7/8-inch paper tape and converted these numbers into electrical pulse trains that carried the machining instructions in each of the space coordinates, x, y, and z. The final function of the director was to decode these pulse trains into angular positions of the command servos, each servo governing the motion of the milling machine in its own particular coordinate. Of course this rough description of the numerically controlled milling machine does not convey how ingenious and sophisticated was the design of the components that performed the various functions; but it does indicate in a general way how this machine could turn out a machined piece automatically according to a program punched on a paper tape.

Progress was rapid on this development, and the original Parsons concepts were supplemented by many inputs from the MIT community. Available resources included not only the Servo Lab staff but also personnel and facilities of the Whirlwind computer project, the Center of Analysis, the Research Laboratory of Electronics, and other research centers. Alfred K. Susskind, S.M. '50, who had been a research assistant in Project Whirlwind, was transferred to Servo Lab in the summer of 1950 and was especially helpful in the design and computer aspects of numerical con-

Gordon Brown's illustration of feedback control, about 1950.

trol. Jay Forrester and R. R. Everett of Whirlwind were consultants, and by February 1951 the machine was about 30 percent complete.

The scope of the project had been broadened to such an extent, however, that the Parsons Corporation terminated its sponsorship, and the machine was completed in March 1952 under a new MIT contract directly with the Air Force. In April 1951 McDonough became project engineer as Pease moved up to the directorship of Servo Lab, Brown having relinquished this post to take on his duties as associate head of the Electrical Engineering Department. Two patent applications covering the inventions were filed in 1952: the first by John T. Parsons and Frank L. Stulen, entitled "Motor Controlled Apparatus for Positioning Machine Tool"; the second by J. W. Forrester, William M. Pease, James O. McDonough, and Alfred K. Susskind, entitled "Numerical Control Servo System." The Parsons patent was issued in 1958 and the Forrester patent in 1962; Parsons acquired title to both and made the Bendix Corporation an exclusive licensee.

In the summer of 1952 a demonstration and discussion of the numerically controlled milling machine was announced for September. The announcement said, in part, "It is now possible to make a preliminary assessment of the economic potential of this new form of machine-tool control. It is therefore appropriate to acquaint industry with this development and to initiate an interchange of information in order to guide the work now in progress at MIT along the lines which will result in maximum usefulness to industry."

Pease, after his resignation from MIT, was succeeded in September 1953 as director of the Servomechanisms Laboratory by Professor J. Francis Reintjes, who carried forward the later phases of numerical control. The industries were not immediately convinced of its practicality. "From time to time," says Brown, "we had visitors from major machine-tool manufacturers,

The beginnings of a major revolution in the machine tool industry can be traced to this numerically controlled milling machine, developed at the MIT Servomechanisms Laboratory.

POST, AUGUST 3, 1952

MILLING OF PARTS DONE BY A ROBOT

M.I.T. Now Operates Mathematical Wizard

A mathematical wizard is in operation at the Massachusetts Institute of Technology which performs in minutes many of the tasks that in current practice take hours.

The numerically controlled milling machine, believed to be the first of its kind in the world, represents a new and pioneering step in the automatic control of machine tools.

RESPONDS TO NUMBERS

The M. I. T. robot responds to instructions transmitted to it as numbers, and substitutes mathematical operations, which are performed on modern computing devices for many of the hand operations now required in industrial production.

The numbers which direct the M. I. T. machine are derived directly from the drawings and specifications of the part to be worked. This digital information is coded and punched on a paper tape similar to that used in the automatic transmission of teletyped messages.

The information on the tape is interpreted by the machine director which employs standard electronic information-processing techniques. The commands are then carried out by power amplifying devices called servomechanisms.

Servomechanisms, or servos, came into prominence in World war II when they were used in the control of radar systems, shipboard guns and secondary equipment for military needs.

This robot has been developed at the Servomechanisms Research Laboratory at the Massachusetts Institute of Technology under contract to the Air Materiel Command at the Wright-Patterson Air Force Base. Its development was supervised by James O. McDonough, project engineer, under the direction of Professor William M. Pease.

He points out that once a tape is properly punched, it provides a permanent, compact control record which may be used at any time for the milling of duplicate parts.

"Such a tape," he says, "guarantees a tremendous saving in time to an industry like the aircraft industry where a single machine must mill a variety of parts on a staggered schedule."

This news item conveys the "gee-whiz" response to the development of the numerically controlled milling machine.

some expressing violent opposition to the work. The machine-tool makers who came from Detroit simply brushed it off and said, 'It's a pretty poor way to build a million automobile fenders.' Of course it was never intended to build a million fenders, but to handle the machining of complex, short-run parts with high efficiency and repeatability.'' McDonough and his group set out on a persistent quest for jobs that would demonstrate the superior abilities of numerical control. Operational data, costs, and time schedules were recorded for each of the customer jobs; and beginning in June 1954, two members of the School of Industrial Management, Assistant Professors Robert H. Gregory and Thomas V. Atwater, Jr., made an economic study, comparing the costs of jobs performed by numerical control at the Servo Lab with costs estimated by conventional job shops.

After the 1952 Servo Lab demonstration, although there was initially little obvious enthusiasm for numerical control in industry, a ''fairly systematic public relations campaign'' began to produce results. By 1953 a number of companies were experimenting with numerical-control systems, and a major program by the Air Force, which was pushing for higher productive capability in the aircraft industry, stimulated new developments. By September 1960, at the machine tool show in Chicago, there were about a hundred numerical-control systems on display; it was apparent that this innovation had made its way out of the laboratory and become a real commercial success. On April 9, 1970, the Joseph Marie Jacquard Memorial Award of the Numerical Control Society was presented to Gordon S. Brown, William M. Pease, and James O. McDonough ''for their pioneering work in developing the first practical numerical-control system.''

Toward Computerization of Numerical Control
The Gregory-Atwater economic study, published in March 1956, indicated that Servo Lab jobs performed with the numerically controlled milling machine were roughly competitive with similar work in conventional job shops. A successful attack upon the programming problem would give numerical control an unqualified lead; in fact, some breakthroughs were at that time already under way.

Certain machining operations, such as cutting a circular shape of given center and radius, appeared frequently in many job specifications. A computer program for carrying out such an operation was called a *subroutine*. John N. Runyon, E.E. '55, a research assistant in the Servo Lab, developed as a thesis project on the Whirlwind I computer the first set of subroutines applicable to numerical control. The idea was to assemble in the computer memory a library of subroutines, any one of which could quickly be used as a component of a particular machining program. In 1955 Arnold Siegel demonstrated at the MIT Digital Computer Laboratory (Whirlwind I) the first pilot model of an automatic programming system for numerical control in two dimensions.

In June 1956 the Servo Lab, under its contract with the Air Materiel Command of the Air Force, reorganized its effort to deal with the programming problem. A fundamental system was needed, one built around an incomplete or generally applicable central or skeletal solution, ''fleshed out'' to complete the program for a particular job through the addition of one or two subroutines of simple structure. By fall such a system had been conceived by Douglas T. Ross, S.M. '54, and his Computer Applications Group. Named APT—Automatically Programmed Tools—it was not a single system but a family of systems, capable of growth and further development around a central process and philosophy. In early 1957 an exemplary system was in operation on the Whirlwind I computer.

At this point Reintjes and Ross saw quite clearly that the development of APT on an industry-wide basis was beyond the capacity of the small research group in the Servo Lab, and in May 1957 representatives from a number of aircraft companies were brought together for a week at MIT to promote and organize this further development. Each programmer became responsible for working out a designated aspect of APT in his own plant and on his own computer. The Servo Lab then assembled these component programs into one overall APT program and in April 1958 shipped the field trial version to 17 of the participating companies for shakedown tests. In June 1958 the coordination and further development of the effectively launched APT systems was transferred from the Servo Lab to the APT Project Coordinating Group, an industry group sponsored by the Aircraft Industries Association. A demonstration of APT at an MIT press conference on February 25, 1959, was sponsored jointly by the aircraft industries, the Air Force, and MIT; Reintjes and Ross made a television presentation on March 4, 1959, on Boston's Channel 2. Further expansion and improvement of the APT system were carried on under joint industrial sponsorship, with central coordination at the Research Institute of the Illinois Institute of Technology.

The APT system has been adopted widely in America and in Europe. Examples of systems developed in the late sixties and early seventies are APT/GECENT of the General Electric Company, which accompanies its Mark Century controls; EXAPT, developed by Professor Herwart Opitz and his associates at the Institute for Machine Tools and Production Engineering, Technische Hochschule, Aachen, West Germany; and the NELNC processor of the National Engineering Laboratory of East Kilbride, Scotland.[5]

Computer-Aided Design Project
Having completed the research phase of APT, Ross and his Computer Applications Group turned their attention in 1960 to the theoretical and practical design of mechanical parts through the use of large general-purpose interactive computers, which were then be-

coming available. The new research was called the Computer-Aided Design Project and developed a family of programming systems called Automated Engineering Design (AED), through which the construction of specialized languages and programs and their adaptation to the various computers could be brought about.

Clarence G. Feldman, XVIII S.M. '59, and, from 1968, Jorge E. Rodriguez, S.B. '60, S.M. '61, Sc.D. '68, assisted Ross in heading this project. The AED material was presented to a group of graduate students in the spring term of 1968 as subject 6.687, Software Engineering. Ross and his associates issued a tutorial report, ''Introduction to Software Engineering with the AED-0 Language,'' in the fall of 1969, and published the *AED-0 Programmer's Guide* in December. In the course of these research years, 32 visiting staff members from 22 industrial, educational, and government institutions worked temporarily at the laboratory and contributed to the practical development of the AED systems; upon return to their sponsoring organizations, they were able to apply and expand the new methods in their own fields. The first AED technical meeting in June 1966 attracted 54 representatives from 24 organizations; the second, in January 1967, was attended by about 350 people from 90 organizations. Air Force funding of the MIT project ended in January 1970; Ross and his key staff members had left MIT in July 1969 to form a private company, Softech, in Waltham, Massachusetts.

From Servos to ESL: Expanding Horizons
The name of the Servomechanisms Laboratory was changed in May 1959 to Electronic Systems Laboratory (ESL), in recognition of its broader base of operations. The director, Professor Reintjes, was at that time continuing his wartime interest in radar; his re-

search group was introducing magnetic and semiconductor devices into radar circuitry and conducting investigations leading to improved airborne radar systems. His classroom subjects were attracting about 50 seniors and a special group of Navy and Air Force officers. Professor Susskind supervised a computer technology group and his class in digital systems engineering attracted about 80 senior and graduate students. Others of the ESL staff—Dertouzos, Liu, Ward, and Connelly, besides Ross and Feldman—showed strong leadership in computer technology. The ascendancy of the digital computer in the 1950s gave rise to the problem of preparing data in suitable form for digital processing and the inverse problem of converting the digital output of a computer into analog form. In the summer of 1956 Professor Susskind and a group of his Servo Lab associates offered a special one-week course in the theory and design of analog-digital conversion devices for systems and design engineers. No well-organized text material was available, and a set of notes was prepared for this course. The widespread interest in the subject led to a repetition of the course in the summer of 1957 with a large registration. The earlier notes were revised and a printed edition made available to the 1957 class by the Technology Press. This material, with minor corrections, was jointly published in April 1958 by the Technology Press and John Wiley and Sons as *Analog-Digital Conversion Techniques*, edited by Susskind.

Even though the Electronic Systems Laboratory extended its activities into these new fields, it maintained a strong emphasis on automatic control. Charles W. Merriam III, S.M. '55, Sc.D. '58, wrote his doctoral dissertation, "Synthesis of Adaptive Controls," under Reintjes's supervision and became a leader in this new aspect of control. At the time of the name change Professor Gould was deeply involved in optimal and adaptive control, and Professor Newton was working chiefly in control-system research but was becoming interested in measurements and instrumentation such as pressure-sensing and gyroscopic devices. Professor Merriam and James F. Kaiser, S.M. '54, Sc.D. '59, left MIT in 1961 for industrial posts. George Zames, Sc.D. '60, a thesis student of Professor Y. W. Lee, was research assistant, instructor, and assistant professor at the laboratory during the years 1956–1966; his specialty was nonlinear and time-varying control systems. Michael Athans (Berkeley Ph.D. '61) and Roger W. Brockett (Purdue Ph.D '64) joined the ESL activities in 1963, bringing new strength to the teaching and research. Professor Athans's book *Optimum Control*, coauthored with Peter L. Falb, appeared in 1966. In June 1969 Athans received the first Frederick Emmons Terman Award of the American Society for Engineering Education, sponsored by the Hewlett-Packard Company.

Dr. Fred C. Schweppe, whose special interest was the control of electric power systems, was first a visiting associate professor, joining the regular faculty in the summer of 1968. Jan C. Willems, Ph.D. '68, was appointed assistant professor upon completion of his doctoral thesis, "Nonlinear Harmonic Analysis"; parts of this thesis were included in his book *The Analysis of Feedback Systems*, published by the MIT Press in 1971. Dr. Sanjoy K. Mitter, a specialist in control systems, came to ESL in 1969, first as a visiting professor from Case Western Reserve University and then as associate professor of electrical engineering. Before Ross had completed the laboratory's Computer-Aided Design Project in 1969, another major project, INTREX, was well under way (see chapter 20).

During the eight years from 1963 to 1971 the annual reports show that thesis research in ESL contributed to 377 degrees—an average of 47 a year—not only in electrical engineering but in chemical engineering, mechanical engineering, physics, mathematics, biology, nutrition and food science, and aeronautical and astronautical engineering. Only a few of the laboratory's major projects have been mentioned here; the total volume and quality of ESL research can be appreciated through a study of the hundreds of project reports assembled on the shelves of the Barker Library.

ESL was an EE Department laboratory until March 1, 1978, when it became an interdepartmental laboratory reporting to the provost. In September 1978 its name was again changed to Laboratory for Information and Decision Systems (LIDS). On September 1, 1981, Professor Athans, who had succeeded Reintjes as director in January 1974, resumed his full-time teaching and research, and Sanjoy K. Mitter became director of LIDS.

Dynamic Analysis and Control Laboratory

During the war the Servo Lab developed automatic control systems for the stabilization of the Navy's Pelican and Bat guided missiles. The successful design of these controls encouraged the Navy to support a postwar continuation of this work. Project supervisor Albert C. Hall submitted, on October 8, 1945, a proposal for a research and development program for a dynamic analysis and control group of the guided missiles project. In January 1946 this group formed the nucleus of the Dynamic Analysis and Control Laboratory (DACL), recognized at that time as an entity separate from the Servo Lab, with Professor Hall as its director. Hall's associates in the early months of DACL were Leonard C. Dozier and Emory St. George, Jr., who had been engineers on the Pelican and Bat projects. William W. Seifert, S.M. '47, Sc.D. '51, was a member of the DACL team from its beginning and was soon involved in the design of a computer. His master's thesis, supervised by Hall, was entitled "An Electronic Multiplier for Use in an Analog Computer."

The showpiece of DACL was its flight simulator. This consisted of a flight table, mounted inside four suc-

cessive gimbal frames, on which the control elements for a guided missile could be mounted and subjected to all the angular rotation phenomena to which these elements would be subjected in the flight of such a missile. Many simulation studies were carried out on the analog AC and DC computers and the flight table. The facilities of DACL were used by other MIT laboratories and by industrial organizations.

After more than four years as director of DACL, Hall left to become associate technical director of the Bendix Research Laboratories in Detroit, Michigan. In 1952 he became technical director of these laboratories, and the associate technical director's position was filled by another DACL member, Charles M. Edwards, VI-A S.B. '40, S.M. '41. When Hall resigned from MIT, DACL became an interdepartmental laboratory of the Departments of Electrical Engineering and Mechanical Engineering. Its new director was Professor John A. Hrones, II S.B. '34, II S.M. '36, Sc.D. '42, of the Mechanical Engineering Department. The joint operation of DACL was highly successful, with increasing numbers of graduate students doing DACL research. In the years 1947–1957, 5 doctoral, 65 master's, and 11 bachelor's theses were written by EE students, and 8 doctoral, 22 master's, and 17 bachelor's theses by ME students.

The educational contributions of DACL staff members resulted in several new classroom subjects. Two large books came out of DACL research, published in 1960. These were *Control Systems Engineering*, edited by William W. Seifert (EE Department) and Carl W. Steeg, Jr. (Math Department), and *Fluid Power Control*, edited by John F. Blackburn (ME Department).

The Dynamic Analysis and Control Laboratory closed in 1958 when major government and industrial support contracts were discontinued.

Chapter 15

The Origins of Whirlwind: Wartime Digital Computer Research

The penetration of high-speed digital computers into every aspect of industrialized society has been so extraordinary in the past few decades that it is difficult to appreciate the fact that these machines simply did not exist before World War II. In the field of electronic calculations, just as in many other fields, the war effort demanded greater power and rapidity. MIT's Center of Analysis, which had been pushing research in electronic digital computers, was performing ballistic computations round the clock; thus, the advanced research work was taken up at other centers, principally the University of Pennsylvania. This chapter, in tracing the early electronic computation research going on at MIT and elsewhere, suggests that had MIT continued its digital computer activities between 1942 and 1945, it could, like the University of Pennsylvania, have developed an effective machine by the end of the war. But wartime pressures and the allocation of tasks nationwide determined otherwise.

Nonetheless, at the close of the war, with the department's thrust of maintaining excellence in all important aspects of electrical engineering—as well as with its keen competitiveness—two programs at MIT began racing to develop a reliable digital computer. One was started by the Center of Analysis, the other by Jay Forrester in the Servomechanisms Laboratory (see chapter 17). How these competing projects evolved, and how Forrester's Whirlwind I project took preeminence by 1947, form part of the story of the origins of digital computation.[1]

Though the digital computer was a new kind of machine, MIT had long been involved in the development of machine computation. The differential analyzers conceived and built by Bush and his colleagues in the 1930s marked MIT as the primary source of sophisticated mechanical aids to computation. These machines were copied and improved, notably at the

Ballistic Research Laboratory in Aberdeen, Maryland, at the University of Pennsylvania, and at the General Electric Company; they enabled mathematicians, physicists, chemists, and engineers to solve systems of integrodifferential equations too complex or too time-consuming for manual solutions.

During World War II the differential analyzers at the Ballistic Research Laboratory, the University of Pennsylvania, and MIT were pressed into overtime service in the study of ballistic trajectories and the preparation of firing, bombing, and range tables. The analyzers remained the fastest means for ballistic computations throughout the war period, not being eclipsed in speed by either the Bell Telephone Laboratories relay computers (1940–1944) or the Harvard-IBM Mark I Automatic Sequence Controlled Calculator, finished in 1944.

Late 1945 and early 1946 saw the testing, completion, and dedication of the first large American electronic digital computer at the University of Pennsylvania, signaling the advent of really rapid computation. (Outside the U.S.A., the British ''Colossus'' was operational in 1943.) This set in motion an avalanche of digital computer technology, which took over the problems formerly handled by differential analyzers, network analyzers, and other relatively slow machines, and went on to stimulate advances in many kinds of human endeavor.

The Rapid Arithmetical Machine, 1937–1942

At MIT, even while the Rockefeller Differential Analyzer was under development and construction during the late 1930s, Vannevar Bush had been thinking about a possible rapid electronic calculator. Between January 1937 and November 1938 he wrote several memoranda, as was his custom when ideas were incubating in his mind. Some of those early ideas were summarized in December 1938 by William H. Rad-

Jay W. Forrester, whose Whirlwind project produced MIT's first digital computer.

ford, then a research assistant working under the supervision of Samuel H. Caldwell:

The Rapid Arithmetical Machine is intended to be an extremely flexible device for automatically and quickly performing extensive and complicated computations involving all combinations of the four fundamental arithmetical operations. . . . In executing a series of computations with numbers supplied to it, the machine will automatically perform all required intermediate operations of selection, transfer, totalization, and storage. Final or intermediate numerical results will be automatically recorded in one or more convenient forms. An essential feature of the machine will be its rapid operation. It is anticipated that it will be capable of performing elaborate arithmetical calculations one hundred or more times as fast as can be done with existing methods and equipment.

It is realized that the first model of the machine will necessarily be a decidedly bulky and complex affair. However, it is not intended to be a portable device, but is designed rather as an extremely powerful and rapid aid for use in centers devoted to computation and analysis—places where enormous amounts of computation must be carried on. The speed of the machine and its ability to deal with complex arithmetical problems will easily justify its size and cost.

As it is now visualized, the ultimate machine will comprise the following essential elements:

1. One or more keyboard units *for entering in suitable form (such as punched tapes or cards or other appropriate devices) the numbers to be operated upon and any constants to be used therewith. There will also be prepared in suitable form, a complete set of operating instructions for the machine.*

2. An input unit *for automatically inserting the prepared numbers and constants as they are required in the course of the calculations.*

3. A control unit *for automatically directing and coordinating all machine operations so as to obtain the desired numerical results.*

4. One or more function units *for automatically introducing required functional relationships.*

5. A computing unit *for automatically performing, on numbers supplied to it by other units, the four arithmetical operations as dictated by the control unit.*

6. Numerous intermediate storage units *for temporarily storing numbers obtained from inserted data and required at one or more points in the sequence of computations.*

7. One or more recording units *for automatically recording either in print or other desirable form such intermediate and final results as may be required.*

In a simplified machine the keyboard, input, and control units might advantageously be merged.

The heart of the machine would be the computing unit, and this was slated to be among the first problems attacked. "One electrical method contemplates the utilization of cascaded electric counting rings comprising either vacuum or gas-filled tubes. In this system all arithmetical operations are reduced to the rapid counting of electrical pulses. A possible alternative method employs a modified form of counting ring in conjunction with a positional procedure. This system obviates the necessity of pulse counting and is hence inherently faster. In some cases a suitable combination of the pulsing and positional methods may prove advantageous."

Radford's year of investigations on the practicability of developing a rapid computing machine was fully reported in October 1939, with diagrams and photographs; at that time he reviewed the literature on electronic counting circuits used mainly in nuclear and cosmic-ray research.[2]

At the beginning of the fall term, 1939, Radford discontinued his research on the Rapid Arithmetical Machine project and began a program of teaching in Bowles's Communication Option. Radford recommended that two men be employed on this research during the ensuing year and that a matrix switch be designed and built. The program called for the design of "all necessary auxiliary equipment, such as pulsers, delayed pass circuits, shifting switches, subtraction controls, failure alarms, routine testing facilities, and all other parts required in a complete calculating unit capable of performing any arithmetical operation." At a conference with the sponsor, National Cash Register Company, one full-time man was authorized for the immediate future and details of the program were modified to include investigation of tube characteristics.

The new researcher was Wilcox P. Overbeck, S.B. '34, who had been employed by the Raytheon Manufacturing Company since his graduation. The earlier research showed that a digital computer utilizing circuits investigated by Radford would require thousands of electronic tubes. Caldwell's experience with the Rockefeller Differential Analyzer, with its 2,000 tubes, had impressed upon him the importance of cutting down the prospective number of tubes. Overbeck's two and a half years of research on the Rapid Arithmetical Machine were therefore directed at tube design.[3]

The computer research program was brought to a sudden halt in the spring of 1942, when Overbeck was transferred to Chicago to work with Dr. Arthur H. Compton on a very secret and important war problem, which everyone would learn later was the atomic bomb. Wartime demands—Caldwell to NDRC, Overbeck to atomic research, Herbert Harris to Sperry, round-the-clock operation of the Center of Analysis on ballistic computations—temporarily stopped the MIT development of electronic digital computers.

Following the cessation of work on the Rapid Arithmetical Machine at MIT, Colonel Gillon of the U.S. Ordnance Department sent to Harold Hazen (then chief of Division 7, NDRC) an outline of a computer project at the University of Pennsylvania—the development of a machine that Gillon called the Electronic Numerical Integrator and Computer (ENIAC). This outline was referred to Caldwell, chief of Section 7.2, for comment. Caldwell's memorandum to Hazen contains the following remarks:

It would appear that the specifications of the equipment under development by the Ordnance Department substantially duplicate the specifications of the Rapid Arithmetical Machine program. . . . We devoted about three years to research in basic circuits and components and were in the process of studying designs for the combination of such equipment into an overall machine when the work was suspended for the duration. As far back as 1939, we realized that we could build a machine for electronic computation. But, although it was possible to build such a machine and possible to make it work, we did not consider it practical. The reliability of electronic equipment required great improvement before it could be used with confidence for computation purposes. . . . Another thing which caused us to discard the idea of building a complete machine immediately was the very large number of small parts required. Colonel Gillon speaks of the machine containing thousands of tubes. In addition it contains thousands of resistors, condensers, and wiring terminals. One of the basic objectives of our grass roots program was to simplify the elements of the machine. I make the above comments . . . to indicate our active interest in the Ordnance Department program because of its similarity to the one which we have at present packed away in camphor balls.

A study of the Radford-Overbeck research indicates that had MIT continued its digital computer activities between 1942 and 1945, with an adequate team under the ingenious and driving leadership of a man such as Overbeck, an effective machine with simplified components would have emerged.

During the war years Warren Weaver of the Rockefeller Foundation was able to view computer developments from several different standpoints. He had been involved with the Bush-Caldwell Rockefeller Differential Analyzer built in the years 1935–46. As chairman of Section D-2, NDRC, and later member of Division 7 (Fire Control), he was intimately associated with Caldwell and came to regard him as a man of outstanding technical ability. Weaver was chief of NDRC's Applied Mathematics Panel, which worked on mathematical problems for all the military services. At the close of the war he became chairman of the Naval Research Advisory Panel. In late 1945 he was closely in touch with the ENIAC development, which was approaching operational status. He also knew about the EDVAC (Electronic Discrete Variable Calculator) at the University of Pennsylvania and had received a proposal for a machine from John von Neumann of the Institute for Advanced Study in Princeton; von Neumann was hoping for financial aid from the Rockefeller Foundation. Seeking help in his own evaluation of the electronic computer situation, and knowing of MIT's interest, Weaver sent the von Neumann proposal to Caldwell for a review. Caldwell gave Weaver a six-page, closely typed reply.

Caldwell discussed the von Neumann proposal under three categories: as it related to the general art of electronic computation; as it related to high-speed digital computation needs during the next decade; and as it related to the MIT program. First, von Neumann's time estimate of two to three years of building and then two years of experimental testing seemed too optimistic: he did not seem to appreciate the engineering problems and the probability that the "large puddles of money" available in the war years would be dried up. Second, Caldwell disagreed with von Neumann's belief that digital computers would quickly displace relay computers, differential analyzers, and the Harvard Mark I. Third,

He may not have been aware of how far in detail we [at MIT] had gone before the war and I hence do not expect him to know that most of the details of his proposal are not new to us. For example, the Rapid Arithmetic Machine was planned to have an even more sophisticated control procedure than that which von Neumann describes. . . . As another example of our early appreciation of von Neumann's basic thesis, I shall be glad to show you or provide photostats of my personal notebook entries of August 23 and 24, 1939, in which I record my initial conception of what I then called "dynamic storage" of digits. You will find it describes the basic idea contained in von Neumann's storage columns.

We now contemplate a balanced program based on cooperative action of the Department of Mathematics, the new Research Laboratory of Electronics, and the Center of Analysis in the Department of Electrical Engineering. . . . The Department of Mathematics is in the midst of a program to improve its position in applied mathematics, and we intend to make use of our computation machinery to that end. There is a strong feeling that mathematics is entering a period of great new accomplishment by way of the mass production of numerical solutions, and that we are potentially situated to operate among the leaders in that advance. . . .

Before the war our developments in computation were entirely within the Center of Analysis. Now we are in a position to exploit the fact that MIT has found many of the best men in the Radiation Laboratory eager to stay here. From these men, and particularly those most skilled in electronics, we have been able to select an exceptionally able group to form the nucleus of the Research Laboratory of Electronics staff.

With such an asset available it would be sheer nonsense to form another group within the Center of Analysis in order to resume electronic computation developments. Accordingly, Dr. Stratton and I have agreed that the basic electronic developments will be the responsibility of his group and that we in the Center of Analysis will retain the responsibility for furnishing development specifications to his laboratory, for carrying out all non-electronic component developments, and later for engineering and building the component groups which become the final machine.

For the electronic machine itself, we would like to proceed now with a program in the Center of Analysis and in the Research Laboratory of Electronics to extend for about two years and to cost about $100,000. This program would do the following:

(a) develop reliable and well-engineered designs of those components for which the general characteristics are known;

(b) study the related problems of switching, control, and recording, and carry out research and development to provide equipment to perform these functions reliably and at speeds comparable with computation speeds;

(c) study the overall content and structure of the final machine in detail, and provide a prospectus of its form, size, ability, and cost, so that any decision regarding support of the ultimate construction program can be made on the basis of a substantial quantity of detailed information.

The above program will require the efforts of a large and varied staff. We have the key men required for the theoretical, developmental, and engineering aspects of the problem, all under one roof, and organized under an exceptionally powerful cooperative arrangement.

In March 1946 Caldwell discussed with Jay Forrester and others the possibility of duplication between the Forrester project in the Servo Lab and the computer project contemplated by the Center of Analysis. By this time a digital computer was clearly part of Forrester's project, but it was thought to have limited application and probably limited availability to the MIT community, since it was sponsored by the Navy and therefore classified. To quote from Caldwell's memorandum covering that meeting,

The more important duplication of effort will be between our program and those of other groups at the Institute for Advanced Study and the University of Pennsylvania. This situation is completely known in the Rockefeller Foundation. But it should be emphasized that we are not in the position of "getting aboard a band wagon." We started our work before the war and we had made real progress in component studies. . . . The exploitation of this opportunity should in no way be contingent upon a program of the Navy. If duplication in details is the price which must be paid for independence of thought and action, then the writer believes that the objective is worth that price.

In a letter to the staff and employees of the Center of Analysis, Caldwell assessed MIT's position vis-a-vis the electronic digital computer as follows:

While we have emerged from the war stronger than we were when we entered it, we have not come out entirely unscathed. Most of our loss lies in one field— that of electronic computation. Many of you know that Mr. W. P. Overbeck had a research program going on this subject before the war, but the atomic bomb took Mr. Overbeck from us and still claims him. During the war other organizations were assigned programs in this field which have resulted in considerable progress, some of which is available to us. But the net overall result is that in the field of electronic computation we entered the war among the leaders and emerged in a much less favorable position.

The importance of the electronic computer to the work of the Center of Analysis, and through the Center to almost every department of the Institute and to a host of outside agencies, need hardly be argued. Those of you who attended the conference sessions here last fall will well remember the analysis presented by Dr. von Neumann. We can only say at this time that resumption of our work in electronic computation development, at a greatly increased rate, stands as the largest and most important single item on our future development program. While the final dimensions of the project cannot be predicted on the basis of present knowledge, we know that functionally we are setting out to do something which is at least of the order of magnitude of the new differential analyzer, but which involves many more hazards and difficulties because we have no original version to guide us.

In April 1946 President Compton submitted an application to the Rockefeller Foundation for a grant of $100,000 to support electronic digital computer research for two years. In his letter of support for this application, Professor Stratton of the Research Laboratory of Electronics said,

I believe that the strength of this proposal from MIT rests on the collaboration that may be anticipated between the Department of Mathematics, the Research Laboratory of Electronics, and the Center of Analysis. The interest in mathematics focuses, of course, on Professor Norbert Wiener and his students. Professor Wiener's current ideas about computing processes, as well as certain physiological analogues that he has studied, are enormously stimulating. The part of the Electronics Laboratory in this program is that of translating ideas such as those of Wiener and his group into physical reality, and of supplying an intermediary stage of basic physical research between the purely mathematical concepts and the ultimate specific application to a computing machine in the Center of Analysis.

The Rockefeller Foundation accepted MIT's proposal. In a progress report issued on February 8, 1947, Professor Richard Taylor of the Center of Analysis pointed out that the computer research had been resumed in September 1945 on a "rather limited scale" and had about doubled by the time the Rockefeller funds became available in July 1946. Taylor envisioned the computer as "a machine designed not only for the mass-production of routine problem solutions, but also for use by the mathematician who is conducting a program of research in the field of applied mathematics. A calculator used in this latter fashion becomes the laboratory instrument for applied mathematics, making possible the . . . same opportunities for experimentation in mathematics as are enjoyed in other fields of science." Studies under way involved, according to Taylor,

the mathematics of problems to find out whether or not there are methods of solution better suited to machine use than the methods now in force for manual computation; we are not adopting the attitude that, because electronic digital computing speeds are high, we are justified in making use of wasteful and inefficient computing routines. . . . Along with this mathematical study, we are conducting laboratory research on some of the more fundamental types of apparatus which we feel will be needed regardless of the ultimate structure of the calculator. Our work along these lines is centered on three items at the moment, high-speed counting and whole-number adding circuits, high-speed storage of numbers for internal memory purposes, and low-speed magnetic storage for external memory purposes.

Considerable study was given to the design of a digital integrator and to methods of solving large systems of algebraic equations.

Warren Weaver's comments on the report centered on his disappointment that Wiener had not been involved during this six-month period. "I think the question, 'What does one want a computer to do?' requires great knowledge of mathematics, great imagination, great sweep and depth of mind. One could get some inspiration, doubtless, by talking to a dozen leaders in applied mathematics. But this is where I had hoped and expected we would have the genius of Norbert Wiener."

In April 1947 Weaver wrote to Hazen. "I continue to worry about Wiener. Is he really putting any time and energy into the computing study? If he is not, the situation is getting worse at an exponential rate, so far as I am concerned; for his contributions obviously ought to be made at the early stages." Hazen replied, "At the last [meeting of the computer group] we began to get into some of the mathematical problems, approaching for example the pile-up errors. I was much interested to see that Wiener got into this wholeheartedly and volunteered out of a clear sky to give a summary three weeks hence of present knowledge on the subject of errors in computation."

By early June Caldwell and his group had made an appraisal of the project, especially in relation to Forrester's Whirlwind development. He wrote,

Competition on equal terms is normal, but when two projects such as, for example, Whirlwind and the Rockefeller Electronic Computer (REC) find themselves on parallel tracks and with a disparity in resources which is obviously great, it is evident that the smaller one has only a poor chance to survive. A reduction of morale within the REC project is already beginning to show. This is not so serious that it would not respond to a strong move to reinforce the project, but it is becoming clearer and clearer to the men on the job that they are not in a position to offer strong competition. The series of staff seminar papers by Forrester and his group made that abundantly clear. The Institute as a whole is placed in an awkward situation by these developments. Through the gradual shift in the objectives and relative emphasis of Project Whirlwind, MIT has come to be in the position of doing a big job for the U.S. Navy plus a watered-down version of a part of that job for the Rockefeller Foundation. This we cannot defend at all, once we have recognized it. Since we cannot defend the position, we cannot occupy it. One of two possible courses has been explored—that is, to stop the REC project. . . . This move has the obvious advantage of concentrating our forces on a united research front. The aims of Project Whirlwind will probably satisfy the ultimate purpose of REC. Mr. Forrester believes his final machine will be able to meet both Navy and MIT needs.

In a letter dated June 25, 1947, President Compton announced to Weaver MIT's decision:

At the end of our first year certain conclusions have become evident that alter our original program in a major way and which I wish to report to you at this time. The major conclusion is that the Navy Whirlwind project here under the direction of Mr. Jay Forrester has so evolved in objectives, scope, and security aspects that it is doing what we had expected to do under the Rockefeller Grant. It is doing this on a substantially larger scale than we had planned under your grant and with such vigor that the purpose of your grant to us is being met as fully as we know how to meet it. This being the case, we think the expenditure of Foundation funds for competitive work in the same institution is not justified and is in fact foolish. Our conclusion therefore is to terminate the work under your grant and to return to the Foundation the monies that have been sent to us by the Foundation.

In his reply of July 9 Weaver remarked, "I think that the conclusion you have reached is entirely characteristic of the intelligence and, if I may use so Victorian an expression, 'the scrupulous honor' which we all confidently expect from MIT. . . . After all, our common object was to forward this very important field of work. It is being forwarded at MIT, and in a very distinguished way."

Electrical engineering students, 1950. In the new generation of students in the postwar era, when curriculum changes were taking place, there was a high degree of optimism about what engineering research promised for their future.

Part IV

The Turning Point

Introduction
A Time
of Reassessment

As the Electrical Engineering Department at MIT had grown in size, the realization had also grown that its faculty and students had the capacity to pursue more avenues of activity in research and teaching, and that it was not necessary to maintain a narrow range of specialization in order to maintain excellence. At the end of the war the EE faculty was still relatively small, but the research horizons had widened almost immeasurably. There seemed almost nothing that couldn't be done. As the pressures of war-related projects receded, new choices had to be made.

These choices involved the entire Institute, whose educational program had been somewhat distorted by the deep involvement of the faculty with weapons research and development and now needed reassessing. Among the issues being debated in the immediate postwar period were the following: the management and balance of new sponsored programs—which were expected to diminish, but which in fact increased in number; the size and composition of the postwar student body, especially at the graduate level; the recruitment of new faculty, mainly from the fine laboratories that had been built up around the nation during the war; the special role of laboratories at MIT; and, particularly, the avenues of scientific and engineering research that should be pursued in conjunction with the educational program. In short, it was an examination even more profound than that which had taken place after World War I.

In 1920 MIT had developed the "Technology Plan," under which it mobilized its research, consultative, scientific, and industrial experience to serve industry needs in return for retainers. Although the program was set up principally to solve the Institute's financial difficulties at that time, and although it failed in certain respects, it was important in MIT's later development inasmuch as it established the Division of Industrial Cooperation, which we have described in earlier chapters.

By the time of World War II, the Division of Industrial Cooperation, which was responsible for handling all the details of contractual arrangements with industry and government, had twenty years' experience; this made the Institute uniquely equipped to handle the huge government-sponsored programs and projects of the war. As an indication of the enormous increase in such research and development, the division was managing about $135,000 worth of sponsored projects in 1940, spread out among 36 projects, whereas by the war's end it had managed $100 million in Radiation Laboratory work alone, plus about another $5 million a year in other projects. Then, despite the closing of the Rad Lab, the level of sponsored work rose to more than $15 million a year by 1949, most of it for the government. The number of sponsored projects went from 123 in 1946, to 153 in 1947, to 176 in 1948, to 190 in 1949. If this high level of support for on-site research and development were to be a continuing feature of MIT's educational environment—not a wartime phenomenon alone—its role had to be carefully evaluated, not only for its potential educational benefits but for its dangers as well.

The benefits of new projects and large research facilities were obvious: they brought more faculty and students into engagement with real problems; they exposed students to the process of research in a way not possible in the classroom; they stimulated new directions, and hence new laboratories,

in American industry; they provided new equipment and research facilities; they provided employment support for students who otherwise might not have been able to continue graduate studies; they brought greater income to the Institute.

Among the many dangers were these: an overcommitment of faculty and facilities would tend to pervert and downgrade the educational mission; a university could gradually be transformed into a commercial enterprise—a role for which it was not specifically structured—and drift into direct competition with industrial companies; and, in the case of near-exclusive sponsorship by government, a university could become identified in the public mind with a weapons mission; moreover, in this latter case, an educational institution could jeopardize its long-term financial stability and health by making itself vulnerable to abrupt changes in government policies. This is a danger always faced by an organization that has only one client, one sponsor, or one product.

Some of these dangers have been with alternative education from the very beginning, stemming from its aspirations to serve the advancement of the industrial arts. There has always been the question of just how closely an educational institution should collaborate with industry. In a special postwar study of this situation carried out by MIT, one conclusion was the following: "As one surveys the current state of contract research in connection with colleges throughout the country, one cannot escape the suspicion that many schools are perilously close to engaging in purely commercial enterprises. In our opinion any school that follows such a course will ultimately suffer serious damage to its standards as an institution of higher learning."

As we suggested in an earlier chapter, in alluding to the reincarnation of the Radiation Laboratory in the form of the Research Laboratory of Electronics, there were many issues that had to be addressed in taking such a "logical" step. The same kinds of considerations would come up later, in an even more serious form, when the Air Force was urging MIT to set up what was to become the Lincoln Laboratory; and, again, when MIT decided to separate itself from Stark Draper's Instrumentation Laboratory in 1973, that undertaking having become a $50 million operation.

Because of the special role it had won during World War II, MIT experienced in a keen way the ups and downs of government agency enthusiasms for research. There was a leveling off in the early 1950s, then a sudden spurt as the United States reacted to the explosion of an atomic bomb by the Soviet Union in 1952. The post-Sputnik reaction in 1957, when President Killian of MIT was called to Washington by President Eisenhower, was another period of intensified research reminiscent of the wartime experience. The establishment of the National Science Foundation in 1951 had something of a stabilizing effect, but its programs and projects too have come under increasing political pressures in recent years, and university research programs across the nation have suffered in terms of support.

These would all be issues for the future; already by the war's end, however, it had become clear that universities in general, and MIT in particular, needed to hammer out a research policy if they were to continue to devote large laboratories and diverse talents to national defense problems. At the same time it had also become clear that MIT must continue the special relation of its educational

programs to American industries; more than that, it had a special obligation, as Dr. James Killian asserts, to perform "public services." World War II had firmly established the policy that any MIT staff members called to public service were to be supported and encouraged, and welcomed back to the Institute without loss of standing when the public service task was accomplished.

The establishment of the Research Laboratory of Electronics (RLE) in 1946, as a joint undertaking of the Departments of Electrical Engineering and Physics (and sponsored jointly by three military services), was part of a much larger tide of change that was affecting MIT and the nation after the war. These changes led MIT in 1947 to establish a special committee to make a total survey of the educational situation at the Institute—to reexamine "the principles of education that have served as a guide to academic policy at MIT for almost ninety years, and to determine whether they are applicable to the conditions of a new era emerging from social upheaval and the disasters of war." This study, initiated by Dr. James Killian, made a major contribution.

The report of this committee, called the Lewis Report, after committee chairman Warren K. Lewis, set up influential guidelines for the period to come. It recapitulated the goals of the Institute since its founding, expressed uneasiness about the forces that had brought "subtle and profound alterations in the very character of the Institute," and, among other things, recommended that there must be an ever greater role for the humanities at MIT. It was evident that not only must students "be trained for a vocation, but they must be made to recognize that the growth of science and technology has a profound impact on society at large." This idea had been at the heart of Rogers's concept for a polytechnic school more than a century earlier, but had never been fully implemented. Moreover, conditions had changed dramatically. In Rogers's day technology was beginning to take off, but needed all the support it could get. Schools like Rensselaer Polytechnic Institute and MIT were then virtual loners in technological education. By 1950 easily a hundred universities had strong scientific and engineering programs. In 1850, technology was a promise; in 1950 it was an accomplished fact, and society was totally and inescapably embedded in it. MIT needed to remember its broader educational mission, to return to the concept of education as ''a training in values,'' not just competence in a field of technology or a profession.

The Lewis Report also took into account the influence of President Compton, who had begun to restore the balance between science and technology. Of Compton's work, the committee observed, ''These changes lifted the morale of the MIT community and resulted in renewed confidence in the ability of the Institute to develop leadership in science as well as engineering. The Institute was no longer dominated by the engineering point of view, and the engineering departments themselves became less narrowly vocational. . . . The appointment of deans of science, engineering, and humanities reflects this altered intellectual emphasis and marks the beginning of a recognition that the Institute has an important mission in fields other than engineering and architecture.'' These broader aims would take time to work out, and the work still goes on; but they did reflect a fundamental shift following the war on many levels—in science, in technology, and in educational purpose.

Chapter 16

The Research Laboratory of Electronics

When the Research Laboratory of Electronics (RLE) was started in 1946, many lines of investigation were already set, their origins in World War II or, in some cases, much earlier. Several RLE projects were continuations of programs that had been going on in the Radiation Laboratory, with staff recruited from those who had contributed to war research. RLE's activities were to expand dramatically, however, and in quite unexpected directions. A most unusual and talented collection of people was brought together at that time, with the beginnings of revolutionary ideas in their heads—ideas in the fields of control, communications, computation—and with the physical and financial means of implementing them. Moreover, it was during this period that the EE Department would become one of the world's leaders in the field of information theory.

The first of the so-called interdepartmental laboratories at MIT, RLE initially brought electrical engineering and physics more closely together, but very soon brought together other specialties. Over the years RLE's researches would reach into many fields; we present in this chapter a somewhat arbitrary selection of persons and researches to give the flavor of RLE development as it relates particularly to the history of the EE Department.

Many of the initial staff were newcomers, young graduate students cutting their way into the domain of information theory. They were learning how to translate the ideas of Norbert Wiener and Claude Shannon into new forms of equipment. One of the important mediators of this translation process was Professor Yuk Wing Lee, who came from China just after the war and opened the eyes of the new generation to the significance of Wiener's difficult ideas on statistical communications. How to detect order in disorder—

how to improve the signal-to-noise ratio—was at the core of these new efforts. Rather than looking exclusively at the physical parameters of devices and systems, this generation of scientists and engineers looked more closely at the phenomena being detected and perceived and at the character of the perceivers, both animate and inanimate. Here was a fundamental shift of viewpoint. Thus, it was not long, in terms of the history of science and engineering, before electronics and communications engineers were working with psychologists, and fledgling computer scientists and mathematicians were working with neurophysiologists. It would not be very long either before the science of acoustics would phase into psychoacoustics.

During this early period, when Wiener's statistical communications was taking hold, he would wander from office to office every day, keeping in touch with everything that was going on. He was also participating in the Macy Meetings, between 1945 and 1952, where the primal ideas of circulatory feedback were being attacked jointly by physical and life scientists, by engineers and philosophers, by people from different professional disciplines who could barely understand one another through the ambiguous "noise" of their specialized jargons. The Macy Meetings, chaired by Professor Warren S. McCulloch, who came to RLE in 1952, were an important prelude to the interdisciplinary ideas that became dominant after the war. A synthesis of many of these ideas would appear in Wiener's 1948 book *Cybernetics*, an exposition of the science dealing with "communication and control in animal and machine," which carried the attack on the classic body-mind duality a stage further.

Equipped now with a rudimentary understanding that living neurons either "fired" or didn't, and were therefore "binary," these scientists and engineers began to look further into the analogies between living intelligent systems and digital computers. Thought and cognitive processes were somehow based on an extraordinary feltwork of neurons, and the first (then large-scale, but in retrospect small) digital computers were also based on a computational system of *yeses* and *noes*, of *fired* or *not-fired*, of ones and zeros. It was tantalizing. Could machines be made to think as people do?

Far more was discovered than was expected, but a high level of inadvertency—of the possibility of accidental discovery—had been allowed for. Electrical engineering was transformed: it could thereafter encompass cognitive science as well as semiconductor devices or energy transformation systems.

Planning and Early Years of RLE

In 1966 a 52-page historical booklet was compiled by the directors and some of the senior members of RLE. A comprehensive story of RLE and its contribution to science and engineering education is still to be written, but the booklet "RLE: 1946 + 20," on which we draw in part for the account in this chapter, presents some of the highlights of those early days.

Postwar planning was encouraged by the Normandy invasion of June 6, 1944, and the recapture of Paris by the French on August 23, 1944. Professor John C. Slater, head of MIT's Physics Department, on leave to do research on microwaves at the Bell Telephone Laboratories, envisioned an electronics laboratory at MIT to be organized and operated by the Departments of Physics and Electrical Engineering. To develop a definite plan, a conference was set up for August 28, 1944, and was attended by President Compton, Dean of Science George R. Harrison, and Professors Slater, Hazen, and Stratton. A memorandum prepared by Slater summarized the discussions and conclusions of that conference.

The top organization might consist of the director, who would be a regular member of one or the other department, as well as holding his position in the Electronics Laboratory; and a steering committee, consisting of approximately equal numbers of representatives of each department. Associated with the laboratory would be such senior staff members, instructors, assistants, teaching fellows, and graduate students of each department as were interested. At the same time the laboratory would have a budget of its own, and could have personnel of its own, such as research associates, assistants, and technicians. As very rough guesses, it was estimated that the total personnel of the laboratory, including staff members and graduate students of both departments associated with it, and scientific personnel of the laboratory itself, might be of the order of 60 to 75. . . . The general feeling of the conference was that the proposed laboratory provided an unusually favorable chance to develop harmonious cooperation between two departments in a field which clearly overlapped the interests of both, and that this cooperation might be hoped to attract industrial support from outside the Institute, to attract good students, and to furnish a pattern for similar cooperative projects in other fields.

Professor Julius A. Stratton, on leave as expert consultant to the secretary of war, was recommended by Slater as director of the new laboratory. Stratton was interested in this prospect, but was also considering an offer from the Bell Telephone Laboratories, where he would supervise research. He was assured of strong support from the Physics Department, since he had worked closely with Slater in the planning for the electronics laboratory; Hazen, then head of the Electrical Engineering Department, pledged similar support and made a special trip to talk with Stratton at the Pentagon to clear up any lingering doubts. In October 1944 Stratton accepted the directorship of the new interdepartmental laboratory.

Stratton saw the laboratory as the expression of ideas that had begun to be formulated in the 1930s. There was in those years, Stratton later wrote, an intense and widening interest in communications and electromagnetic theory among a substantial number of faculty members. They were

preoccupied with a host of practical and theoretical questions ranging from the generation of microwaves and the design of high-frequency circuits to the properties of electron gases and the study of electromagnetic radiation and boundary value problems. Out of these interests there grew a remarkable spirit of collaboration between the physicists and the electrical engineers. A whole new field of effort began to emerge, focusing our attention upon a closely knit set of problems which we now identify as electronics. Indeed, in his President's Report for 1939, Karl Compton made special mention of the related research projects which were coalescing into wide-ranging programs breaking across departmental boundaries and involving cooperation with outside agencies. . . . The founding of the new electronics laboratory in 1946 represented a major new departure in the organization of academic research at MIT and was destined to influence the development of interdepartmental centers at the Institute over the next two decades. These centers have been designed to supplement rather than to replace the traditional departmental structure. They take account of the fact that newly emerging fields of science commonly cut across the conventional disciplinary lines. And they afford a common meeting ground for science and engineering, for the pure and applied aspects of basic research, to the advantage of both. Perhaps more than any other development in recent years they have contributed to the special intellectual character and environment of MIT.

In December 1944 Hazen sent to certain electronics-oriented staff members of the Electrical Engineering Department two memoranda, one giving Slater's RLE physics program and the other Hazen's own thoughts on a possible electrical engineering program. Hazen said, "It is likely that the interests of individual electrical engineers and physicists will interpenetrate extensively the rather nebulous boundary between the two departments. This is as it should be, assuming of course that personal clashes are not incurred."

Following V-J Day, August 14, 1945, RLE matters began to move fast. The National Defense Research Committee, at its meeting on August 24, 1945, voted to permit continuation of basic research under the Radiation Laboratory contract. On September 8 the MIT administration issued a bulletin announcing "special arrangements to enable men with extended war experience" to become graduate students as research associates. "The Executive Committee of the Corporation has given its sanction to this extension of our research associateship in the belief that the Institute, at war's end, has an unprecedented opportunity to aid 'cream-of-the-crop' graduate students."

A Stratton memorandum of September 17, 1945, presented a plan for the transition from Rad Lab to RLE, the latter to be formally established on January 1, 1946. The research division of the transitional organization would include 17 faculty members of the Physics and Electrical Engineering Departments and 27 former staff members of Rad Lab. The west wing of Building 20 was made available and equipped with machine shop, tube shop, instrument room, and stock room. Of great value to the new laboratory was the acquisition, largely through the herculean efforts of Professor George G. Harvey (Physics), of pioneering electronic equipment developed by the Radiation Laboratory.

A research program under the following five headings was announced: Microwave Electronics, Microwave Physics, Modern Electronic Techniques Applied to Problems of Physics and Engineering, Microwave Communications, and Electronic Aids to Computation.

The first three directors of RLE. *Left to right:* Julius A. Stratton (1946–1949), Albert G. Hill (1949–1952), Jerome B. Wiesner (1952–1961).

Under each heading were projects to be begun at once, some representing continuations from Rad Lab. By the time RLE's first quarterly progress report was issued in March 1946, 22 MIT faculty and instructors and 16 RLE staff members constituted the senior research personnel. Thirty research associates, assistants, and graduate students were listed as junior research personnel, of whom 8 were registered as graduate students in electrical engineering.

Of the projects reported in March 1946, the following involved EE Department personnel:

- Construction of High-Power S-band Magnetron, under the leadership of Stuart T. Martin, Jr., VIII S.B. '34, VIII Sc.D. '38, a new EE faculty member who had taught at Clark University, worked as a research engineer with RCA Manufacturing Company, and been a captain in the Signal Corps during the war.
- Studies Leading to the Design of a Microwave Accelerator, led by Slater and Trump, with five graduate students, including S. J. Mason, S.M. '47, Sc.D. '52, and M. E. Van Valkenburg, S.M. '46, of the EE Department.
- High-Speed Oscilloscope and Magnetron Moding, under O. T. Fundingsland, VIII S.M. '50, and D. F. Winter, S.M. '48, the latter formerly of Rad Lab.
- Measurement and Control of Deionization Time in Thyratrons, under F. M. Verzuh, S.M. '46, Sc.D. '52 (Professor Edgerton, advisor).
- Synchrotron Project, under I. A. Getting, new EE associate professor, formerly of Rad Lab. This project was transferred to the new Laboratory for Nuclear Science and Engineering, directed by Professor J. R. Zacharias, formerly of Rad Lab.
- Noise, Interference, and Distortion in Pulse Modulation, under W. G. Tuller, VI-A S.B. and S.M. '42, Sc.D. '48, formerly of Rad Lab.

- Modulation and Stabilization of Oscillators, under F. P. Zaffarano, formerly of Rad Lab.
- DC Amplifier Problems, under Professor T. S. Gray, at this time in charge of the EE Department's Measurements Laboratory.
- Study of Waveguides with Dissipative Walls, by R. M. Fano, formerly of Rad Lab.
- The Broadbanding of Microwave Networks, by R. M. Fano.
- Circuit Synthesis of Active Networks Involving Vacuum Tubes, by M. Cerrillo, assistant.
- Transients in Waveguides, by M. Cerrillo.

There were 29 projects in all. The physics projects had very strong leadership, including 14 men from Rad Lab at the doctoral level, as well as Professors W. P. Allis, VIII S.B. '23, S.M. '24, S. C. Brown, VIII Ph.D. '44, F. Bitter, C. F. Squire, L. Tisza, and M. S. Livingston. Back in 1944 Slater had envisioned the new laboratory as stronger at first in physics than in electrical engineering, in order to develop new electronics fundamentals. RLE would then "accelerate the transfer of these new methods from the physics laboratory to engineering practice."

A letter of intent had been received by MIT assuring continued armed service support for RLE. The many years of Stratton's participation in the MIT educational program, combined with his wartime experience as expert consultant to the secretary of war, enabled him to work out with the armed services a financing plan that proved to be of value to the students, the faculty, MIT as an institution, and the military services. Vannevar Bush's recommendation for government support of research through a national research foundation was not adopted by the United States Congress until 1951; but in early 1946 an agreement was reached whereby the Navy Office of Research and Inventions, the Army Air Force, and the Signal Corps would fund the MIT research started under OSRD auspices. The participation of the three services was equal, but for contractual convenience the legal arrangement was between MIT and the Signal Corps only. A technical advisory committee to the laboratory was set up, consisting of representatives from all three services.

The march of events quickly dispelled the idea that academic research and development would become disengaged from military developments. Julius Stratton had observed that the war had "demonstrated in a spectacular manner that henceforth military strength and security would be utterly dependent upon a very advanced, highly sophisticated technology." Moreover, massive federal support for basic research was emerging as a national policy. With the Russian explosion of an atomic bomb in 1949, and with the beginning of the Korean War in June 1950, the RLE programs would double in the fall of 1950 under strong military support.

Despite the good working relations that had been established between the civilian scientific and engineering communities and the military establishment during the war years, there continued to be apprehension about the influence of military aims in the academic community. Looking back years later, RLE's second director, Albert G. Hill, reflected on the positive aspects of the relationship:

We have seen, in the six-year period from August 1945 to August 1951, a new academic, peacetime laboratory start from the defunct wartime Radiation Laboratory, and in turn the academic laboratory enabled a new applied military laboratory [Lincoln Laboratory] to get off to a running start. From this, all kinds of conclusions can be drawn regarding six-year cycles, military expediency, and the beauties of the ad hoc study. I firmly believe, however, that the only verifiable conclusion is that military support of university research can be extraordinarily fruitful for both parties. Certainly there are dangers, but care and mutual re-spect can obviate most of them. Personally, I feel that basic research can profit from an association with the applied or programmatic research—an association that should be close, but must not be smothering.

One of the first responsibilities of RLE was to participate in a study of missile guidance and telemetry as part of Project Meteor, a U.S. Navy Bureau of Ordnance project involving several MIT departments. This secret RLE project was operated under a steering committee initially headed by Director Stratton and later by Professor H. G. Stever. In January 1947 Henry J. Zimmermann, S.M. '42, joined RLE and became supervisor of the Meteor guidance group. Two systems for causing the missile to home in on a target were under study. In one the missile's homing signal was conventional microwave pulsed radar; in the other, continuous-wave radio.

Although the project was highly classified, many of the student theses were general enough to be conducted without security classification. The following are samples of more than 30 guidance theses supervised by Zimmermann: Raymond A. Glaser, S.M. '47, "Measurement of Small Frequency and Phase Deviation"; Thomas J. Fitzgerald, S.M. '50, "Phase Measurements Applied to Instrument Landing of Aircraft"; Albert E. Cookson, S.M. '51, "High-Frequency Phase Measurement Technique"; Campbell L. Searle, S.M. '51, "Phase-Angle Multiplication by Means of Frequency Multipliers"; James B. Angell, Sc.D. '52, "Errors in Angle Radar Systems Caused by Complex Targets."

The telemetry group was supervised by Professor William H. Radford. Among his 30 theses were these: James K. Hunton, S.M. '48, "An Experimental Study of the Characteristics of a Vibrating Wire Pressure Gauge"; Lester L. Kilpatrick, S.M. '48, "A High-Capacity-Matrix-Commutated Telemetering System"; Harold A. Spuhler, S.M. '50, "A Miniature Pressure

Gauge for Use in Airborne Telemetering''; and William F. Santelmann, Jr., E.E. '52, ''Electronic Scaler for Telemetered Data Reduction.''

RLE's seventh quarterly progress report, dated October 15, 1947, lists Professor Wiesner as assistant director; a review of the projects at that time shows that the Electrical Engineering Department was steadily increasing its involvement in RLE.

Professor Lawrence B. Arguimbau found RLE facilities ideal for his experimental studies in television, phonographic recording, and multipath radio transmission. John Granlund, S.B. '44, S.M. '46, completed his doctoral thesis, ''Interference in Frequency-Modulation Reception,'' under Arguimbau's supervision in 1950.

Professor Lan Jen Chu, S.M. '35, Sc.D. '38, an expert in field and wave theory, was widely regarded as a talented teacher and an important influence in RLE research, starting with his pioneering studies in wave guides. Chu and Wiesner studied traveling-wave amplifier tubes and started Lawrence A. Harris, VI S.M. '48, on his 1950 doctoral thesis, ''Axially Symmetric Electron Beam and Magnetic Field Systems''; Samuel A. Sensiper, VI S.B. '39, on his 1951 doctoral thesis, ''Electromagnetic Wave Propagation on Helical Conductors''; and Richard B. Adler, VI-A S.B. '43, on his 1949 doctoral thesis, ''Propagation of Guided Waves on Inhomogeneous Cylindrical Structures.'' In collaboration with Richard Adler and Robert Fano, Chu later developed two textbooks, *Electromagnetic Fields, Energy, and Forces* and *Electromagnetic Energy Transmission and Radiation,* two of the five books of the so-called green series at MIT (see chapter 19), books that exerted an influence on the teaching of electrical engineering far beyond MIT.

Ernest R. Kretzmer investigated pulse-modulation phenomena, leading to his 1949 doctoral thesis, ''Interference Characteristics of Pulse-Time Modulation,''

and his subsequent career at Bell Laboratories. Professor Wiesner started Edward E. David, Jr. (who would become science advisor to President Nixon), in his studies on locking phenomena in microwave oscillators, leading to his 1950 doctoral thesis, ''Some Aspects of R-F Phase Control in Microwave Oscillators.''

Samuel J. Mason, S.M. '47, Sc.D. '52, was appointed to RLE when it was formed in 1946 and later became head of the Cognitive Information Processing Group. His research focused on reading systems for the blind, computer-based optical character recognition, and the psychophysics of tactile and auditory displays. His 1952 doctoral thesis, ''On the Logic of Feedback,'' enabled him to introduce his ''signal-flow graphs'' into the undergraduate electronics subjects. During the EE Department curriculum revision in the later 1950s he coauthored two textbooks with Henry Zimmermann, *Electronic Circuit Theory* and *Circuits, Signals, and Systems.* In the late 1960s and early 1970s Mason served as associate director of RLE.

Manuel Cerrillo (first Ph.D. in EE, '47) and Professor Guillemin did early studies in active networks. John G. Linvill (later head of Stanford's EE Department) joined this activity to write his 1949 Sc.D. thesis, ''Amplifiers with Arbitrary Amplification-Bandwidth Product and Controlled Frequency Characteristics.''

The January 1948 quarterly progress report (no. 8) listed 19 EE Department members out of RLE's 40 faculty members and instructors, and 42 EE members out of 85 research associates, assistants, and graduate students. This laboratory was proving to be an important new research facility, enabling the department to carry its greatly expanded load of postwar graduate students. Among the early EE projects in RLE were those stimulated by Norbert Wiener of the Mathematics Department. The laboratory was fortunate to have a team of sharp and energetic individuals—Tuller, Fano, Wiesner, Cohen, Kretzmer,

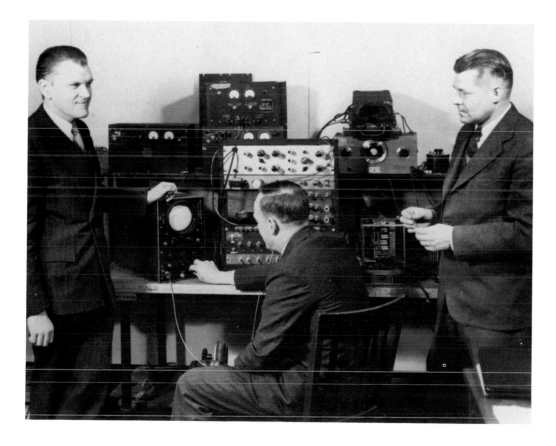

Left to right: Henry J. Zimmermann, J. N. Thurston, and W. H. Radford, examining test equipment related to multichannel telemetering investigations.

**Research Groups Listed in RLE Quarterly Progress Reports
at Five-Year Intervals, 1946–1961**

1946

Microwave and Physical Electronics		Microwave Physics	Communications and Related Projects

1951

Physical Electronics	Microwave Spectroscopy	Low-Temperature Physics	Solid-State Physics	Magnet Laboratory Research

1956

Atomic Beams	Microwave Spectroscopy	Low-Temperature Physics	Solid-State Physics	Microwave Electronics	Nuclear Magnetic Resonance and Hyperfine Structure
Statistical Communication Theory	Multipath Transmission	Processing and Transmission of Information	Speech Analysis	Mechanical Translation	Sensory Replacement

1961

General Physics and Plasma Dynamics

Molecular Beams	Microwave Spectroscopy	Low-Temperature Physics	Solid-State Physics	Microwave Electronics	Thermoelectric Processes and Materials

Communication Sciences and Engineering

Statistical Communication Systems	Modulation Theory and Systems	Processing and Transmission of Information	Process Analysis and Synthesis	Artificial Intelligence	Signal Detection by Human Observers
Computer Components and Systems			Circuit Theory and Design		Network Synthesis

		Modern Electronic Techniques Applied to Physics and Engineering		Aids to Computation		
Microwave Gaseous Discharges	Tube Research and Development	Communication Research	Air Navigation	Analog Computer Research		
Irreversible Thermodynamics	Physical Electronics	Microwave Gaseous Discharges	Semiconductor Noise	Stroboscopic Research		
Communications Biophysics	Transistor Circuits	Neurophysiology	Nonlinear Circuits	Network Synthesis		
Noise in Electron Devices	Nuclear Magnetic Resonance and Hyperfine Structure	Statistical Thermodynamics	Physical Electronics	Physical Acoustics	Plasma Dynamics	
Speech Communication	Linguistics	Mechanical Translation	Sensory Aids Research	Communications Biophysics	Neurology	Neurophysiology
		Computer Study of Dynamics of National Economy				Stroboscopic Research

Cheatham, Davenport, Singleton, Kraft, and Reintjes—who could comprehend Wiener's ideas and translate them into engineering processes and devices.

Wiener's Early Work

It would be difficult to overstate the intellectual impact of Norbert Wiener upon RLE in the postwar era. Wiener joined the MIT Mathematics Department in 1919 and in 1921–22 offered graduate instruction in Fourier series and integral equations. (One of his first two graduate students was Karl Wildes, who wrote his master's thesis under Wiener's supervision.) Wiener was already familiar with the traditional Fourier series of Byerly and other nineteenth-century mathematicians, so that his presentation moved ahead into the study of normal and orthogonal functions other than those employed in trigonometric and exponential series and integrals. He had not studied integral equations in depth, however, so his classroom treatment of this subject was a consolidation and critique of the only works on integral equations that had come to his attention. These were *Integral Equations and Their Applications in Mathematical Physics* by Adolf Kneser of the University of Breslau, a chapter in Goursat's third volume, and a small book by Maxime Bocher of Harvard. Thus began Wiener's intensive study and development of generalized harmonic analysis, culminating in a 141-page treatise in *Acta Mathematica* in 1930.

During these early years Wiener was deeply interested in the work of the Electrical Engineering Department. He and Manuel S. Vallarta helped Bush to establish the Heaviside operational calculus on a sound mathematical basis. Bush's book *Operational Circuit Analysis* was published in 1929 with an appendix by Wiener.

Two influential leaders in the Electrical Engineering Department were Samuel J. Mason (*left*) and Lan Jen Chu, shown here discussing the theory of a new radar antenna developed in RLE.

Wiener's wartime study, *Extrapolation, Interpolation, and Smoothing of Stationary Time Series* (1942), marked his discovery that an important subclass of communication and control problems are best solved through the use of probability and statistical considerations. It was begun in late 1940; the manuscript was essentially complete when, near the end of 1941, Wiener learned that Andrei N. Kolmogoroff had published in 1939 a two-page article in *Comptes Rendus,* "Interpolation and Extrapolation of Stationary Sequences," and an 11-page, similarly titled paper in the *Bulletin of the Academy of Sciences, U.S.S.R.,* in 1941. Wiener's work was independent, however, and was built on his *Generalized Harmonic Analysis,* the calculus of variations, and the Wiener-Hopf integral equation, as well as the statistical mechanics of J. Willard Gibbs of Yale University (published in 1902) and G. I. Taylor in England (1920).

Wiener's early thinking on correlation functions appeared in a doctoral thesis he supervised. Entitled "A New Method of Periodogram Analysis with Illustrative Applications," it was the work of Gleason W. Kenrick, VIII S.B. and S.M. '22, IX-C Sc.D. '27, an instructor in electrical engineering from 1923 to 1927. Kenrick's first application was an investigation of the fading of Station KDKA's radio signal. The second application involved random probability studies suggested by telegraph signals and radio static. In his article "Mathematical Analysis of Random Noise" (*Bell System Technical Journal,* July 1944), S. O. Rice says that Kenrick appears to be one of the first to compute the power spectrum through use of the correlation function. By the time Wiener had published his collected studies in *Generalized Harmonic Analysis* in 1930, he had analyzed in depth the existence and properties of the correlation function. The concept of correlation became so central in Wiener's later studies in prediction, detection, and information processing that the early RLE experimenters chose to investigate correlation functions thoroughly and to develop devices to display these functions automatically.

Information Studies by RLE Associates

One of the first research associates to enter the new RLE in the fall of 1945 was William G. Tuller. He would normally have finished his Course VI-A requirements in June 1940, but his thesis supervisor, Edward Bowles, was at that time setting up a special group for ultrahigh-frequency research in Alfred Loomis's research laboratory in Tuxedo Park, New York (see chapter 13), and Tuller elected to join this group. He worked in the Radiation Laboratory from its beginning in 1940 and changed his master's thesis project to study adjustable-frequency oscillators in the microwave region, finishing his S.B. and S.M. requirements in 1942. He was on leave from Rad Lab to work with the Raytheon Manufacturing Company during the latter part of the war period and headed Raytheon's microwave component development laboratory. In RLE he proposed, for a doctoral thesis, the extension of ideas expressed by Dr. R. V. L. Hartley in his paper "The Transmission of Information," published in the *Bell System Technical Journal* in July 1928. Tuller's thesis was supervised by Guillemin, with the collaboration of Professor Henry Wallman of the Mathematics Department. Tuller showed that earlier investigators (Kupfmuller, Nyquist, Hartley, Gabor) had omitted the effect of noise on the transmission of information, and acknowledged that Wiener's "application of statistics had made possible an adequate treatment." Tuller, who developed a mathematical expression for the amount of information, showed that Hartley incorrectly believed that "the total amount of information which may be transmitted over such a system is proportional to the product of the frequency range which it transmits and the time during which it is available for the transmission." Having concluded that the only limits to the rate of transmission of the information on a noise-free circuit are economic and practical, not theoretical, Tuller went on to show that noise limits the rate of transmission for both coded and uncoded messages, and he discussed the trade-offs among

Norbert Wiener, shown here in
1954 with Yuk Wing Lee (*left*)
and Amar G. Bose, discussing
an aspect of statistical communication theory.

bandwidth, time, and power. The thesis was successfully defended before a department committee in September 1947, although some of Tuller's claims to mathematical rigor had to be modified.

Another important contributor to information studies in RLE was Robert M. Fano, who had been a graduate student and instructor in the department before entering the Radiation Laboratory in June 1944. In the fall of 1945 he returned to academic status as a research associate, teaching subject 6.622, Radio Lines, Antennas, and Propagation, and continuing to prepare his Rad Lab developments for publication in the Rad Lab Technical Series. In March 1946 he entered upon a program of study for the doctorate. His doctoral thesis, entitled "Theoretical Limitations on the Broadband Matching of Arbitrary Impedances," was supervised by Guillemin and completed in June 1947. As he looked for a new research field, his imagination was captured by Wiener's concept that the amount of information in communications is analogous to negative entropy in statistical mechanics. By the spring of 1948 Fano had justified this concept in terms of message encoding and begun the preparation of RLE Technical Report 65, "The Transmission of Information."

Dr. Claude E. Shannon, XVIII S.M., Ph.D. '40, of the Bell Telephone Laboratories had done similar work, published in his classic paper, "A Mathematical Theory of Communication" (*Bell System Technical Journal*, July and October 1948). In this paper Shannon acknowledged that his method of "encoding" was "substantially the same as one found independently by Fano"; he also cited the Tuller thesis and Wiener's forthcoming book *Cybernetics*: "His [Wiener's] classic report contains the first clearcut formulation of communications theory as a statistical problem, the study of operations on time series." Fano and Shannon extended the Hartley concept of *amount of information*

to include systems in which choices among the symbols are not equally probable.

In 1948 Shannon gave a general definition of the *capacity* of a communications channel; in a formula that would become famous, he asserted that for noisy channels the channel capacity can be approached as closely as may be desired through suitable coding. Shannon sketched an incomplete proof of this theorem in 1948; a rigorous proof did not appear until Amiel Feinstein, VIII Ph.D. '54, gave one in his doctoral thesis, supervised by Fano. Shannon followed with a complete proof along the lines of his 1948 ideas.

Shannon, who is most famous for laying the foundations of information theory, joined RLE and the MIT faculty in 1956 as Donner Professor of Science and Professor of Electrical Engineering and Mathematics. As a graduate student at MIT in the late 1930s, Shannon had worked on Vannevar Bush's differential analyzer and had done research that would become important in the development of digital computers. During the war years he was at the Bell Telephone Laboratories, where he worked on cryptography and his statistical theory of communication. His return to MIT in 1956 greatly increased MIT's strength in the area of information theory.

Fano's chief research interest was in the transmission of speech and coded messages. He kept in touch with the work at the Bell Telephone Laboratories and other research in the field of speech intelligibility. Dr. Joseph C. R. Licklider of the Harvard Psychoacoustic Laboratory had found that "clipped speech," produced by limiting the amplitudes of speech waves, was still 70 to 90 percent intelligible. David F. Winter, S.M. '48, of RLE turned his attention to speech-clipping studies after finishing his master's thesis, "High-Frequency Components of Encephalograms." Fano decided to make a study of speech through its auto-

Lee's familiarity with Wiener's early work prepared him for a mastery of Wiener's wartime statistical treatment of stationary time series. Wiener's expositions were usually without graphs and often omitted logical steps that were perfectly obvious to him, but that could be supplied by the student only after considerable thought. Lee's influential achievement was first to understand the Wiener concepts and then to explain them fully with diagrams and illustrations. In the fall of 1947 Lee presented a graduate subject, 6.563, Synthesis of Optimum Linear Systems (later changed to 6.571, Statistical Communication Theory), based on the Wiener treatise of 1942 and concepts in Lee's earlier thesis. Lee's expositions were so lucid that the RLE Communications Group, who listened to the lectures, began to get involved in statistical communications research. Jerome Wiesner also became interested in the new statistical theory and joined Lee in supervising research in this field in addition to his broader field of communications engineering.

Lee's first thesis student was Robert Cohen, S.B. '47, S.M. '48. Cohen discussed in considerable detail the correlation function; among other things, he showed that high-frequency components of the power spectrum correspond strongly to the nature of the correlation function near its central region, and that the power spectrum is best found as the Fourier transform of the correlation function, rather than from experimental data. The extensive manual computations for this thesis were made by the RLE Computation Group.

Lee's first doctoral student was Thomas P. Cheatham, Jr., S.M. '47, Sc.D. '52. Working closely with Lee, Cheatham also had the cooperation of Wiener and Wiesner, who acted as thesis readers. Cheatham and his RLE co-workers designed and built the first electronic correlator, completed and tested in September 1948. He not only described the successful

Richard B. Adler lecturing on transistors, 1963.

Yuk Wing Lee, shown here with a class of graduate students studying statistical communications theory. Amar G. Bose is seated in the center.

J. C. R. Licklider.

machine, but described the false starts and ideas that didn't pan out. The final machine represented the waveform as a sequence of sample values, and multiplied parts of samples by generating a pulse with a height proportional to one sample value and a width proportional to the other. The Cheatham thesis, "Experimental Determination of Correlation Functions and the Application of These Functions in the Statistical Theory of Communications," presented to the department in September 1949, was a superb piece of research and exposition. It included the depth of understanding of the Wiener Theory and its extensions as worked out by Lee and his RLE staff during the first two years following Lee's first lectures.

Thirty years later a correlation function could be produced in a few minutes on any one of a multitude of sophisticated digital computers; but in 1949 MIT's Whirlwind I was not yet reliably operational. Without a correlating machine, researchers had to endure long periods of calculation with desk calculators. The electronic correlator was a useful machine, but Cheatham noted that a typical calculation for one point would require two minutes and the 90 points for a complete curve would require three hours. Cohen had shown that certain correlation values needed much greater precision than was afforded in the Cheatham correlator. A faster and more accurate machine was therefore planned.

Leadership in the design and construction of a digital correlator came from Henry E. Singleton, who joined the RLE staff in September 1948. Singleton, who had completed Course VI-A in 1940, joined Professor Bennett in the degaussing of ships at the Naval Ordnance Laboratory during the war. At the time of his return for a doctoral program at MIT, he was employed as an engineer with the International Telephone and Telegraph Company in New York City. Singleton's correlator, in operation in October 1949, was described in Technical Report 152, issued in

February 1950, after adjustments, improvements, and tests had been made on the machine. The input was sampled data, as in the Cheatham machine; but storage, multiplication, and integration were carried out rapidly by a binary digital process. A complete correlation curve could be produced in less than an hour. Leon G. (Jake) Kraft, Jr., an instructor at the MIT Radar School, joined RLE in January 1949. After the completion of his master's thesis, "A Device for Quantizing, Grouping, and Coding Amplitude-Modulated Pulses," in May 1949 under Wiesner's supervision, Kraft joined forces with Singleton in the completion of the digital correlator and continued with its tests and applications.

With the Singleton correlator Lee and his group developed some striking examples of the detection of periodic signals submerged in overpowering noise. His RLE Technical Report 181 (1950), "Applications of Statistical Methods in Communication Problems," pictured sinusoids, periodic triangular waves, and radar signals recovered from very noisy signals by correlation. In 1951 Lee and Kraft showed the power of the method in a paper, "Detection of Repetitive Signals in Noise by Correlation," presented at the national convention of the Institute of Radio Engineers in New York City. Here two periodic waves, one triangular at a fundamental frequency of 30,000 Hz and the other rectangular at a fundamental frequency of 30,005 Hz, were simultaneously submerged in a strong random noise. Through cross-correlation of the mixed signal with periodic unit impulses of each of these frequencies in turn, both periodic waves were clearly recovered.

By the summer of 1951 this group had built up four years of development and application in the new field, and the time was ripe for a sharing of statistical communications knowledge with interested workers in industry and in other educational institutions. That summer Lee presented a special three-week summer

Norbert Wiener, Jerome
Wiesner, and Yuk Wing Lee.

program, subject 6.629, Modern Communications. Lee himself lectured on messages and noise as random processes, correlation theory, the automatic correlators, detection of repetitive signals in noise, filtering, and prediction. Jerome Wiesner, associate director of RLE, gave special attention to noise spectra, nonlinear processes, and detection and integration of signals. Robert Fano applied the Wiener-Shannon theory of information and his own developments to different modulation schemes, to the rate of transmission in speech communication, and to television, as well as to fields other than electrical communication. Wilbur Davenport discussed probability distributions and their applications. J. C. R. Licklider, whose field was the psychology of communications, lectured on the neurophysiology of vision and hearing, the perception of speech, and the presentation and assimilation of information. Dr. Alex Bavelas, also a psychologist, gave special attention to communication in task-oriented groups and human organizations as information handling systems.

In all his work Lee was a perfectionist. His 6.571 notes absorbed the research contributions of his doctoral students as well as his own research during the 1950s, and the resulting book, *Statistical Theory of Communication* (1960), was judged by his publisher, Wiley, as close to perfection. This book was confined to linear theory and applications; during the 1950s, however, Lee began a study of nonlinear phenomena. Henry Singleton's doctoral thesis, "Theory of Nonlinear Transducers" (1950), was supervised by Wiesner with Fano and Lee as readers.

Lee, assisted by Amar G. Bose, VI-A S.B., S.M. '52, Sc.D. '56, persuaded Norbert Wiener to present his ideas on nonlinear problems in random theory as a series of lectures. As Wiener lectured, Lee took hundreds of photographs of the complex equations with which Wiener covered the blackboard, and he tape-recorded the lectures. Working with Bose, Lee then compiled all the material into the book *Nonlinear Problems in Random Theory* (1958), the first book on this subject in English.

An important new direction for RLE research was established when Walter Rosenblith joined the MIT Electrical Engineering Department in 1951. With a background in communications engineering and classical physics, Rosenblith had first worked at the Harvard Psycho-Acoustic Laboratory. He soon came in contact with Norbert Wiener and cybernetics. At MIT, as a staff member in both RLE and the Acoustics Lab, Rosenblith started research in the field then called communications biophysics. An indication of the relative novelty of research that attempted to link mathematical modeling of electrical brain activity and man-made communication processes can be found in Professor Schmitt's reply when Rosenblith described the direction he wished to pursue. Said Schmitt, "You have come two hundred years too early!" Nonetheless, the new field grew rapidly, and Rosenblith served as professor of communications biophysics in the EE Department from 1957 to 1975. During this period the Center for Communication Sciences in RLE was formed, and Rosenblith was a member of its steering committee from 1958 until 1969, when he became associate provost of MIT.

In the Center for Communication Sciences, electrical communication engineers, computer scientists, neurologists, and neurophysiologists, as well as scientists with research interests in speech, hearing, learning, linguistics, sensory processes, and logic, worked together on a range of problems related to both natural and man-made communications. Rosenblith's research emphasized the use of computers as investigative tools in trying to unravel the meaning of the electrical activity in central nervous system functioning.

As Rosenblith has pointed out, during this early RLE period there evolved numerous groups that in various ways linked engineering and living systems. Among his own collaborations was the formation of the Eaton-Peabody Laboratory for Auditory Physiology at the Massachusetts Eye and Ear Infirmary, which was supported by the National Institutes of Health. That program brought together researchers in medicine, biology, and engineering.

Among the many influential scientists whom Jerome Wiesner brought to RLE during his tenure as director of the laboratory was Warren S. McCulloch. Arriving in 1952, he brought years of experience as a neuroanatomist and neurophysiologist at Yale and at the University of Illinois. McCulloch's career was dedicated to finding out how brains work. In his experimental work he made fundamental and extensive contributions in mapping the cortico-to-cortico connections in the central nervous system; on the theoretical side, working with the brilliant young mathematician Walter Pitts, he formulated the fundamentally computational character of thinking in 1943 in a paper entitled "The Calculus of Ideas Immanent in the Nervous System," which had far-reaching significance and which influenced John von Neumann in his early designs for computers. McCulloch regarded the "Calculus" paper, along with papers by Wiener and Arturo Rosenblueth in the United States and by Craik in England—all published in 1943—as having bridged the mind-body split that had bedeviled scientists since the time of Descartes. McCulloch did not teach formal courses in the EE Department at MIT, but his brain-research activities in RLE excited numerous young electrical engineers, mathematicians, and computer scientists at MIT to work on fundamental problems of brain functioning. One avenue of his influence was evidenced in research on artificial intelligence (AI), especially that of

Marvin Minsky in the AI Laboratory established in the EE Department, research which attempted to build elements of human-like intelligence in computer systems. Another avenue was the neurophysiological work of Jerome Lettvin, who with McCulloch and Pitts demonstrated the functioning of ganglia that served as "property detectors" in the visual system of the frog. Their now-classic paper "What the Frog's Eye Tells the Frog's Brain," published in the *Proceedings of the Institute of Electrical and Electronics Engineers* in 1959, was a landmark piece of research. It represented a culmination of the viewpoint established in the "Calculus" paper years earlier; it demonstrated how much information processing went on in the peripheral (sensory) systems of organisms, and started a rush of new research (notably the work of Hübel and Wiesel at Harvard); in McCulloch's view, the frog's-eye work put epistemology on an experimental basis for the first time in the Western scientific tradition.

The far-reaching influence, on all these scientists, of Wiener's statistical theory is seen here only in its beginnings. A few of the later applications are included in *Theory of Servomechanisms* (1947), by James, Nichols, and Phillips, part of the Radiation Laboratory Series; in *Control System Synthesis* (1953), by James G. Truxal, S.B. '47, Sc.D. '50; and in *Control Systems Engineering* (1956), by J. Halcomb Laning, Jr., X S.B. '40, XVIII Ph.D. '47, and Richard H. Battin, S.B. '45, XVIII Ph.D. '51. Some of Wiener's statistical communication ideas were included in the 6.05 textbook *Electronic Circuits, Signals, and Systems* (1960), by Mason and Zimmermann.

Jerome Wiesner, in an essay written in 1966 for the booklet "RLE: 1946 + 20," reflected on the dramatic growth of the communication sciences during RLE's first two decades. The following excerpts from that essay convey the atmosphere of the period.

Walter Rosenblith (*right, stand-ing*), with William Siebert (*left*) in lab.

Warren S. McCulloch brought a new dimension to electrical engineering at MIT and, among other matters, inspired early research on artificial intelligence. In this photo McCulloch (*left*) is conferring with Betty Goodrich and Jerome Lettvin beside a machine for making electrodes.

We shared the ferment and excitement of discovering for ourselves the universal role of communication processes in man's universe. . . . Fired up by Norbert Wiener's cybernetics, we explored the far-ranging implications of the concepts of information and communication theory; our interests ranged from man-made communication and computing systems to the sciences of man, to inquiries into the structure and development of his unique nervous system, the phenomena of his inner life, and finally his behavior and relation to other men.

All of this took place in that unique scientific incubator, the Research Laboratory of Electronics, which for two decades has provided an almost ideal research environment and has been a model for the structure of other research centers. . . .

When the Research Laboratory of Electronics was created with a charter that agreed only ''to do basic research in the field of electronics,'' no one knew just where this license would lead. Those of us who had spent our war years as plumbers in the Radiation Laboratory and expected somehow to employ our knowledge of the new microwave technology usefully could hardly imagine the excitement and intellectual pleasure that lay ahead of us. In fact, as I look back on twenty years of the communication sciences at MIT, I have the impression of powerful personalities and even more powerful ideas drawing people together from all over the world. I see Wiener, Gabor, Bavelas, Lee, Shannon, Fano, Rosenblith, Tuller, Guillemin, Elias, McCulloch, Pitts, Lettvin, Ikehara, Licklider, Miller, Halle, Jakobson, Chomsky, Bar-Hillel, Kretzmer, Singleton, Mason, Davenport, David, Mandelbrot, Cherry, Schutzenberger, and many others. . . . The two decades of RLE [were] like an instantaneous explosion of knowledge.

Norbert Wiener drew the communication sciences together at MIT. The ideas were in the air. Shannon's classic papers, Wiener's own wartime activities, the interest in signals and noise problems that many of us had developed all contributed; but Wiener was the catalyst. . . . He did his job almost without recognizing his role, for his interest was in ideas. But his action did bring together people with related interests. He did this in many ways—by personal contact, by stimulating meetings, and particularly by writing and speaking. Many can remember Norbert's daily visits around the Institute from office to office and his conversations that always began with "How's it going?" He never waited for the answer before sailing into his latest idea. "By the way," he would say, "have you heard about what Arturo and I have been thinking?" Or, "Have you heard about what Walter Rosenblith has said?" Sometimes he was disturbed: someone had challenged one of his ideas; or he had become convinced that some foolish action of the president or secretary of state was going to catapult the world into oblivion. "Do you think there is going to be a war?" was his standard question on those blue days. Whatever was on his mind, Norbert Wiener's visit was one of the high points of the day at MIT for me and many others. For those of us working in the lonely isolation of Building 20, Norbert's visits were especially welcome, for he was one of the best links with the main building. . . .

Wartime activity also led Claude Shannon to develop his mathematical theory of communication.. . . By defining a unit of information and showing the relationship between signals, noise, and rate of transmission of information, he provided a conceptual basis for the attack on communication problems. Together Shannon and Wiener aroused the interest of people all over the world whose activities brought them into contact with those mysterious "nonphysical" properties of ensembles that stem from their organized complex-

ity, i.e., their symbolic and informational aspects. In the forefront were the more thoughtful communication engineers who had long been troubled by the fact that they did not understand the commodity they were handling. They could not see it or measure it or know why it changed as it was manipulated.

The communication engineers were not alone in their enthusiasm for new ideas. Neurophysiologists and other biologists, linguists, economists, social scientists, and psychologists of the various persuasions recognized almost intuitively the usefulness and relevance of feedback and information theory concepts to their fields. For all of them, the concepts of information theory, coding, feedback, prediction, and filtering provided new pathways to explore, pathways that seemed to wind unendingly. Even before the publication of Cybernetics, Wiener was working hard to stimulate general appreciation of and interest in the broad applicability of these emerging ideas. To this end he organized dinner meetings for a diverse (the label "interdisciplinary" stuck) group of scientists and engineers. Together they explored each other's fields and slowly began to comprehend each other's lingo and exhibit that spirit of mental intoxication that characterizes the pursuit of an exciting new idea. . . .

Any mention of those exciting years would be incomplete without recalling the important role that a sister laboratory, the MIT Acoustics Laboratory, played in some of the developments, largely because of the presence there of J. C. R. Licklider. Licklider—experimental psychologist and electronics experimenter par excellence—like Rosenblith and several other members of the cybernetic marching society, spent his war and postwar years at Harvard's Psycho-Acoustic Laboratory. When he arrived at the Institute (where he was to play a notable role in research and teaching throughout most of the 1950s), Alex Bavelas was already involved in the study of human communication networks. Bavelas and several young mathematicians, among them Duncan Luce, examined the ways in which the patterns of communication facilities linking people influenced their performance of a task and their attitudes toward each other in these task-oriented networks. Licklider's complementary interest in all aspects of speech communication systems was to be an important factor in studies of biological communication for a decade or more. With Bill Locke he organized several speech communication conferences, which resulted ultimately in the formation of a linguistics group. By becoming involved with the engineering aspects of speech communication in the RLE (and the Acoustics Laboratory), the Department of Modern Languages was able to attract the nucleus of what is today sometimes called MIT's Linguistic School: Morris Halle and Noam Chomsky came and gathered numerous students and associates. Roman Jakobson, who had early understood RLE's promise for the future of Linguistics, gradually shifted his intellectual abode from Harvard to the Institute.

Spread-Spectrum Origins

Among the many contributions by Jerome Wiesner to RLE was his participation in what is now known as spread-spectrum communications. Considerable attention has been given to the "spread-spectrum concept" since its recent release from military classification. Two articles on this subject have been published by Professor Robert A. Scholtz in the *Institute of Electrical and Electronics Engineers Transactions on Communications*: the first, entitled "The Spread-Spectrum Concept," appeared in August 1977; the second, "The Origins of Spread-Spectrum Communications," in May 1982. We are indebted to Professor Scholtz's work, and to the sources from which he has drawn his account, for the information that follows.

"Summer Studies" at MIT brought together academic persons of great competence who were relatively free during the summer months. One of these summer studies, Project Lexington, was conducted in 1948 "to determine the technical feasibility of nuclear-powered flight." Project Charles (1951) was another. Project Hartwell was set up at the request of the U.S. Navy for the summer of 1950 to find new ways to protect overseas transportation. The chairman of the Hartwell project was Jerrold Zacharias, Professor of Physics at MIT, a former leader in the Radiation Laboratory and in the Atomic Bomb Laboratory at Los Alamos. Among the many developments reported in the 700-page report delivered to the Office of Naval Research in September was a method of radio communication that could neither be picked up by the enemy nor "jammed."

The concept involved the use of a noise signal as a message carrier. This noise signal was spread so thin that it would not be noticed, as it was only a fraction of the normal transmission noise. In order to operate this system, the noise signal transmitted must be replicated at the receiver, so that the transmitted signal can be cross-correlated with the replicated carrier to extract the message. Two methods of providing the noise carrier at the receiver are described in the Hartwell report. One is to transmit the noise signal to the receiver on a second channel; the other is to store the noise carrier at the receiver. In spread-spectrum terms, the first is called TRSS, or transmitted-reference spread spectrum.

Soon after the Hartwell project Wiesner described to Robert Fano the new idea of using a noise-modulated carrier to provide secure military communications. This made a profound impression on Fano, who discussed it with his recent doctoral student Wilbur B. Davenport; they agreed that Davenport would apply the new technique to communications, while Fano would apply it to radar.

The first thesis in the new field was classified "secret" and was completed by Bennett L. Basore, Sc.D. 1952, under Fano's supervision; Wiesner and Davenport were readers. It compared the transmitted-reference (TR) and stored-reference (SR) systems. Basore coined the term NOMAC (Noise Modulation and Correlation) to describe the systems under study. Under Davenport's supervision a study of TR-NOMAC systems was made by Bernard J. Pankowski in a secret S.M. thesis, also completed in 1952.

When it became evident that SR systems had advantages over TR systems, Davenport turned his Lincoln Laboratory studies to SR-NOMAC. Paul E. Green, Jr., Sc.D. '53, gave special attention in his secret thesis to the generation of the same noise signals simultaneously at the two ends of a transmission link. His success spelled the beginning of truly operational SR-NOMAC systems. Green remained at Lincoln Laboratory and took charge of getting a prototype SR-NOMAC built and tested. Called the F9C, this work was done for the Army Signal Corps. Course-VI alumni worked on this project.

The F9C was tested with a transcontinental link in 1953. Signals arriving over different paths (multipath transmission) presented problems. Robert Price's 1953 Sc.D. thesis, supervised by Davenport, addressed such multipath problems. A solution was found by Price and Paul Green: instead of tuning to only one transmission path, they proposed utilizing the energy of the several paths. This was accomplished through the use of tapped delay devices whose signal outputs were brought into synchronism and added together. To quote Green, "Price and Green polished this mutually arrived-at set of ideas and figured out a way to build an add-on to the F9C system. The modified system was built and called an 'Anti-Multipath System,' the F9C-A." The anti-multipath system, known as RAKE, saw vital service during the Berlin Crisis of the early 1950s. For these early spread-spectrum developments Green and Davenport received 1981 Pioneer Awards from the Institute of Electrical and Electronics Engineers' (IEEE) Aerospace and Electronic Systems Society. For his developments in communications, including the RAKE system, Price received the 1981 Edwin Howard Armstrong Achievement Award of the IEEE Communications Society.

Green remarked that John M. Wozencraft "later came up with the first practical frequency-hopping proposal, the WICS, which some said stood for 'Wozencraft's Incredible Communication System,' but in fact is patented as the 'Wozencraft Iterated Coding System.'"

There were many other significant researches in RLE's early years that cannot be described at length here. For instance, David A. Huffman's 1953 Sc.D. thesis, "The Synthesis of Sequential Switching Circuits," under Samuel H. Caldwell, led into Huffman's outstanding leadership in this field. New avenues of research were established by Louis Smullin's work on microwave tubes, Richard Adler's on solid-state devices, Morris Halle's on the mechanical translation of languages and on speech analysis, and Ernst Guillemin's on network synthesis.

The list of important doctoral theses is so large that it cannot be included here. The increase in the total numbers of doctoral theses produced in RLE under EE faculty supervision—from the first 2 in 1947 to an extraordinary 21 in 1965 and a more representative 14 in 1975—reflects the growing volume of EE research. We observe that 16 of the 1981 faculty did their doctoral theses in RLE and that several of these professors have established RLE research groups in the laboratory.

Administrative Changes

The year 1949 was one of reorganization, not only for RLE but for MIT as a whole. James Rhyne Killian was inaugurated as president in April as part of the Mid-century Convocation. In his annual report to the corporation Dr. Killian said, "The fact that an institute of technology held a convocation that inquired so widely and deeply into the social problems of the midcentury is evidence of a vigorous humanism that recognizes the interdependence, unity, and social value of all useful learning." Quoting from Rashdall, *The Universities of Europe in the Middle Ages*, he saw the true university as "a place where the different branches of knowledge are brought into contact and harmonious combination with one another." Killian remarked, "In our university organization we must devise new organizational methods, such as interdepartmental laboratories and programs, in order to provide an integrated approach to fields such as nuclear science, international relations, electronics, and public administration." Killian announced the creation of a new senior administrative post, that of provost: "His primary concern is the administration and coordination of educational and research activities which do not fall within the jurisdiction of any single School. He thus has cognizance over the interdepartmental laboratories and the research projects of the Division of Industrial Cooperation. At its meeting in March, the Corporation of the Institute elected to this new post Professor Julius A. Stratton of the Department of Physics. Professor Stratton's background, both in electrical engineering and in physics, his brilliant direction of the Research Laboratory of Electronics, and his membership on the Committee on Educational Survey have provided him with an extraordinarily rich background for his new responsibilities."

Albert G. Hill succeeded Stratton as director of RLE; in 1952 he became director of Lincoln Laboratory and was succeeded at RLE by Jerome Wiesner. Henry Zimmermann became director in 1961; his was the longest of the directorships, ending, at his request, in June 1976, when Professor Peter A. Wolff of the Physics Department took over. Professor Jonathan Allen of the EE Department succeeded Wolff in 1981 when the latter became director of the Francis Bitter National Magnet Laboratory.

RLE Research, 1960 to the Present

For the account of recent RLE activity that follows, the authors are indebted to Professor Jonathan Allen, director of the laboratory since 1981.

In the 1950s, as a heritage from Rad Lab's wartime research in gaseous T-R (Transmit-Receive) tubes for the protection of sensitive radar echo receivers from the powerful radar-transmitter pulses, Professors of Physics William P. Allis, Sanborn C. Brown, and Malcolm W. P. Strandberg established RLE as one of the major centers of research in gaseous electronics, with special emphasis on radio-frequency and microwave breakdown, and microwave spectroscopy. In 1958, soon after thermonuclear fusion was declassified at the First International Conference on the Peaceful Uses of Atomic Energy, held in Geneva, the group's interests began to turn in the direction of hot dense plasmas. There followed a diversity of studies in a field called plasma dynamics, which has grown substantially in RLE over the last 25 years. In the early days radiation from plasmas was little understood, and because of that an intense program of studies was undertaken, leading to the first quantitative measurements of cyclotron emission and bremsstrahlung, under the guidance of Professor George Bekefi (Physics), and the first complete theoretical analysis, by Professor Tom Dupree (Physics), VIII S.B. '55, Ph.D. '60. Perceiving that wave instabilities represented one of the major stumbling blocks to thermonuclear fusion, Professors Abraham Bers, VI S.M. '55, Sc.D. '59, Richard Briggs, VI S.B. '59, S.M. '61, Ph.D. '64, and

Peter A. Wolff of the Physics
Department, RLE director from
1976 to 1981.

Jonathan Allen of the EE Department, RLE director since
1981.

Louis Smullin, VI S.M. '39, undertook research in this area. This has led to the first classification of instabilities by Briggs and Bers.

The same instabilities that are detrimental to thermonuclear fusion can often be employed in the generation of coherent electromagnetic radiation and in the development of novel microwave and millimeter-wave devices. Some of the earliest work in this area was undertaken by Bekefi and Bers in the early 1960s but was largely abandoned because of the lack of interest at that time. In the late 1970s, however, Bekefi resumed this work and formed a group that is now devoted entirely to novel devices such as the relativistic magnetron, the gyrotron, and the free-electron laser. The group has produced the first narrow-band free-electron laser capable of being tuned over a very wide range of frequencies.

Work in the 1970s focused on several difficult plasma phenomena, including nonlinear wave-plasma interactions and the characterization of plasma turbulence. Dupree and his collaborators have done pioneering research in this area, achieving the first fundamental understanding of nonlinear theory and the prediction of "clumps." Professor Bruno Coppi (Physics) introduced the study of the magnetohydrodynamics of hot fusion plasmas; he developed the exciting concept of the high-field tokamak, the first model of which, the Alcator A tokamak, was built at the National Magnet Laboratory under the guidance of Professor Ronald Parker, VI S.M. '63, Sc.D. '67. This work continues and is embodied in a second-generation machine, the Alcator C tokamak, now at the Plasma Fusion Center.

Supplemental plasma heating is one of the major concerns in thermonuclear fusion research. RLE has been particularly strong in this area, with an emphasis on radio-frequency heating. RF-driven current in tokamaks was predicted for the first time by Nathaniel Fisch, VI S.B. '72, S.M. '75, Ph.D. '78, a graduate student in Bers's group, and the first definitive demonstration has been performed by Professor Miklos Porkolab (Physics) and Stanley Luckhardt (Plasma Fusion Center) on a small experimental device called the Versator II tokamak, housed in RLE.

Research in the plasma area has always involved a close coupling of interests in both theory and experiment, cutting across the EECS and Physics Departments. While RLE interest has always focused on basic plasma phenomena and their theoretical characterization, increased government interest in plasma fusion led to the forming in 1980 of the Plasma Fusion Center, where large mission-oriented experiments involving both toroidal and mirror-confinement machines have been carried out with the goal of developing practical means for generating electromagnetic energy from nuclear fusion. Thus, RLE continues in a diverse pattern to explore fundamental plasma issues, whether or not they are related to fusion, while a strong interaction with the Plasma Fusion Center nevertheless provides an important motivation for much of the RLE work.

Early involvements with the general problem of sensitive microwave amplifiers led naturally to interests in radio astronomy instrumentation, pursued by groups involving Professors Alan H. Barrett (first in the EE Department, later in Physics), David H. Staelin, VI S.B. '60, S.M. '61, Sc.D. '65, Paul L. Penfield, Jr., VI Sc.D. '60, and Robert P. Rafuse, VI S.M. '57, Sc.D. '60. This line of investigation broadened eventually into studies of the remote sensing of the terrestrial atmosphere and ground resources by electromagnetic waves from either satellites high above the earth or probes inserted deeply into it, pursued by professors Staelin and Jin Au Kong.

Indeed, the RLE interest in electromagnetic wave propagation under a wide variety of circumstances has continued in a virtually unbroken, though convoluted, line from RLE's postwar inception to the present time. It has involved, on the one hand, fundamental theoretical work by Professors Haus, Penfield, Kong, and Frederic R. Morgenthaler, VI-A S.B., S.M. '56, Ph.D. '60, on such topics as the electromagnetism of moving matter, microwave-acoustic-magnetic wave interactions in magnetic materials, coupled-mode theory, and propagation in stratified and random media. On the other hand it has extended into optical propagation systems and detection methods, investigated by Professors Robert S. Kennedy, VI S.M. '59, Sc.D. '63, and Jeffery H. Shapiro, VI S.B. '67, S.M. '68, Ph.D. '70, leading to recent studies of quantum optics and associated systems.

A related trend in RLE, building from the interest in electromagnetic propagation and high-frequency amplifiers, has been the study and characterization of electrical noise. It has long been recognized that amplifiers add noise of their own to any incoming signals and noise, and that the ultimate source of amplifier noise is the graininess of the electrical charge. Fundamental work along these lines was carried out by Professor Hermann A. Haus, VI Sc.D. '54, on the underlying noise limitation of traveling-wave microwave amplifiers, which is attributable directly to fluctuations of the very electron beam that produces the amplification. Haus, suspecting that greater generality was hidden in this result, then enlisted the collaboration of Richard Adler to bring to bear some results of the circuit theory of solid-state amplifiers originally developed by Samuel Mason and Richard Thornton, VI S.M. '54, Sc.D. '57. There resulted a new general concept of "noise measure" to characterize the fundamental signal-to-noise ratio limitation of *any* linear amplifier, however constructed. Unfortunately, it turns out that the optimum noise measure has a lower bound that is proportional to the frequency of operation of the amplifier. Accordingly, it tends to be very large in optical systems, particularly in lasers, in which the noise is therefore actually measurable experimentally, even though it is of quantum-mechanical origin. Haus realized that optical communications systems operate at the limit set by the Heisenberg uncertainty principle. This natural transition from electronic amplifiers to the study of optical systems, where the quantum-mechanical noise limit could be detected, has led to a new emphasis by Haus and Professor Erich Ippen, VI S.B. '62, on the high-speed properties of optical systems. Thus, while the noise properties of coherent laser optics are not encouraging for practical application to analog-based communications, there is a benefit that outstrips any competitive physical mechanism, namely, the ability to produce optical pulses of ultra-short duration. The work in RLE in the late 1970s and early 1980s has led to a stunning series of experimental demonstrations, first a picosecond pulse train and more recently pulses as short as 16 femtoseconds. The availability of such pulses may be extremely important for digital optical communication, as well as for the continuing investigation of integrated optical systems, where electronic integrated circuits fabricated in direct band-gap semiconductors can be directly interfaced to optical communication and computation systems.

The trend by which research interests in RLE have moved from the characterization of noise and amplifiers to the study of the high-speed properties of optical systems is continuing in the direction of device implementation using modern epitaxial techniques that are at the forefront of integrated-circuit technology. Likewise, the fundamental investigations of noise have had an important interaction with RLE studies in radio astronomy, where very faint signals must be recovered through the use of large antenna arrays. Here the work, led by Professor of Physics Bernard Burke,

VIII S.B. '50, Ph.D. '53, focuses on the use of sophisticated computation to extract signals provided by the Very Large Array, a tool that is providing a continuing stream of astronomical discoveries.

Through the earlier work of Mason and Adler, RLE had a strong influence in the characterization of semiconductor devices, particularly in device modeling and the utilization of these models in nonlinear circuits. More recently, however, there has been an increasing effort within the condensed-matter physics part of RLE to understand semiconductor fabrication and defects. Here again RLE has provided the means for interaction between an experimental group—including Professors of Physics Robert Birgeneau, David Litster, VIII Ph.D. '65, and Mark Kastner—and a theoretical effort, led by Professors of Physics Nihat Berker, John Joannopoulos, and Patrick Lee, XVIII S.B. '66, VIII Ph.D. '70. By means of high-energy synchrotron radiation it has been possible to study the way in which semiconductor surfaces are modified when other materials are adsorbed onto them. These phenomena are fundamental to such processes as epitaxial growth, where failure to provide strong bonding to the semiconductor lattice has led to unreliable devices and circuits. Over the last 20 years work in this area has built up to the point where an accurate model for surface reconstruction has been provided, showing that the underlying lattice of the semiconductor is not invariant during such processes, but actually reconstructs itself in such a way as to adapt to the material being adsorbed. It has even become possible, using modern pseudopotential theory and computational methods, to predict how individual atoms will bond to a semiconductor surface. Again, fundamental research turns out to be of the utmost practical value. Moreover, while the underlying physical properties have not changed, the means to exploit our knowledge of these properties in a computationally realistic framework is an important contribution. A continuing trend in RLE is thus the application of fundamental physics to increasingly practical electronic systems, where the ability to produce ultra-small and high-performance components becomes ever more significant. RLE provides an environment in which semiconductor fabrication interests in both physics and electrical engineering are coming together to offer scientific support in areas where technology often leads science.

Another example of this trend is the work in physical chemistry by Professors Sylvia Ceyer and Keith Nelson. Ceyer has concentrated on the detailed experimental study of chemical reactions when accelerated streams of neutral species impinge on a semiconductor surface. This work is of fundamental benefit to the understanding of such important processes as plasma etching in integrated-circuit fabrication. Nelson, using coupled lasers, has introduced acoustic phonons into a semiconductor lattice, providing controllable strain and hence the ability to introduce defects in a controlled manner. As semiconductor devices scale down in size, the ability to characterize defects accurately is of increasing importance, and technological invention alone cannot be counted on for further progress without adequate support from fundamental scientific investigations. RLE, through its Joint Services Electronics Program, provides the means not only to support this basic work but also to bring it together with important application areas in an appropriate engineering context.

The engineering side of the fabrication of very small semiconductor structures is illustrated by the introduction to RLE in the late 1970s of new efforts in creating submicron structures, led by Professor Henry I. Smith. Structures with a resolution as small as a tenth of a micron have been reliably produced through the use of X-ray lithographic techniques, which are also an invention of Professor Smith. Not only is this work of great practical importance, but its facilities have been used to support many other projects within RLE as well. The careful growth of thin semiconductor and metal films has also been an emphasis of Smith's activities and can be seen as an important new complement to conventional epitaxial techniques.

One of the interesting attributes of RLE from the beginning is the way in which small projects have grown into large efforts, sometimes even leading to new research laboratories. This phenomenon can be seen in the case of computing, which grew up at MIT within RLE and the neighboring Computation Center. Early work in artificial intelligence by Professor Minsky and in experimental time sharing by Professor Dennis took place in RLE before the formation in 1963 of Project MAC. Indeed, the first director of Project MAC, Professor Fano (see chapter 22), has already been identified with the early history of RLE. Artificial intelligence subsequently split off from Project MAC, illustrating once again the tendency of groups to divide and form new enterprises when the size of the research activity reaches some critical mass. In this way research in RLE can be seen in a biological analogy, wherein occasionally a cell splits, producing new cells beyond the parent organism. Thus, the structure and fabric of RLE have been continually changing over the years, but there has always been a consistent and strong focus on electronics and high-frequency electromagnetics as well as on language, speech, and hearing.

With the growth of computing in RLE in the early 1960s, it was natural that many areas of research would exploit this new capability, and in many cases new research groups evolved through coupling traditional disciplines with a new computational emphasis. One example is the extension of the earlier study of continuous systems, both linear and nonlinear, to a study of discrete systems, thus forming a new area now called digital signal processing. Professor Alan

Oppenheim, VI-A S.B. and S.M. '61, Sc.D. '64, built up a very strong group within RLE, initially addressing problems in speech such as spectral estimation and coding techniques, but in more recent years becoming involved in two-dimensional image-processing techniques. Professor William Schreiber also built up a group, starting in the early 1960s, focused on image coding and transmission. This group developed the first laser-based wirephoto machine, now in standard use, and the group's extensive experience with picture coding has led to advances in the study of high-resolution television.

Image coding and processing have been studied by other RLE researchers over the years, in particular the Cognitive Information Processing Group, led originally by Professors Samuel Mason and Murray Eden. Eden used techniques borrowed from phonology to characterize handwritten script in terms of a set of features that were then used for the recognition of such script. Mason, desiring to build a reading machine for the blind, concentrated on the development of character recognition techniques for a wide range of fonts. This work was furthered by Professor Donald Troxel, VI S.M. '60, Ph.D. '62: a scanning machine coupled to a computer provided the first realistic demonstration of input scanning capability for a reading machine. The scanner was then coupled to a computer-driven system, initially developed by Professor Francis F. Lee, VI S.B. '50, S.M. '51, Ph.D. '66, for synthesizing speech by rule; there followed a 15-year effort toward unrestricted text-to-speech conversion, resulting in the MITalk system, which forms the conceptual basis for all commercially available text-to-speech systems. The MITalk system was developed under the direction of Professor Jonathan Allen, VI Ph.D. '68. It included a wide variety of linguistic capabilities, such as morphemic analysis, syntactic parsing, letter-to-sound rules, intonation analysis, and phonemic synthesis, the latter developed by Dr. Dennis Klatt. While this work was motivated by the need to develop sensory aids for the blind, it was obviously strongly linked to the earlier speech communication work involving the acoustic modeling of the human vocal tract by Professor Kenneth Stevens, VI Sc.D. '52, and his colleagues in the Acoustics Laboratory.

The speech synthesis work went on to involve the study of articulation, acoustic correlates of vowels and consonants, intonation and timing, developmental concerns, and the relationship of phonetics to phonological theory, particularly at the phonemic feature level. Thus, within RLE important instrumental investigations in phonetics were coupled with the burgeoning work in linguistics. In 1968 Professors Morris Halle and Noam Chomsky published their famous book *The Sound Pattern of English*, which gave a comprehensive view of the segmental phonology known at that time. The contributions of linguistic theory to the determination of lexical stress were particularly important in helping to provide the high quality of speech output of the MITalk system.

In addition to the work in linguistics and speech communication, the 1960s saw increasing research in both auditory psychophysics and auditory physiology within RLE's communications biophysics laboratory, led by Professor Walter Rosenblith. From initial correlation studies, already mentioned, experimental techniques developed until it was possible to measure signals directly from single auditory nerve fibers in laboratory animals such as cats. During the last 25 years there has been intensive work in all aspects of auditory physiology, including studies by Professor William Peake, VI S.B. '51, S.M. '53, Sc.D. '60, on mechanical transduction in the middle ear; studies by Thomas Weiss, VI S.M. '59, Ph.D. '63, on the transduction process within the cochlea, involving the excitation of hair cells; and the related program of comprehensive

recording from auditory nerve fibers at the Eaton-Peabody Laboratory of the Massachusetts Eye and Ear Infirmary conducted by Nelson Kiang, now a professor of psychology at MIT. Important work in the theoretical modeling of these processes has been carried out by Professor William Siebert, VI S.B. '46, Sc.D. '52; and recently Dr. Donald Eddington of Eaton-Peabody, in research aimed at producing a new prosthetic device, has developed experimental techniques for the direct electrical stimulation of the bony exterior of the cochlea. In the area of auditory psychophysics Professor Louis Braida, VI S.M. '65, Ph.D. '69, and Mr. Nathaniel Durlach have studied auditory intensity perception in great detail, and Professor Steve Colburn, VI S.B. '63, S.M. '64, Ph.D. '69, and Campbell Searle, VI S.M. '51, have studied auditory localization and other features of auditory perception. A continuing interest has been the study of pitch perception as a psychophysical process, and much attention has been given to the transformation of speech stimuli in forms that could be used as the basis for hearing aids, as well as the study of tactile stimulation used by "listeners" who are both deaf and blind. RLE has provided the means to integrate interests in linguistics, speech communication, auditory physiology, and auditory psychophysics. Each of these areas provides important support for the others. For example, in the early 1980s models based on physiological understanding of the auditory periphery are being developed for use in speech-recognition algorithms; these represent a significant improvement over earlier simplistic filtering models.

Although there was a great deal of activity in RLE in the early 1960s in characterizing semiconductor devices and circuits, MIT did not participate strongly in the early development of integrated circuits as opposed to the study of discrete devices. In 1978, however, Professors Jonathan Allen and Paul Penfield recognized that important new design techniques,

**Research Groups Listed in RLE Quarterly Progress Reports
at Five-Year Intervals, 1966–1976**

1966

General Physics

Molecular Beams	Microwave Spectroscopy	Radio Astronomy	Optical and Infrared Spectroscopy	Ultrasonic Properties of Solids	Geophysical Research	Gravitation Research	Noise in Electron Devices

Plasma Dynamics

Gaseous Electronics	Plasma Physics	Plasmas and Controlled Nuclear Fusion	Energy Conversion Research

Communication Sciences and Engineering

Statistical Communication Theory	Processing and Transmission of Information	Artificial Intelligence	Speech Communication	Linguistics	Cognitive Information Processing	Communications Biophysics

1971

General Physics

Molecular Beams	Microwave Spectroscopy	Atomic Resonance and Scattering	Radio Astronomy	Solid-State Microwave Electronics	Cooperative Phenomena in Solids and Fluids

Plasma Dynamics

Gaseous Electronics	Plasmas and Controlled Nuclear Fusion

Communication Sciences and Engineering

Processing and Transmission of Information	Detection and Estimation Theory	Speech Communication	Linguistics	Cognitive Information Processing

1976

General Physics

Molecule Microscopy	Electron Optics	Semiconductor Surface Studies	Atomic Resonance and Scattering	Quantum Electronics	Infrared Instrumentation and Astronomy

Plasma Dynamics

Plasma Dynamics

Communication Sciences and Engineering

Optical Propagation and Communication	Detection Estimation and Modulation Theory	Digital Signal Processing	Speech

Magnetic Resonance	X-Ray Diffraction Studies	Soft X-Ray Spectroscopy	Physical Electronics and Surface Physics	Physical Acoustics	Electrodynamics of Moving Media
		Spontaneous Radio-Frequency Emission from Hot Electron Plasmas		Laser-Plasma Interactions and Magnetic Scattering	
Neurophysiology	Computer Research	Network Synthesis	Computation Research	Advanced Computation Systems	Stroboscopic Light Research

Gravitation Research	Magnetic Resonance	Physical Electronics and Surface Physics	Physical Acoustics	Electrodynamics of Media	Laser Applications
Applied Plasma Research			Relativistic Beams		
Communications Biophysics		Neurophysiology	Signal Processing	Digital System Design Automation	

Microwave and Millimeter Wave Techniques	Radio Astronomy	Electrodynamics of Media	Physical Acoustics	Gravitation Research
Linguistics	Cognitive Information Processing		Communications Biophysics	Neurophysiology

focused on MOS (Metal-Oxide-Semiconductor) circuits, had opened up the possibility of designing large circuits at a system level with impressive performance. Building on a semester visit by Ms. Lynn Conway of the Xerox Palo Alto Research Center, a robust new activity in integrated-circuit design built up, spanning interests all the way from the device level to large complex computing systems. The emphasis in RLE has been on the building of a number of software design tools, many of which are used extensively in industry, and on the development of new circuit-analysis techniques and high-performance architectures for digital signal processing. Once again, RLE research can be seen as spanning the entire spectrum of concern in electronics and high-frequency electromagnetics, from materials and fabrication, through device invention and characterization, to optics and the study of semiconductor surfaces and interfaces, and hence to circuits, logic, and architecture. Indeed, just as in the language, speech, and hearing area, the laboratory is committed to the integration of all the disciplines within the electronics and optics area. The ability to bring together the study of fundamental phenomena, theoretical modeling, and the design of important applications remains the hallmark of research in RLE.

The complexities of the evolution of ideas in RLE have been such that it is difficult to present a concise or balanced picture of its activities over the years. The fact that thousands of MIT degrees have come out of its research and that many departments beyond Physics and Electrical Engineering have participated in RLE programs is statistical testimony to the vitality of the laboratory, but hardly explains how the microwave heritage of the Radiation Laboratory could produce such a diversity of interests. Yet there is a singular center to a phenomenon like RLE, which is explained by the fusion of basic research and education. Henry

Zimmermann describes it as follows: "Much of the motivation for learning comes from the element of discovery inherent in the acquisition of new facts or the understanding of new ideas. At a research frontier where the unknown becomes knowable for the first time, the excitement that accompanies learning is that of the explorer in scholar's clothing. Research leads naturally to education because new knowledge demands to be disseminated, and education leads inevitably to more research because the gaps in our knowledge are never more apparent than when we try to teach."

Chapter 17

From Whirlwind to SAGE

Some of the early work on a project in the Servomechanisms Laboratory that would grow into the Whirlwind computer was first mentioned in chapter 14. That project had many stages: it started as an effort to devise a real-time, universal, analog-type flight simulator for training aircraft pilots; it outgrew its original mission and became redirected into the development of a reliable general-purpose digital computer; it then looked to real-world applications, such as traffic control of fast planes; and finally it was conscripted into the service of a new American air defense system (SAGE) at a critical juncture in the international Cold War. The later stages of the Whirlwind story are highlighted in this chapter.

Because it involves an invention that would become famous in computer history, the story of Whirlwind is one that belongs to both engineering and science. It embodies what some might nowadays call synchronicity; however, to historians of science and engineering, it suggests a pattern or set of laws underlying the development of man-made artifacts. Though there is planning involved in research and development, there is also the element of the accidental and peripheral, which creative people take seriously, and which in the postwar era has been raised to a high art in some fields.

The Whirlwind story also demonstrates the importance of giving a good research person the freedom to pursue the ideas he believes in, and protecting him while he does so—a proposition probably honored more often in the breach than in practice, despite generations of experience with the character of research. Gordon Brown, in whose Servo Lab the Whirlwind project originated, goes so far as to say that the widely used Forrester version of magnetic memory for computers might never have been invented had not

Jay Forrester been allowed the freedom to pursue his ideas and been given the support to do so. Given the history of invention—in which the same idea tends to erupt at almost the same moment in different people—this is unlikely; but Brown's statement does make a strong point about why it happened first in Forrester's work.

Forrester did his undergraduate work in electrical engineering at the University of Nebraska and had the highest record of the 70 engineering graduates of 1939. At a regional meeting of nine student chapters of AIEE, he took first prize for his paper and demonstration entitled ''An Electrostatic Dust Precipitator.'' The University of Nebraska was to confer upon him in 1954 an honorary Doctor of Engineering degree. In the fall of 1939 he began his graduate study program at MIT as a research assistant in Trump's High-Voltage Laboratory; he joined Gordon Brown's new Servomechanisms Laboratory in the middle of the 1940–41 academic year. The pressure and excitement of research in the Servo Lab during the war years led him to interrupt his degree program for three years, from the spring of 1942 to the spring of 1945; during that period he was an employee of the Division of Industrial Cooperation. His master's thesis, ''Hydraulic Servomechanisms Developments,'' was supervised by Brown and completed in 1945.

One day during the war, on the steps of MIT, Jay Forrester fell into a discussion with Perry Crawford, who was about to begin his work with the Special Devices Division of the Navy's Office of Research and Inventions. Crawford suggested that Forrester look into the digital method of doing computation. It was, according to Brown, ''the turning point in Forrester's vision and perception of the problem he was then working on.'' Digital computation went far beyond the capabilities of an analog simulator for a four-engine bomber, which was the formal task assignment in the

Jay W. Forrester with magnetic-core plane.

lab, and promised the development of a general-purpose computer that would be sufficiently fast and have the "word length" to operate in real time—that is, do calculations in synchrony with actual events. It was an unusual concept at that time, and it led Forrester to visit von Neumann and the ENIAC and EDVAC builders at the University of Pennsylvania. There, Forrester concluded that those machines were so unreliable that they would never make an impact on the kinds of applications he imagined. For instance, in tracking, controlling, and guiding a fast plane coming in for a landing in a crowded airspace, one does not have the leeway for making a mistake and then going back and correcting it that one has, say, in the solution of mathematical or scientific problems. Real time was simply beyond the capability of current analog computers, and reliability was not inherent in the current digital computer architecture and technology. Early postwar computer development could have taken another interesting turn had not fate thrown in one more factor. Upon Norbert Wiener's urging, a major architect of computers, John von Neumann, considered the idea of leaving the Institute for Advanced Study at Princeton and coming to MIT. Moreover, he was actually offered an appointment (about 1945).[1] However, von Neumann stayed at Princeton, where he managed to obtain support for the building of a general-purpose computer, despite considerable opposition to such a move. Apparently, writes one historian of that event, "members of the [Princeton] Institute faculty . . . regarded the computer—hardware, engineers, technicians, large government contracts—as the unwelcome machine in the idyllic garden."[2]

Forrester and his chief partner, Robert Everett, had no less of a struggle at MIT for enough money and time to build the Whirlwind as a reliable machine. In fact, when Whirlwind finally became operational in 1951 or

so, it functioned robustly and reliably for many successive users until the late 1970s—an astounding feat for a machine invented at the beginning of the computer age, when nearly everything was new and untried.

Brown asserts that Forrester should get more recognition than is generally accorded him for the work he did in the improvement of the life of vacuum tubes. With the assistance of numerous graduate students, Forrester pushed the research on electrostatic storage tubes—then seemingly the most promising memory devices for computers—to its absolute limits in the search for something that would behave reliably. In itself, this search proved an impressive educational contribution of Whirlwind.

Having gone as far as he could with such tubes, Forrester, in the spring of 1949, seized on the idea of using assemblies of small magnetic cores—elements like miniature doughnuts—which could be "switched" individually into one state or another and thus might serve as the ones and zeros of binary calculations. Working with graduate student William Papian, who was doing his master's thesis, Forrester pushed through to his "great invention," a magnetic storage system, which became the basis for a generation of computational machinery. For that invention Forrester has recently (among other recognitions) been elected to the Inventors' Hall of Fame.

However, at the time Forrester was pushing for reliability and real-time capacity, Whirlwind was arousing controversy in the MIT community. According to Dr. James Killian, then president of MIT, the steadily growing need for funds made the Navy, which was supporting the project, and the MIT administration "very restive. . . . I think Project Whirlwind would probably have been canceled out had not George Valley, as chairman of the Scientific Advisory Board of the U.S. Air Force, come up with the pressure to use

Jay Forrester (*standing*),
Patrick Youtz (*center*), and
Stephen Dodd with the electrostatic storage tube, the predecessor of the magnetic-core
memory.

Whirlwind as part of the SAGE system"—the proposed national air defense system, Semi-Automatic
Ground Environment.[3]

The proposal for SAGE came as the result of the Russian atomic bomb explosion of August 1949. The
prospect of jet bombers and missiles armed with nuclear warheads coming over the polar region into the
continental United States galvanized air defense proponents. Alarm was further exacerbated by the beginning of the Korean War in June 1950. A proposal for
an extensive network of radars—an "electronic
fence," monitoring the polar airspace and sending
data to a central computerized command center—
was urgently pushed. MIT was importuned by the Air
Force to set up an electronics laboratory to develop
such an air defense system, in which Whirlwind I
would serve as the prototype computer. Here, then,
Whirlwind I found its most significant real-world
application.

MIT's administration, Killian reports, had extended discussions about whether or not to undertake the running of a new large laboratory. Finally, only because
the Air Force was "unequivocal" in its determination,
MIT agreed to undertake the management of the Lincoln Laboratory. Though both Lincoln Laboratory and
Draper's Instrumentation Laboratory proved enormously creative and productive, says Killian, "there
were periods when both labs posed problems for us
that gave us concern about their future impact on
MIT. . . . These big labs are not without their compli-
cations. For one thing, they tend to have an inflation-
ary effect on the institution. And when, for instance,
the Instrumentation Laboratory philosophy of cradle-to-
grave development began to involve the Institute in
contracting with industry for many elements, we
thought these presented inappropriate kinds of decisions for an academic institution."[4]

Here, then, in the Whirlwind story we see decisions arising out of different perspectives and pressures, needs and desires, which have had a profound influence on MIT, on education, on the nation, and on the evolving role of science and technology.

The Beginnings of Whirlwind

During the Second World War the armed services had brought into use training devices of various kinds in order to familiarize prospective operators with such new equipment as antiaircraft fire-control directors, radar sets, and bomb sights. In particular the Navy was using a flight trainer, designed by the Bell Telephone Laboratories, in which a pilot could get the feel of operating a new airplane before taking it into the air. Toward the close of the war Captain Luis de Florez, II S.B. '11, himself an aviator for many years, was director of the Special Devices Division of the Navy's Office of Research and Inventions. He and his engineers evolved the idea that rather than building a trainer for each new airplane, it might be possible to build a universal trainer that could simulate any existing aircraft or even one still in the design stages. He sought advice from MIT's Department of Aeronautical Engineering, where the Wright Brothers Wind Tunnel was working round the clock to supply aerodynamic analysis data to many manufacturers.

After a study of the universal trainer concept, Professors John R. Markham, Otto C. Koppen, and Joseph Bicknell reported in April 1944 that such a machine was feasible. Captain de Florez's plan at this point was to set up a working arrangement whereby MIT would supply wind-tunnel data and the Bell Telephone Laboratories would design the machine to be manufactured by the Western Electric Company.

When the Navy decided in the fall of 1944, after much internal discussion, to place a contract for an aircraft stability and control analyzer (ASCA), the Bell Telephone Laboratories were not in a position to take it on. Captain de Florez then turned to MIT, where the feasibility study had been made; a contract was awarded to Gordon Brown's Servomechanisms Laboratory. The ASCA project was officially launched in December 1944 under the direction of Jay Forrester.

Forrester's first attack on the ASCA problem was to become thoroughly acquainted with the Bell Telephone Laboratories' trainers and with the aerodynamic and engine equations necessary for the design of a cockpit and instrument panel. In the spring of 1945 it became clear to him that in order to include the pilot's control reactions and make the solution time correspond to the airplane's actual flight time, very high speed servos having extremely short response times (high natural frequencies) and other computer elements must be developed.

The summer of 1945 brought strong doubts as to whether analog devices could be made rapid enough for real-time computation, and the group turned its attention toward computing processes other than analog. It was at this point that Perry Crawford of the MIT Center of Analysis urged Forrester to give consideration to digital methods. After meeting J. G. Brainerd and J. Presper Eckert, Jr., at MIT in late October, Forrester and Crawford arranged to visit the Pennsylvania project, where they discussed the ENIAC and EDVAC computers with von Neumann, Goldstine, and others. Although the Pennsylvania machines were not yet operating reliably, Forrester became convinced that digital methods would be the best solution to the ASCA problem. He persuaded Gordon Brown of this, and a digital computer development program was set up in the Servo Lab in January 1946. In April 1946 the contract specifications were changed to include the new digital computer development in place of the former analog computer program. In this revised contract the Navy designated the new task as "Project RF-12, Known as Whirlwind."[5] It was understood that the digital computer would eventually become an essential component of the ASCA and would also "have

Robert R. Everett, Forrester's chief partner in the development of Whirlwind, was a Duke University EE graduate who finished his S.M. program at MIT in June 1943. Here he is seen testing the initial analog-type flight simulator, later abandoned.

the required capacity for many other families of scientific problems.'' This redirection of the project needed an extension in time and funds.

Scaling Up the Project

Forrester saw this heavy commitment as demanding the most creative theoretical and experimental talent available, and he sought to build up a top-level staff with experienced personnel and with graduate students in electrical engineering, mechanical engineering, physics, and mathematics. By October 1946 there were seventeen engineers on Project Whirlwind, including Forrester, who had become associate director of the Servo Lab. Harris Fahnestock, an aeronautical engineer with a master of arts degree in communications from Harvard, had come from the Radiation Laboratory, where he had worked on the 1-cm airborne radar H_2K. Professor Philip Franklin of MIT's Mathematics Department headed the Mathematics Group. In 1946 five graduate students in electrical engineering were working on the project. By the fall of 1947 the number of engineers had grown to 33, and by September 1948 to 63 after the project had occupied its new quarters in the Barta Building at 211 Massachusetts Avenue, just north of campus.

Early in 1946 Forrester and Everett were considering an EDVAC-type computer having binary arithmetic, a 1-MHz pulse repetition rate, a mercury-delay-line memory, serial operation, and a three-address order code. Because the ASCA demanded calculations in real time, it soon became evident that an EDVAC-type computer would be too slow. By late 1946 the Whirlwind thinking had evolved toward a general-purpose computer with a (tentative) 2-MHz pulse repetition rate, a random-access memory—in view of the RCA Selectron, the British Williams Tube, and another electrostatic tube developed by the MIT Radiation Laboratory—parallel operation, and a single-address order code. A program under Steve Dodd and Pat Youtz

The Barta Building, where
Project Whirlwind was housed
after 1948.

Norman H. Taylor, shown here
with the assembly of the 5-digit
multiplier, an important development of the Whirlwind
computer.

was set up to modify the Rad Lab electrostatic tube as a storage device, and during the early months of 1947 the various components, such as flip-flops, gates, pulse generators, and circuits for adding, complementing, shifting, and carrying, were designed and tested. An important development at this point was the 5-digit multiplier, which was built to bridge the gap between a block diagram and the organization of components into a large-scale working system. The 5-digit multiplier, designed by Norman H. Taylor (a 1937 Bates College graduate and a 1939 MIT Course VI graduate), multiplied two 5-binary-digit numbers in 5 microseconds and displayed the product as a 10-digit number on an indicator panel.

The first 840 hours of operation, during which multiplications were in process only as initiated by a hand-operated push button, caused the machine to idle most of the time. On December 22, 1947, a "periodic program control" was added, which allowed the multiplier to carry out a multiplication of two 5-binary-digit numbers every 1,000 microseconds. During the fall of 1949 several runs as long as 20 days (almost two billion multiplications of two 5-binary-digit numbers) with no errors were obtained. Life tests on tubes and other components under pulsed conditions were greatly speeded up by this rapid repetition process.

The successful operation of the 5-digit multiplier was a giant step forward. It demonstrated an accurate performance at the 2-MHz pulse repetition rate, which was considered fast enough for real-time computations, and it showed that a parallel method of computation was practical. The experience with this system formed the basis for many decisions in the Whirlwind design. Problems with vacuum tubes led to considerable vacuum-tube research, not only at MIT, but in the plants and laboratories of tube manufacturers like RCA as well. This work eventually yielded a supply of reliable vacuum tubes, and contributed significantly to the vacuum-tube art.

Two major objectives in Forrester's Whirlwind plans had been high speed and reliability. By early 1948 high-speed computations had been achieved.

In November 1947 Forrester, aware that the deterioration of vacuum tubes and crystal diodes could shut down the Whirlwind computer several times each day, conceived the idea of *marginal checking*. This was a scheme of checking tubes, diodes, and other components to discover those that were about to fail, so that they could be replaced before making a computational run. A marginal checking system was developed and made automatic. Subsequent experience showed marginal checking to be a very significant factor in the reliability of the Whirlwind computer.

The term *binary digit* was shortened to *bit*, following the suggestion in 1948 of John W. Tukey, a professor of mathematics at Princeton University and a member of the technical staff at the Bell Telephone Laboratories. In present-day language the 5-binary-digit multiplier would be called the 5-bit multiplier.

Whirlwind Seminars

In the spring of 1947 Forrester began the digital-computer education of the MIT community. A series of five lectures was part of the Electrical Engineering Department's weekly staff seminar program. These followed a series on the differential analyzer. At this time Caldwell, in his June 1947 review of the relative stages of development of Whirlwind and the competing Rockefeller electronic computer, pointed out the strong lead of Whirlwind.

Two Whirlwind computers were envisioned. Whirlwind I was to have a word length of 16 bits (about 5 decimal digits), to be transmitted simultaneously over a parallel transmission line of 16 coaxial cables. Because most mathematical problems would require a longer word length, Whirlwind I could double it by combining two 16-bit registers; multiplication would require more time. The Whirlwind I computer was used for the study and demonstration of computing circuits and troubleshooting methods, for mathematical research, and for the study of engineering problems. Whirlwind II, the final, large-scale objective of this project, was to have 40-bit words. Both computers were to have electrostatic internal storage and some form of external storage. Growing technical and administrative problems in the development of Whirlwind I pushed the planning of Whirlwind II farther and farther into the future, and from 1948 to 1952 very little thought was given to a second machine.

Forrester closed the first series of lectures to the MIT community with remarks that showed not only his appreciation of the large and complex problems soluble on the large-scale computer, but also his foresight into the development of libraries of control programs, long-distance teletype connections, and a national network of computing facilities, all of which have since become a reality.

The Whirlwind seminars of 1948 were opened to a larger audience and were attended by members of several MIT departments as well as by representatives from military services and manufacturing companies. The number of graduate students in electrical engineering associated with Whirlwind I rose from about 15 in 1948 to a level of 25 to 30 continuously between 1951 and 1954.

Beyond the walls of MIT Forrester spent a great amount of thought and energy in preparing the nation for the oncoming use of high-speed computers. In a talk at a UCLA symposium in July 1948, for instance, he said, "I believe that if a high-speed computer capable of 1,000 to 20,000 arithmetic operations per second were sitting here today, it would be nearly two years before the machine were in effective and efficient operation. One would be caught totally unprepared for feeding to this equipment problems at its

high acceptance rate." Not only must new advances be made in adapting mathematical problems to computer solution, but "it will be necessary to develop some auxiliary equipment and in particular the administrative procedures required to maintain a schedule of high-speed traffic flow. In other words, it should be possible to pass such problems through a machine just as the telegraph company transmits messages. The machine would be available to a large number of groups through a wire communication facility, probably teletype initially." In forecasting the time span and cost of the computer development, he said,

I believe that, barring an all-out emergency effort such as went into the development of radar, large-scale computers will for several years need the sympathetic care of a laboratory crew. During these several years personnel must be trained in the proper use of the machine. Trouble location methods must be designed and suitable facilities provided in the machines for rapid and probably automatic location of faulty parts. . . . One must remember that by comparison a radar set is a very simple and straightforward device, yet hundreds of millions of dollars went into the development of this equipment. True, we are building on the results of that work, yet I believe that corresponding millions of dollars will be spent in computer development before achieving even a representative sampling of the objectives which this audience visualizes. . . . We must therefore look upon it as a new field in its earliest stages of physical development; it will require the same growing period as any other branch of engineering.

Postwar Funding

The ASCA project started during the war while funding was large. The year 1946 saw the cessation of OSRD activities and the planning for heavy government support for basic science research. President

Roosevelt had written to Vannevar Bush on November 17, 1944, asking how "the information, the techniques, and the research experience developed by the OSRD and by the thousands of scientists in the universities and in private industry should be used in the days of peace ahead for the improvement of the national health, the creation of new enterprises bringing new jobs, and the betterment of the national standard of living." In this letter the president had said, "New frontiers of the mind are before us." Bush's reply, dated July 5, 1945, embodied the recommendations of four special committees set up to study these questions. The Bush report, *Science the Endless Frontier,* was widely recognized for its recommendations for government funding of basic research in colleges, universities, and research institutes. The MIT Research Laboratory of Electronics, established in 1944, was one of the earliest postwar research projects set up under armed-service funding.

In August 1946 the federal government created the Atomic Energy Commission and the Office of Naval Research (ONR), which replaced the Navy's Office of Research and Inventions (ORI); the Special Devices Division of ORI at Sands Point, Port Washington, New York, became the Special Devices Center (SDC) of ONR. The Whirlwind contract between the Navy and MIT continued to have the interest and attention of Perry Crawford, who headed the computer section of SDC under ONR, where Dr. Alan T. Waterman had become deputy chief and chief scientist. (Waterman was an able and understanding civilian administrator in ONR, having been vice-chairman of Division D, NDRC, and in 1945 Compton's successor as chief of the Office of Field Service, OSRD. In 1951 he was chosen as the first director of the National Science Foundation, serving until 1963.) Changes in SDC funding between 1947 and 1949 led to the placement of the Whirlwind contract under the Physical Sciences Division of ONR, where Dr. Mina Rees headed the Mathematics Branch.

There were new problems in financing, however. The demands of Whirlwind caused ONR to reconsider the distribution of its research funds. By July 1948 Forrester had spent $1.5 million of Navy funding and had entered the new fiscal year with a program requiring over $1 million more. In 1948 design work on aircraft cockpit simulation equipment was indefinitely postponed, thus removing the real-time application that had called the digital computer into being in late 1945.

In the view of the Office of Naval Research, the Whirlwind general-purpose computer program seemed to be growing beyond the bounds of justifiable ONR funding, and it seemed advisable to direct the project toward the completion of a usefully operating computer at the earliest date. On the MIT side, Forrester's administrative superiors maintained a high regard for Project Whirlwind. Gordon Brown, director of the Servomechanisms Laboratory, Harold Hazen, head of the Electrical Engineering Department, and Nathaniel Sage, director of the Division of Industrial Cooperation, who managed the contractual arrangements, were in close contact with the work and saw Whirlwind as an innovative and important technical development that was contributing strongly to the education of young engineers and scientists.

Bush was chairman of the Electrical Engineering Department's Visiting Committee, which was to meet in the spring of 1948. In January he discussed Project Whirlwind with Vice-President Killian and suggested that a member of the Visiting Committee, perhaps Ralph D. Booth, S.B. '20, of the Jackson and Moreland firm, be asked to make a critical appraisal of the project for the guidance of the MIT administration. Forrester welcomed the proposal, provided the examination were done thoroughly; he mentioned the rather numerous "semi-investigations" he had experienced over recent months. Booth did in fact make a thorough study of Project Whirlwind during the summer of 1948 and concluded that Forrester's plans for the successful completion of the high-speed computer showed every promise of fulfillment. This solidarity of MIT support was of course very heartening to Forrester, but frustrating to the ONR officials, who had a strong conviction that the Navy ought to bring its Whirlwind funding to an early conclusion.

Over the months as Whirlwind approached operational status in 1949, numerous proposals, studies by investigating committees, and counterproposals were made. For instance, as first chairman of the Research and Development Board of what was soon to become the Department of Defense, Vannevar Bush early in 1948 developed plans for a national committee on digital computer programs. These plans were carried out after Karl Compton became chairman of the Research and Development Board in late 1948. An ad hoc committee of the board visited the several government-sponsored computer projects under way in 1949 and wrote a comprehensive and critical report, recommending that unless Whirlwind I could find a suitable application, its financing should be discontinued. Late 1949 was certainly a low point for Forrester, Everett, and their organization.

Computer Applications

Jay Forrester was not at this time in any doubt about suitable missions for Whirlwind I. Back in 1948 he had said, "In control applications the computer is an operating part of a larger system. By way of illustration we might mention the control of industrial processes, military gunnery and fire control, and the centralized control of air traffic." Earlier an application had been suggested for antisubmarine warfare. In 1949 the Mathematics Group worked out computer programs for solving various types of difficult differential equations. Phyllis A. Fox, S.M. '49, XVIII Sc.D. '54, showed in her master's thesis how to use the digital computer to solve power-system problems such as

were at that time solved on the network analyzer. James E. Pierson, S.M. '49, showed in his master's thesis how Whirlwind I would make a point-by-point solution in the study of a nonlinear servomechanism. Charles W. Adams, VIII S.B. '48, XVIII S.M. '49, worked out a Whirlwind procedure for the calculation of autocorrelation and cross-correlation functions, although he noted that the cost efficiency of such calculations is greater when performed on a special-purpose machine such as the analog and digital correlators of RLE. It was during this period that William K. Linvill became interested in the analysis of servo systems containing sampled-data devices such as a digital computer. His thesis proposal was submitted in October 1948; the final thesis document, with supervision by Guillemin, was dated May 13, 1949. Linvill, in turn, supervised the master's thesis of Roger L. Sisson and Alfred K. Susskind, entitled "Devices for Conversion between Analog Quantities and Binary Pulse-Coded Numbers," submitted in January 1950; and the doctoral thesis of John M. Saltzer, entitled "Treatment of Digital Control Systems and Numerical Processes in the Frequency Domain," submitted in August 1951.

With this kind of experience and methodology Forrester and Everett were prepared to tackle a real-time application of considerable magnitude and significance. An investigation of the civilian air-traffic control problem showed that although radio, radar, and various kinds of computers had been applied to increase the safety and efficiency of air travel, the anticipated fast jet commercial airplanes would need fast computers like Whirlwind I as part of the control system. A contract, effective March 1, 1949, was arranged between MIT and the Watson Laboratories, Air Materiel Command, Red Bank, New Jersey, with two objectives, "the first, to determine the contributions which high-speed digital computers might make to the long-term air-traffic control problem for the conditions likely to exist after 1960; and the second, to extract from the long-range work those phases which can be put into service after 1952 as part of an interim air-traffic control system leading toward the final system." This Air Force contract supplied additional funds to replace part of the announced decrease in ONR support. Existing air-traffic control operations and plans for the future were first thoroughly reviewed, and then studies were made to show how a high-speed digital computer could be integrated into the system to form its nerve center.

The Whirlwind computer was almost, but not quite, in reliable operation in early 1950. In analyzing the project's large budget needs, Forrester and his colleagues argued the importance of reliability.

This has accounted for a sizable part of our time and cost. To one who really understands this field (and they are very few), principles are not so much at stake as their realization. The theory and potential usefulness of a digital computer can be established on the philosophical and paper-study level. Its ultimate utility depends on demonstrated successful operation. . . . Any contemplation of digital computers for air-traffic control, fire control, or aircraft interception must be supported by a plan for getting a machine of the dependability required. Reliability is by far the most important unknown quantity. . . . An experiment in air intercept control, plagued by digital computer failures, could easily do much more harm than good. In ultimate real-time applications one has not the opportunity to start over and repeat a run as in scientific computation.

The budget included $84,000 to make the electrostatic storage tubes reliable, but also a modest amount ($9,400) to continue research on the new magnetic storage, on the basis that "one can expect that all forms of storage tubes now in use will become

obsolete within a decade." The Office of Naval Research had no intention of supporting the Whirlwind project on a high financial level.

Survey of the MIT Computer Situation

The wide discrepancy between ONR and MIT was one of the problems of the new MIT administration following the inauguration of Killian as president in April 1949. Julius Stratton terminated his directorship of RLE to become MIT's first provost; together with Dean of Engineering Thomas K. Sherwood he made a thorough investigation of the computer situation.

Stratton was already acquainted with the RLE computers. Professor Henry Wallman of the Mathematics Department had articulated the need for a differential analyzer of high speed, moderate accuracy, and low cost for solving all types of ordinary differential equations, and in 1945 had proposed the development of such a machine to Alan B. Macnee (VI-A S.B. '42, S.M. '43, Sc.D. '48) as a doctoral project. The Macnee machine was all-electronic, containing well-known components such as electronic adders and integrators, but also new function generators and multipliers that Macnee had originated. In one of the examples described in the thesis, 200 trial-and-error runs with varying initial conditions were made in about an hour, in comparison with an estimated three eight-hour working days if this differential equation were solved on a Bush-Caldwell mechanical differential analyzer (RDA). Accuracy was 1 to 5 percent and the cost of the machine was less than $20,000. Other RLE computers were the analog and digital correlators previously mentioned and special-purpose devices such as Ronald E. Scott's (Sc.D. '50) network synthesizer. The Bush-Caldwell machine was in financial difficulties, a $40,000 surplus left over from war work having been spent, and Stratton believed the time had come to discontinue this obsolescent machine. Hazen made

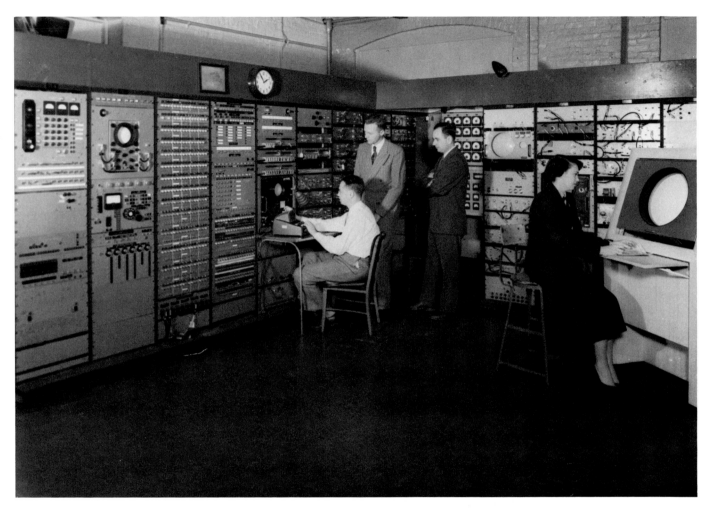

The Whirlwind I Operational
Control Center in 1950. *Left to
right*: Stephen H. Dodd, Jr.,
Jay Forrester, Robert Everett,
and Ramona Ferenz.

plans to dismantle it in June 1950, but Stark Draper found it useful until 1954 in his design of new types of lead-computing gun sights, tracking systems, and aircraft components. The Center of Analysis had completely disintegrated by 1950, and the various computing facilities of MIT needed some kind of organization and overall direction.

Two computation groups that had originated in the Center of Analysis during the war had been moved to Building 20-C in 1946 when Professor Caldwell acquired the Building 20 space for the Rockefeller digital computer research. The first group, under Dr. Zdenek Kopal, an astrophysicist, was engaged in work on Army and Navy ballistics, continuing after the war with such problems as the trajectories of yawing cones at supersonic velocities, the analysis of meteor-track photographs to obtain ballistic properties of the upper atmosphere, and the propagation of shock waves. Kopal developed new and efficient methods of numerical analysis and directed a group of computing personnel using desk calculators. By 1950 he had been appointed to an associate professorship in electrical engineering, and groups of graduate students were studying numerical analysis in his classes, laboratory, and seminar.

The other group, called the Punched-Card Section, under Frank Verzuh, worked with IBM calculators—the older 602 and 602A and the new, faster 604—performing various kinds of calculations as arranged with clients inside and outside MIT. The loss of a principal client, the A. D. Little Company, threatened such an operating deficit that a decision was made to discontinue the contract for rent of the IBM machines. Stratton learned of the situation in time to prevent dissolution of the group. At his direction the IBM punched-card facility was reactivated in July 1949 as a separate computing facility under Division of Industrial Cooperation sponsorship. A year later the IBM Punched-Card Section was set up as the Office of Statistical Services, furnishing leadership in the mechanization of the Institute's administrative offices. The Dynamic Analysis and Control Laboratory (see chapter 14) was developing a flight simulator that included a large analog computer and several smaller computing devices. Departments other than Electrical Engineering, such as Civil Engineering, Aeronautical Engineering, and the latter's Instrumentation Laboratory, had built up computing groups using all kinds of computing machinery, ranging from Marchant, Monroe, and Friden desk calculators to the sophisticated analog and digital machines.

Stratton's wartime association with United States government officials, especially those in the military services, and his longtime involvement with MIT affairs gave him an advantageous position in resolving the tension between ONR and Project Whirlwind. ONR had already established several centers of numerical analysis in the United States and now offered financial support for such a center at MIT. The ONR-MIT group was aware that something like $500,000 would be available from the Air Force for the use of Whirlwind facilities as part of an air defense system. With this prospect in view ONR proposed that its 1950–51 support for Whirlwind be reduced to about $280,000, but that it subsequently fund at MIT a center for machine computation and numerical analysis. This proposal was accepted by MIT, and in the following months a Committee on Machine Methods of Computation was formed "to integrate the efforts of all the departments and groups at MIT who are working with modern computing machines and their applications." Philip Morse of the Physics Department was named chairman of this committee; the other members were Herman Feshbach of the Physics Department, Eric Reissner and Chia-Chiao Lin from

Mathematics, and Jay Forrester and Samuel Caldwell from Electrical Engineering. Provost Stratton's directive of November 1950 said, ''The Committee will inquire into and advise on the education of students in the use of newly developed computing machines and on the most effective methods of exploiting these devices by the staff and students of the Institute in the furtherance of education and research.'' A new task order was spelled out in the old ONR contract, and the first funds for the MIT committee became available in July 1951.

In September the Barta Building operation was separated from the Servomechanisms Laboratory and became the Digital Computer Laboratory of MIT under Forrester's direction. Four research assistants were appointed for the summer, one to work with Whirlwind and the other three with the IBM punched-card group. One of these was Fernando J. Corbató, VIII Ph.D. '56, later an associate head of the department with special responsibilities in the computer science program. Corbató's research assistantship was continued in subsequent years; he worked with the various MIT computing facilities, including Whirlwind I. By the second quarter of 1953 Forrester's Digital Computer Laboratory had 27 EE research assistants.

Air Defense: A New Factor
In August 1949 the Russians exploded an atomic bomb; three months later Professor George E. Valley of the MIT Physics Department, a member of the Air Force Scientific Advisory Board (SAB), proposed that a working group be set up under SAB for the purpose of improving America's air defense. A committee was promptly formed and held the first of its weekly meetings late in December 1949 in the Air Force Cambridge Research Laboratory building at 230 Albany Street, Cambridge. This committee, chaired by Valley, was called the Air Defense System Engineering Committee (ADSEC); it defined its task as ''(1) to bring

the presently planned Air Defense System to its maximum inherent effectiveness, and (2) to design a model of the best air defense system conceivable, under only the limitations of basic natural and economic laws.'' (This and the following quotation are from the Lincoln Laboratory Quarterly Progress Report of June 1, 1952.) Early in 1950 ''the conception of a centralized computer system with many small radars, each of which functioned only as an aircraft detecting device, began to take form. At this time only the haziest notions were current of what the central computer would be like, or what it would actually have to do. After a nationwide search and getting familiar with computing devices of all sorts, the committee learned that the MIT Whirlwind computer existed, and that moreover its design parameters ideally suited it for the purpose in mind. As a result, a permanent invitation to sit with the committee was extended to Forrester in March 1950.'' At this point the Whirlwind air-traffic control study was interrupted.

The urgency of building an air defense was heightened by the outbreak of the Korean War in June 1950; in the fall of 1950, after about a year of studies, ADSEC was ready to begin research and development on its emerging program. The Air Staff requested that MIT set up and administer a laboratory for this purpose, asserting that the Institute was uniquely equipped to carry out this mission. MIT took on the new responsibility, as had been done in 1940 when the Radiation Laboratory was established to develop microwave radar. In order to assure sound and adequate planning for an air-defense laboratory, Project Charles was organized as an ad hoc committee to work out a set of definite recommendations leading to a program for the projected laboratory. The Charles study got under way in February 1951 and produced a report the following August.

Soon after Project Charles concluded its meetings, the Lincoln Laboratory was organized in August 1951 under F. Wheeler Loomis, director, and Jerrold Zacharias, associate director, with headquarters in Building 22, a temporary World War II structure just south of Building 20. The Lincoln Laboratory organization was patterned after that of the earlier Radiation Laboratory, with *divisions* and their component *groups*. Division 3 (Communications and Components) took over an RLE activity already doing important work for the Army, Navy, and Air Force. Albert Hill was its head, with Jerome Wiesner as associate head and George Harvey as assistant head. Group 31 (Presentation) of Division 3 was led by J. C. R. Licklider and Herbert Weiss; Group 32 (Continuous-Wave Components and Tubes), by Robert Fano and Louis Smullin; Group 33 (Long-Range Communications), by William Radford and J. T. de Bettencourt; and Group 34 (Secure Communications), by Wilbur Davenport and Yuk Wing Lee. Division 2 (Aircraft Control and Warning) was set up under George Valley, with Ragnar Rollefson as associate head. It worked in cooperation with Division 6 (the Digital Computer Laboratory), under Jay Forrester, head, Robert Everett, associate head, and Harris Fahnestock, executive officer. There were six groups within Division 6.

Division 6 remained in the Barta Building during its early years. A new set of buildings in Lexington, adjacent to the Bedford Airport, was soon under construction, and by late 1952 Albert Hill had been made director of Lincoln Laboratory and was occupying his new office in the Lexington complex. The air-defense activity was called Project Lincoln and operated in the eastern part of Massachusetts.

The Whirlwind I computer now began its air-defense application. The first specific goal in this work, according to a Lincoln Laboratory report of June 1952, was to use the computer to

track-while-scan (TWS) multiple targets and to generate a filtered display. . . . At this time the computer was operating with only five electronic (flip-flop) registers of test storage; the electrostatic storage system was not yet installed. . . . Although [Whirlwind I] was designed for 2,048 electrostatic storage registers, the first increment planned was 256 registers. As the program developed, it became evident that interception could be accomplished with 256 registers. . . . During the spring of 1951, programs for velocity tracking, guiding an aircraft to a fixed point, and collision-course interception were tested by recorded data flight tests. A flight test held in April resulted in a successful interception from an initial target-interceptor separation of about 40 miles.

Both ADSEC and Project Charles advocated a network of many short-range radars with overlapping coverage in order to assure the protection of the defended area against low-flying enemy planes. Quoting again from the June 1952 report, "The experimental system, consisting of a network of radars, remote transmission of radar data, and a digital computer, became known as the *Cape Cod System.* . . . The system might be thought of as a proving ground, and it should be developed to the point where it is a complete model air defense system suitable for tactical evaluation." Insertion of data from the radars of the Cape Cod network would require two major additions to the Whirlwind I computer: "(1) a buffer storage system to collect and store radar data so that the computer may obtain it by blocks under the computer control, and (2) an auxiliary computer memory to hold the large amount of aircraft track data needed in the radar network." After several months of planning, two magnetic-drum storage systems were ordered in December 1951 from Engineering Research Associates. The auxiliary storage drum was supplementary to the high-speed electrostatic storage and was delivered in late January 1953.

A Special-Purpose Machine

In June 1952 the logical design of the specialized Whirlwind II machine, now aimed at fitting air defense requirements, was still unsettled, but it would obviously build upon the six years of experience with Whirlwind I. There was a hope that transistors and magnetic-core memory banks could be included. The invention of the magnetic-core memory banks had grown out of the intensive research program to produce a reliable electrostatic storage tube. The story of that research, which led to a steady stream of master's theses in the EE Department, is an impressive example of the educational contribution of Whirlwind. Although Forrester believed that the succession of problems that held back the reliable operation of the electrostatic storage tube would eventually be solved, he was in quest of a more compact storage device. In an April 1947 memorandum he described a three-dimensional array of glow-discharge tubes as an embodiment of this idea, but the tubes did not turn out to be reliable.

In the spring of 1949 Forrester saw a possibility of using a three-dimensional assembly of magnetic cores. William N. Papian, S.B. '48, in quest of a master's thesis project, entered upon an investigation of such small toroidal cores under Forrester's supervision. In his thesis, entitled "A Coincident-Current Magnetic Memory Unit," completed in August 1950, Papian described the initial concept of magnetic-core memories and showed how the cores could be combined in planar arrays, which could in turn be combined into three-dimensional assemblies.

By the time Papian had finished his master's thesis, the concept of a magnetic-core storage looked promising, but the realization of a reliable working device was to require much thoughtful experimentation. Of paramount importance was the selection of a suitable material for the magnetic cores. Two kinds of cores were pictured in the Papian thesis. One was a metallic

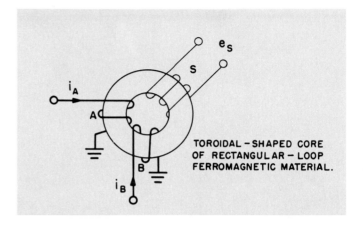

TOROIDAL – SHAPED CORE OF RECTANGULAR – LOOP FERROMAGNETIC MATERIAL.

Figure 1 from William N. Papian's master's thesis, "A Coincident-Current Magnetic Memory Unit."

Figure 2 from Papian's thesis: a planar storage array with nine magnetic cores.

Figure 3 from Papian's thesis: a three-dimensional assembly of magnetic cores.

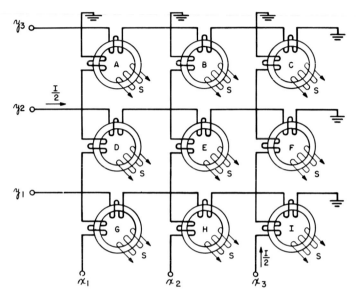

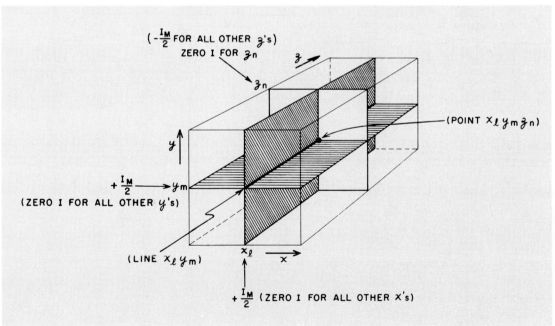

ring having about a three-inch outside diameter, formed like a roll of one-mil-thick Scotch tape. The other was a much smaller doughnut of powdered magnetic ferrite, pressed into toroidal form and heat-treated. These cores were wound with test coils of insulated copper wire having 25 or more turns each. The manufacturers of these core types, Allegheny Ludlum Steel Corporation for the metallic cores and General Ceramics and Steatite Corporation for the ferritic cores, showed much interest and cooperation in the development of materials with the appropriate magnetic and hysteresis (B-H) characteristics suitable for magnetic cores.

Professor Arthur R. von Hippel of the MIT Laboratory for Insulation Research was interested in the ceramics problem; he visited General Ceramics in August 1951 to aid in the improvement of B-H characteristics and the fabrication of small-size samples for test and development. The Laboratory for Insulation Research then undertook to conduct fundamental investigations for the improvement of ferrites—to synthesize ferrites and to analyze and measure their composition and properties.

A Magnetic-Materials Group was organized within the Digital Computer Laboratory to study and develop materials for use in computers. Among those doing the research was Kenneth H. Olsen, S.B. '50, S.M. '52, whose master's thesis was entitled ''A Magnetic Matrix Switch and Its Incorporation into a Coincident-Current Memory.'' This switch, which selected a specified core among the cores in a ceramic memory plane, had a much shorter access time than most other random-access memories.

The Cape Cod study and increasing computer demands by the MIT departments ruled out the use of Whirlwind I to test the new magnetic-core memories. In the summer of 1952 a Memory Test Computer Section was organized to design and construct a small computer for testing the magnetic memory

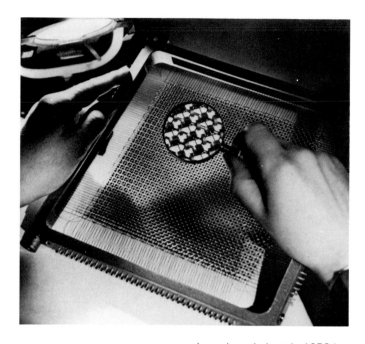

In work carried out in 1952 by the Whirlwind team, an important simplification in the memory planes was the reduction of both magnetizing and sensing coils to one turn per core. In this photograph the magnifying lens enlarges a section of a core plane so that the winding structure is visible.

being designed by the Magnetic-Core Memory Section. This computer was built from the standard Whirlwind circuits available in plug-in packaged form. At first the memory was to be a 32 x 32 x 17–core bank of metallic units, but by winter 1952–53, progress in ferrite materials was so encouraging that the memory test computer design switched to ceramic cores. The 1,024-register ceramic memory was running well in the computer by May 1953, one of its memory planes having been built by IBM.

The collaboration with IBM was begun in the fall of 1952 under an MIT subcontract. In the spring of 1953 IBM was given a contract by the Air Force Cambridge Research Center to work with Lincoln Laboratory on the design and manufacture of Whirlwind II, now designated by the Air Force as AN/FSQ-7.

The year 1953 was a very significant one for Project Whirlwind. The earlier dream of a fast, reliable computer was realized as Whirlwind I got its first bank of 32 x 32 x 17 coincident-current magnetic memory cores from the memory test computer on August 8, and its second bank, quickly assembled from the available supply of cores, on September 7. Computing speed was doubled and useful operating time was increased to more than 90 percent. The 1953 Cape Cod Model Air Defense System was in successful operation with live target and interceptor aircraft. The design of the air-defense computer, AN/FSQ-7, had been completed by the combined efforts of IBM and Lincoln engineers, and early 1954 was to see the manufacture of components for the first prototype, XD-1, in the Poughkeepsie plant of IBM. Project ADES—Air Defense Engineering Service—was organized by the Bell System to work with Lincoln Laboratory and IBM on the communications aspects of the Cape Cod and AN/FSQ systems.

To replace the 1,024-register magnetic memory given to Whirlwind I in August 1953, a new 64 x 64 x 17–core memory was installed in the memory test computer in January 1954. Simultaneously, IBM at Poughkeepsie was beginning the construction of the 64 x 64 x 36–core memory for XD-1 with cores supplied by General Ceramics. The 1954 Cape Cod system, using the Whirlwind I computer, was the experimental model of a subsector of the SAGE (Semi-Automatic Ground Environment) system, a new, extensive phase of the air defense program. In 1954 the Cape Cod system had an expanded radar network of two long-range radars (at South Truro and Montauk), seven gap-filler radars, and a Mark X IFF radar. It had a sophisticated computer program for automatic tracking and used greatly improved cathode-ray-tube displays. The SAGE system was to have several *subsectors*; the first experimental subsector was a square approximately 400 nautical miles on a side and centered at South Truro, Massachusetts. A new building was constructed at Lincoln Laboratory to house the XD-1 computer, brain center of the experimental SAGE subsector.

In Division 6's first quarterly progress report of 1955, Forrester and Everett say, ''The central computer equipment of AN/FSQ-7 (XD-1) was received from IBM in January and installed in the Experimental SAGE Subsector Direction Center building at Lexington. Approximately three weeks after the disassembly of the equipment at Poughkeepsie it was again operating with the same reliability and margins observed prior to shipment. As additional elements are installed, testing is continuing, largely by engineers from IBM stationed at the Lincoln Laboratory.'' The memory test computer was moved from Cambridge to Lexington to participate in the testing of prototype XD-1. The production model of AN/FSQ-7 was to be a duplex system, that is, two intercommunicating computers, the second a standby machine to be brought quickly into service when, for any reason, the active machine was out of service.

When military operations crossed the boundary between two SAGE subsectors, it was necessary to transmit information between direction centers of the sectors. This information transmission was called *cross telling*, and an experimental cross-telling arrangement was set up between XD-1 in Lexington and Whirlwind I in Cambridge. In December 1955 Stephen Dodd reported that messages were transmitted from Whirlwind I to XD-1 and back without error. "The Whirlwind I computer has continued its 168-hour-per-week schedule at a high level of reliability. The average good time over the period September 1 through November 30 has been 97.4 percent, with a high of 99.3 percent over one two-week period." The core memory had an access time of 10 microseconds.

In addition to the joint design and construction of the AN/FSQ-7 system by IBM and Lincoln Laboratory, there was intensive cooperation by the Bell Telephone Laboratories, the Western Electric Company, the General Ceramics and Steatite Corporation, the General Electric Company, Convair, Hughes Aircraft, Raytheon, Jackson and Moreland, and several others. The entire Air Force Defense System was developed under contracts placed by the Air Force Cambridge Research Center and the Rome (New York) Air Development Center. The first full sector of the SAGE system went into operation in June 1958 for the protection of the New York–Philadelphia area. As a measure of the magnitude of the task assigned to SAGE, it had to screen potentially hostile aircraft from tens of thousands of daily routine flights and to direct interceptors and missiles in case of intrusion.

The programming for the SAGE experimental subsector and for the first three operational SAGE sites was carried out by Lincoln Laboratory. Programming for subsequent sites was taken over by the RAND Corporation, after much cooperative planning and recruitment of new personnel for both organizations. The MIT Lincoln Laboratory was relieved of SAGE system responsibilities through the establishment of the MITRE Corporation.

The MIT policy relating to the Lincoln Laboratory and the MITRE Corporation is clearly set forth in the following quotation from President Stratton's 1958 report:

The Institute has always accepted sponsored research projects as an integral part of its educational system. Such work, whether on the campus or in the defense laboratories, provides unusual opportunities for both graduate students and faculty to participate in research at the frontiers of their respective fields. We recognize also that urgent demands will be made upon our resources in times, such as the present, when the safety and strength of the free world depend so greatly on advanced science and technology. Such responsibilities we continue to accept on a selective basis, when the work is in fields in which the Institute has particular competence and which will benefit our educational mission.

A major organizational move during the year, designed to promote this trend, has been the cooperative effort of MIT and the Department of Defense in the formation of MITRE, a new, independent, nonprofit corporation, to undertake a major advisory role in the systems engineering of the country's air defense. Much of this responsibility, relating to the electronic ground environment of air defense, has previously rested on the Lincoln Laboratory. One objective of this organizational move is to permit the Lincoln Laboratory to devote more of its time to the type of advanced research which fits most naturally and productively into the scheme of an educational institution.

Lincoln Laboratory continued to conduct extensive research programs in the digital computer field, including the development of the well-known TX-0 and TX-2
computers and special computers for dealing with biological and medical problems. The LINK (Laboratory
Instrument Computer), for instance, was conceived by
Lincoln's Computer Group in cooperation with the
Communications Biophysics Group of RLE to satisfy
such needs.

Jay Forrester, who had led the successful Whirlwind
work, joined the MIT faculty in September 1956 as
Professor of Industrial Management. He initiated a
program in industrial dynamics and worked to bring
high-speed computing techniques to managerial
decision-making problems.

Chapter 18

Growth of the Acoustical Sciences: The Acoustics Laboratory

The first acoustics laboratory in electrical engineering at MIT began in a small basement room just west of the Building 10 elevator. Professor Bowles called this "Com Lab IV," and here Professor Richard D. Fay and his assistant William M. Hall inaugurated, in the spring of 1931, the first student experiments for the new graduate subject 6.69, Sound in Electrical Communications. The laboratory work comprised measurements in the testing of electroacoustical apparatus.

From this modest beginning there emerged a strong laboratory and a revival of the interdisciplinary interest in the art of acoustics that had been instilled in the Institute at the time of Charles Cross and Alexander Graham Bell. In the evolution of the MIT Acoustics Laboratory between 1931 and 1958, when it was formally closed, we see how a laboratory in an educational context can stimulate unusual directions in research, how it can spawn new fields of endeavor and new companies and then, through its very success, bring on its natural dissolution. During this same period of about three decades, the acoustical arts—ranging from the study of physical phenomena and architectural problems to psychoacoustical disciplines such as speech recognition and speech synthesis systems—gradually acquired a solid scientific basis. In all, faculty members and students from five different MIT departments participated in work in the Acoustics Laboratory during this period.

Toward a New Theoretical Basis
When Philip M. Morse joined the Physics Department in the fall of 1931, he soon discovered Dick Fay—a quiet, friendly person, well educated and experienced in acoustics—and finding that they had common interests in acoustics, they became friends. Morse had completed his doctoral studies at Princeton in 1929 and, after a year at the Bell Telephone Laboratories, spent a year in Munich studying with Arnold Sommer-

feld. Julius Stratton relates how William P. Allis, Morse, and Stratton spent the summer of 1931 together in Cambridge, England, often thinking and talking together about the prospects for physics at MIT under the new regime of Compton and Slater.[1]

Morse inherited subject 8.05, Sound, Speech, and Audition, which Professor William R. Barss had designed and taught from 1926 to 1931 in the last term of Bowles's communications curriculum. Morse saw that his new understanding of wave phenomena, together with the new electronic measuring devices that had become available, could move acoustics out of its empiricism into the realm of a precise and useful science.[2] Morse changed the name of subject 8.05 to Vibration and Sound and developed a set of class notes that was published in 1936 as his first book.[3]

By early 1934 Fay and Hall had acquired new space for their laboratory. On May 9, 1934, Fay, as chairman of the Joint Committee on Acoustics, circulated a report dealing with acoustics instruction, facilities for research, and allocation of research projects for MIT as a whole. The instruction comprised Morse's 8.05, Vibration and Sound, chiefly for communication students; 8.06, Acoustics, Illumination, and Color, for students in architecture; and Fay's 6.69, Sound in Electrical Communications. The committee recommended that acoustics experiments related to 8.05 be provided in the senior communications laboratory subjects and that 8.05 be moved to the first term, so that a two-term sequence in acoustics would be available. Included in the report were the following descriptions of research projects:

An application of the theories of sound vibrations is being made in the problem of detonation in internal combustion engines, investigated by C. S. Draper under the supervision of Professor Morse. The explosion in an engine cylinder can be studied by means of the sound waves it generates inside the cylinder, and a large number of important facts can be obtained by means of this analysis. . . .

Since the theory of sound radiation is intimately connected with the theory of all sorts of wave motion—light waves, radio waves, and electric waves—the research is at present concerned with the general mathematical theory of wave motion. This problem is being studied by Professors Stratton and Morse in a joint project.

Draper's work was published in 1939.[4]

The acoustics team at MIT was further strengthened in the fall of 1939 when Richard H. Bolt, who had just completed his doctorate at the University of California, came to spend his National Research Fellowship year with Morse. Although on the staff at the University of Illinois for a short period in 1940, he was soon back as a research associate in the MIT Physics Department. Morse characterized Bolt as "an imaginative experimentalist, an excellent teacher and thesis supervisor." They worked cooperatively with Professor Frederick V. Hunt's Harvard group and especially with Hunt's student Leo L. Beranek, who, after achieving his doctorate in 1940, continued to direct acoustics research at Harvard. With the leadership of Morse and Bolt in physics and of Fay and Hall in electrical engineering, MIT had built up a strength in acoustical research and teaching. But war responsibilities interrupted these activities temporarily. Morse took organizational and technical leadership in antisubmarine warfare problems, eventually setting up and heading the Operations Research Group (ORG), so effective in the closing weeks of World War II.[5] Bolt was in London on a mission for the Office of Scientific Research and Development in 1943, returning to the United States to become a technical aide in Division 6 (Subsurface Warfare) of the National Defense Research Committee until the end of the war. Fay was engaged in underwater sound problems and Hall was working on radar problems at Raytheon Manufacturing Company.

Planning the New Acoustics Laboratory

In spite of Morse's intensive wartime activities, he was giving thought in the fall of 1944 to postwar MIT developments. In a memorandum entitled "Proposed Program in Acoustics at MIT after the War," he pointed out that "the Institute had the opportunity of becoming one of the two or three outstanding institutions in acoustics." In research "the work divides itself naturally into three parts: the acoustical properties of materials, the acoustical properties of rooms, and the study of electroacoustical coupling." Morse envisioned a laboratory layout for student experiments, for research, and for studies in room acoustics. He suggested that the new facility be named the MIT Acoustics Laboratory and that it be run by the Departments of Physics, Electrical Engineering, and Architecture—perhaps also by Mechanical Engineering.

The Morse program was set in motion through a recommendation submitted in November 1945 to the MIT administration by John C. Slater, head of the Department of Physics; Harold L. Hazen, head of the Department of Electrical Engineering; and William W. Wurster, Dean of Architecture and Planning. As a result of a substantial contract with the Navy through the Division of Industrial Cooperation and a $15,000 appropriation by MIT, activities were actually already under way. The recommended administration of the laboratory was as follows: director, Richard H. Bolt; chairman of the Supervisory Committee, Philip M. Morse, representing the Department of Physics; committee members, Richard D. Fay of the Department of Electrical Engineering, Lawrence B. Anderson of the School of Architecture, and Julius A. Stratton of the Research Laboratory of Electronics. A formal announcement by President Compton came in January 1946, giving lists of equipment and some of the fields of investigation. A brochure, "Acoustics Laboratory of the Massachusetts Institute of Technology," was prepared for the information of prospective industrial sponsors. Acoustics seminars were held during the spring of 1946. The northeast corner of Building 20 was reconstructed in the summer of 1946 to provide space for the new laboratory, which Bolt, with several of his staff, occupied in the fall of 1946.

Research Activities, 1946–1958

The first quarterly progress report of the new Acoustics Laboratory, issued in January 1947, observed: "The principal objective of the work of this laboratory is to understand the factors affecting sound fields. The portion of this work under Navy sponsorship is primarily concerned with sound fields in water, but the similarities between sound in air and sound in water are strong enough to offer the possibility of adapting techniques from one medium to the other. The mathematical treatment of sound fields in the two media is identical; practical results in air and water appear dissimilar, however, when the differences in density and sound velocity are taken into account." Since acoustical materials are used to control sound fields, the acoustic properties of these materials were studied intensively, and during the year 1946 about 11,000 measurements of the acoustic impedance (a quantity that describes the degree of difficulty involved in producing mechanical motion of the material) of underwater materials were made. Among the researchers were six students, four in electrical engineering and two in physics.

William Hall did not return to MIT after the war; however, fortunately for the Electrical Engineering Department, Leo L. Beranek became interested in the acoustics developments of Morse and Bolt and joined them in 1946–47 as a research associate in physics. He was made technical director of the Acoustics Laboratory in the spring of 1947 and was appointed to an associate professorship in electrical engineering. During the war Beranek had been engaged in acoustics problems in the armed services. Even before the organization of Bush's National Defense Research Committee in 1940, a Committee on Sound Control in Combat Vehicles was set up in the National Research Council, with Morse as chairman and Beranek as secretary.[6] Research was done during the war years in this field by Beranek's Cruft Laboratory group, known as the Electro-Acoustic Laboratory, and by S. Smith Stevens's Psycho-Acoustic Laboratory group. In one of these studies Beranek designed an echoless chamber, known as "the world's quietest room," and originated the term *anechoic* for such a chamber.

Bolt, Beranek, and Fay attracted a succession of able graduate students to do research in the new laboratory. Samuel Labate, for example, completed his master's thesis in 1948 under Professor Bolt, beginning an association that eventuated in Labate's presidency of the Bolt, Beranek, and Newman firm. Jerome R. Cox, Jr., did a master's thesis, "Direct Reading Device for Measuring Acoustic Impedance," in 1949 and in 1954 an Sc.D. thesis, "Physical Limitations on Free-Field Microphone Calibration," both supervised by Beranek. James L. Flanagan did his master's thesis, "Effects of Phase Distortion on Speech," under Beranek in 1950 and his Sc.D. thesis, "Speech Analyzer for a Formant-Coding Compression System," under K. N. Stevens in 1955.

The science of architectural acoustics has turned out not to be a simple one, as shown by adverse experiences in New York's Avery Fisher Hall and MIT's Kresge Auditorium, both of which had to go through extensive redressing before satisfactory sound qualities were achieved. A pioneer in this field was Wallace C. Sabine, who in 1898, while an instructor in physics at Harvard University, took on the acoustical design for the new Symphony Hall in Boston. Beranek reports that on October 29, 1898, Sabine first expressed his formula for the "reverberation time" for a concert hall, which he considered to be the principal acoustical design criterion for such an auditorium.[7] Asserts Beranek, "His formula marked the beginning of the science of room acoustics and remains a cornerstone of calculation today."

For the study of acoustical problems in broadcasting studios, an experimental studio was constructed. Bolt and Anderson directed this project with participating students in physics, electrical engineering, and architecture. Robert B. Newman, a student in architecture and planning, was a member of the team. Newman's longtime association with Bolt and Beranek began in 1949.

In the spring of 1948 plans were made for an extensive study of the acoustic properties of materials and structures suitable for walls, floors, and other surfaces. Anderson, Bolt, Beranek, and Newman were involved in this project, and Jordan J. Baruch entered it in the summer of 1948, after completing his VI-A program with the General Radio Company. In 1950 Baruch completed his doctoral thesis, "Instrumentation for a Transmission Research Program," under Beranek's supervision.

The design of auditorium acoustics is complicated by the fact that a short reverberation time, suitable for the clear understanding of speech, does not provide a rich musical quality. An auditorium designed for speech clarity with a reverberation time of slightly more than one second can be made suitable for music by the creation of an artificial reverberation in the form of a machine that records the music on a tape

In an echoless chamber designed for acoustical experiments are three of the principals of the Acoustics Laboratory, *(left to right)* Professors Richard H. Bolt, Jordan J. Baruch, and Leo L. Beranek.

loop, with pickup heads placed at later positions along the tape. These delayed sounds are then played from loudspeakers along the sides of the auditorium, thus simulating the effect of a reverberant room. Such a machine was the subject of a 1951 master's thesis, "Design and Construction of an Artificial Reverberation Producer," by William Guy Redmond, Jr., supervised by Beranek. Later a reverberation system was designed for Kresge Auditorium at MIT, one of many reverberation machines designed by acoustical engineers throughout the world.

Another Beranek-supervised doctoral thesis was "Perception of Sounds Shaped by Resonant Circuits," completed by Kenneth N. Stevens in 1952. Stevens had taught mathematics and physics at his alma mater, the University of Toronto, before entering MIT in 1948 for a doctoral program. His first staff duty at MIT was to assist Beranek in the teaching of undergraduate and graduate students in acoustics. He became involved with the work of the Acoustics Laboratory in the technical and psychological aspects of speech analysis and synthesis. The primary purpose of his thesis was to advance the understanding of speech communication. In 1955 Stevens, now an assistant professor, supervised the doctoral thesis, "Speech Analyzer for a Formant-Coding Compression System," of James L. Flanagan, who became head of the Acoustics Department of Bell Laboratories and in 1978–79 president of the Acoustical Society of America. Jordan J. Baruch, as a member of the EE Department faculty, continued his interest in the Acoustics Laboratory and supervised several theses, such as that of Kenneth W. Goff in 1954, "The Application of Correlation Techniques to Some Acoustic Measurements." Other EE faculty on the Acoustics Laboratory staff in 1953 were Joseph C. R. Licklider, Osman K. Mawardi, Dennis U. Noiseux, and Walter A. Rosenblith.

By 1953 the MIT Acoustics Laboratory had earned a reputation throughout America and Europe, but there were problems. In June Gordon Brown, who had become head of the Electrical Engineering Department the preceding year, wrote Stratton, then vice-president and provost, that although he recognized the outstanding qualities of Beranek, Rosenblith, and Mawardi as EE Department faculty members, he saw the administrative setup of the Acoustics Laboratory as untenable. Bolt was nominally giving three-fourths of his time as director; Beranek was giving half of his as technical director. But the organizational problems of the firm of Bolt, Beranek, and Newman (BB&N) were requiring more and more attention as demand grew for the company's services. Professor Nathaniel H. Frank, head of the Physics Department, did not see the Acoustics Laboratory as a major part of the physics program.

Brown's remedy for this situation was to make the Acoustics Laboratory a division of the Research Laboratory of Electronics, with Beranek full-time professor of electrical engineering and perhaps an associate director of RLE. Bolt and Baruch would presumably give full time to BB&N. This remedy was not acceptable to the Acoustics Laboratory staff, especially Bolt and Beranek. After consultations with Stratton and Dean of Engineering Edward L. Cochrane, Bolt drafted, on October 20, 1953, an "Acoustics Plan for MIT," which would establish a department of acoustical engineering, reporting to the dean of engineering. The existing laboratory facilities would become the Research Laboratory of Acoustics, administered by the new department but serving as an interschool research center for students and staff from the several interested departments. This plan had the concurrence of most of the Acoustics Laboratory staff, but Beranek and Licklider raised questions and Mawardi and Rosenblith were not in favor. Rosenblith, who had joined the EE faculty in 1951 from the Harvard Psycho-Acoustic Laboratory, pointed out that his principal loyalty and research activity were in the Research Laboratory of Electronics, but that he hoped to continue his studies of the electrophysiology of hearing in the Acoustics Laboratory.

When in the spring of 1954 Dr. C. Richard Soderberg was taking over the deanship of engineering from Dean Cochrane, he and Brown worked out another solution. Bolt was made a full professor (Professor of Acoustics) in the Department of Electrical Engineering and continued as director of the Acoustics Laboratory. Dean Soderberg's recommendation for Bolt's promotion listed his accomplishments, including his contributions to the acoustical design of the Royal Festival Hall in London and the United Nations Building in New York, and acclaimed him as "the unquestioned leader in the field of acoustics, not only in the United States, but also internationally. He has achieved this position by a combination of talents in science and unusual qualities of personal leadership." Beranek, in view of his presidency of the Acoustical Society of America, asked to be relieved of his duties as technical director of the Acoustics Laboratory, but Fay continued as research supervisor. John A. Kessler had become executive officer in 1953. Research and teaching by the laboratory staff continued. Members of the staff were in great demand as speakers at meetings of various organizations. Rosenblith, for example, was involved in several problems in these years: the effect of airplane noise on residents near airports, the effect of noise in industrial plants, noise in the home, noise in human behavior.

Then, in the fall of 1956, Bolt requested a year's leave of absence. He was to become a principal consultant to the Study Section on Biophysics and Biophysical Chemistry of the National Institutes of Health, responsible for the planning and coordination of a new program aimed at the stimulation and development of the rapidly growing field of biophysics on a national scale. After Bolt's departure administrative responsibility for the laboratory was vested in the Department of Electrical Engineering. Gordon Brown, now responsible for the Acoustics Laboratory, put the research supervision and administration in the hands of Professors Fay, Halle, Ingard, Rosenblith, and Stevens and Dr. Heuter, with John Kessler continuing as executive officer. The principal academic offerings at that time were subjects 4.41, Architectural Acoustics, taught by Newman; 6.35, Fundamentals of Acoustics, and 6.36, Engineering Acoustics, a two-term sequence taught by Beranek; and 8.13, Vibration and Sound, taught by Ingard. Graduate subjects were 6.696, Hearing, Speech, and Language, taught by Rosenblith and Halle; and 8.14, Sound Propagation in Moving Media, taught by Ingard.

The very success of acoustical engineering and of its leaders, such as Bolt, Beranek, and Baruch of the Electrical Engineering Department, led to the termination of the Acoustics Laboratory as such. Baruch became a lecturer in 1955, resigning his assistant professorship and investing most of his professional interest in BB&N. Beranek retained his associate professorship until 1958, then continued as a lecturer. His outstanding contribution to MIT in the later years was a series of summer courses on noise reduction in buildings, machines, and enclosures, beginning in 1953 and repeating with new material at intervals of two or three years. Beranek organized the courses, choosing participants from BB&N, MIT, and other organizations. Bolt was not to return to his MIT professorship.

The final quarterly report of the Acoustics Laboratory was issued in September 1957. Listed in the report were 9 doctoral and 58 master's theses in electrical engineering, 19 doctoral and 11 master's theses in physics, and 1 to 3 theses in each of four other departments. On January 31, 1958, the MIT Acoustics Laboratory was officially closed.

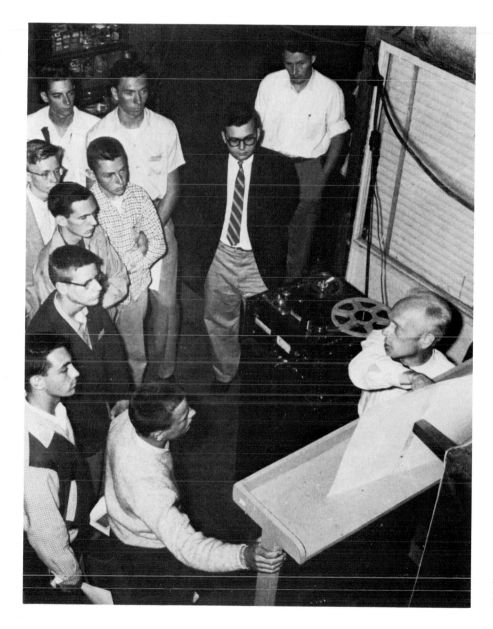

First-year MIT students in 1956 being guided through the Acoustics Laboratory by its executive officer, John A. Kessler.

Chapter 19

Innovations in Engineering Education: The Gordon Brown Era

Hardly any fundamental transition comes about without struggle. The turnabout at MIT in the decade following the war, part of a larger pattern of change in engineering education in midcentury, brought with it a lot of turmoil. It flowed out of the wartime experiences of physicists and engineers, many of whom returned to academia with the conviction that engineering education before World War II had not prepared the engineers sufficiently for participation in enterprises like the Radiation Laboratory and the Manhattan Project. As a result, there began to be a kind of general agreement that more physics and mathematics were needed in the engineering curriculum.

"There began to be stirrings in the EE Department," says Gordon Brown, "led largely by men like Ivan Getting, who was quite convinced we could do better by the educational process." By that time Brown had been promoted to professor, and had begun to direct more and more of his own energies toward strengthening the department's teaching programs. Concurrently, both the student enrollment and the number of professional specialties were increasing, with each faculty group of specialists seeking autonomy. The administrative problem was becoming very much one of providing money, space, and order for a growing cluster of fiefdoms. Hazen sought administrative assistance and began to lean more and more heavily on Gordon Brown, having him appointed associate department head in 1950. Brown was one of the electrical engineers (the others were Julius Stratton, Harold Hazen, and Jerome Wiesner) who were asked in 1947 to serve on the Lewis Committee, reevaluating MIT's educational programs.

But Gordon Brown was also perceived by many people as an engineer, an "engineer par excellence," which meant, in some quarters, that he was not theoretical enough. In the new RLE, for instance, which still carried much of the physics ethos of the Rad Lab, there were rumblings that the computer work (Whirlwind) then going on in the Servo Lab was not proper academic research and should be kept off the campus. In general, the new area of what would become known as computer science—the first science of the artificial, a science of man-made artifacts rather than of the phenomena of nature—only slowly began to become acceptable on college campuses. And when the rumor began to spread that Gordon Brown might become head of the Electrical Engineering Department, apparently some people in RLE vowed they would resign if that came to be.

The resignations never happened, and Gordon Brown lost only one electrical machinery professor when he began to revamp the department; for by 1952, when Brown became department head, he had won a lot of important support for the direction in which he was proceeding. Moreover, in spite of supposedly being a gut-level engineer, Brown had, as one later department chairman would put it, "bought lock, stock, and barrel the idea that engineering education before World War II had failed to prepare them for the things coming, and that what was needed was a science-based engineering education."

In highlighting some of the changes brought about in the EE educational program in the Brown period, one might see them as part of a progressive series of changes running through the entire MIT structure. These changes are reflected in the way in which research has been organized over the decades. The organization of research at MIT has, in fact, altered drastically over the years, going from small projects, run by staff members and involving students at the graduate level, to large interdisciplinary projects. In recent years the concept of "centers" of research has been taking shape, dynamically involved with both the

Gordon S. Brown.

As head of MIT's Department of Electrical Engineering (1952–1959) and subsequently as dean of engineering, Gordon Brown (*second from right, at head of table*) launched a major program of educational innovation; he succeeded in radically changing the department as well as exercising a widespread influence in engineering education throughout the country. *Clockwise from upper left*: Peter Elias, Truman Gray, Karl Wildes, Murray Gardner, Carlton Tucker, Brown, Jerome Wiesner, and (*with backs to camera*) an unidentified participant and Francis Reintjes.

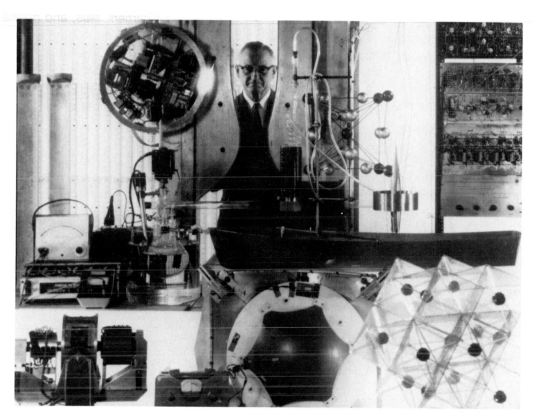

Gordon S. Brown.

undergraduate and the graduate educational programs. Many people in many departments and in the institutional structure participated in the evolution of these ideas, but it is useful to see Brown's involvement in this evolution during his tenure as department head and later as dean of engineering.

Brown had been preparing for such changes for a long time; he had started very gradually soon after the war. "I made my first approach to getting a little emancipation into the electrical machinery area," he states, "by inviting Leo Beranek, then one of the key people in the Acoustics Laboratory, to begin quietly collaborating with me on an idea that struck me as pertinent, namely, that a loudspeaker is in principle the same thing as a motor. It was certainly an energy conversion device, with a dynamic performance far in excess of any rotating machine; nonetheless, it could be described by the same equations. I thought that Leo could be of great help in revising the machinery curricula." It was one of Brown's first steps in a long process of developing a new, unified view of all the things that were going on under the name of electrical engineering.

He says of those first steps, "I had to learn that you approach such changes very discreetly in a place like this. The acoustics people objected because they thought that Leo was selling himself out to the machinery people, and they thought I was intruding in their areas. So we didn't move very fast." Nonetheless, Brown's search for a unified vision of engineering was to lead, in interesting ways, to a resurgence of activity in electric power systems at MIT.

About the time he became department head, Brown had begun to pull together the nucleus of a simplified and clarified idea for revising the curriculum. "All the electrical engineer has to work with," he would say, "are charge carriers . . . either ions, electrons, or holes, their associated electric and magnetic fields, and their interaction with materials. Out of this basic nucleus of carriers, fields, and material, the engineer builds configurations of devices, machines, or systems that do either of two things—they process energy or they process information. That's all!"

Armed with this simple model, Brown began to persuade others—department members, visiting committees, industrialists who looked to MIT for the next generation of engineers—and, as described in this chapter, he brought about sweeping changes in the electrical engineering curriculum. Earlier, as associate head of the department, he had already recruited some important allies. As early as 1947 he was announcing, "I can foresee a whole new era in the approach to machines . . . not simply as motors, generators, and the like, but as *energy processing systems.*"[1] One of Brown's early recruits, David White, who joined the Servo Lab, would later (in 1972) become head of the MIT Energy Laboratory.

As the new EE Department head, Gordon Brown also traveled around the United States seeking financial and other support for his new program. At the California Institute of Technology he met with the man from Rad Lab who had become Cal Tech's new president, Lee DuBridge, who also offered him support in his new direction. At Cal Tech Brown crystallized further his thinking about the EE curriculum. A week or so later, while visiting the University of California at Berkeley, he was asked to give a lecture about his ideas on energy and information processing. In the middle of that lecture, quite suddenly, he reports, "I got my own ideas straightened out, and in that moment, saw the arguments I needed." To his audience he said aloud, "The more I think about this, I believe I'm going to go back to MIT and ask President Killian to scrap the machinery laboratory." Sitting in the front row of that audience was a distinguished engineer, Everett Lee of General Electric, a man of the traditional school. No sooner had Brown uttered this thought than Lee "shot up out of his chair." Brown, the iconoclast, the politician, the promoter, was committing "absolute heresy, unadulterated treason." But wherever Brown met with electrical engineers who had had the wartime experience of radar and electronics, he reports, "I was amazed at how much support I got."

He returned to MIT, saw Killian, presented his argument. All Killian said was, "How quickly do you want to do it?" The hour of revolution had arrived. The great, revered machinery lab, housed in a huge space in Building 10, in which generations of electrical engineering students had got black to the armpits wiring up generators and motors and making traditional measurements, would be thrown out. It was more than a symbolic gesture—more space, both physical and mental, was won for new kinds of teaching and research.

The changes that Brown wrought in the EE Department, and some of their consequences, as well as the resurgence in electric power systems that continues up to the present day, are sketched in the following pages. The chronology at the end of this volume (see Appendix) lists some of the other events of this period.

Launching the New Curriculum

In large part, it was the extent and depth of Gordon Brown's department experience that enabled him to embark so quickly on a new undergraduate curriculum and to announce his evolving philosophy of engineering education. The new scientific demands of electrical engineering had, as we have already seen, convinced him that the time had come for another deepening of scientific foundations. In the fall of 1952 he presented a paper, "The Modern Engineer Should Be Educated as a Scientist," before the American Society of Engineering Education, spelling out his views.[2] Some Course-VI alumni were offended, com-

plaining that the department had gone out of the engineering business. Not so, replied Brown, the plea was for enough science to support the new demands of engineering. As he had put it in the paper, "It is more to the point to say that engineers should be educated with depth in the sciences."

These matters were thoroughly threshed out with the Corporation Visiting Committee, which met in February 1953. This was a strong committee of eight members under the chairmanship of Harold B. Richmond of the General Radio Company. The Brown recommendations were endorsed. A paragraph from the committee report is enlightening:

Much of the traditional instruction in machinery analysis will be discontinued. A whole new concept for laboratory instruction on energy conversion devices and systems will emerge as a result of this change. Laboratory instruction will coordinate closely with classroom instruction as an integral part of the subject. Much of the equipment in the present electrical machinery laboratory is inadequate and even a deterrent to progress in instruction in the electric power field. Much new equipment will be needed. The new subject— Control and Conversion of Energy—will aim to bring about a closer coupling between science and engineering than has been characteristic of the past, since engineering today is everywhere vastly more scientific than it was a decade or so ago—in fact, science dominates engineering.

Working committees were organized and announced early in the fall term of 1953. A large and carefully selected Committee on Educational Policy worked devotedly under Brown's chairmanship to develop the new curriculum. The introductory sophomore subjects, 6.00 and 6.01, in circuit theory under Ernst A. Guillemin's leadership were strengthened to 12 hours per week, including a closely integrated laboratory. Junior and senior EE subjects were proposed on the same lecture-recitation-laboratory pattern. The 1953–54 year was a transitional one in the energy conversion area. Professors David C. White and Alexander Kusko, with their teaching staffs, presented a two-term sequence, 6.03 and 6.04, on control and conversion of energy. In 1954–55 Professor Robert M. Fano began to present 6.03 under the title Fields, Materials, and Components, which evolved into Fields, Energy, and Forces.[3] This junior subject was studied in parallel with Professor Samuel J. Mason's new synthesis of elementary electronics, 6.02.[4] Kusko's 6.04 was studied in the second junior term in parallel with Professor Truman S. Gray's Applied Electronics.[5] During the same academic year, 1954–55, Professor Richard B. Adler began to present his Energy Transmission and Radiation, subject 6.07;[6] and in parallel for first-term seniors, Professor White presented Electric Power Modulators, 6.06, which evolved into Electromechanical Energy Conversion.[7] An important later addition to the core curriculum was Professor David J. Epstein's Molecular Engineering, 6.08, given for the first time in the fall term of 1958. This dealt with materials in their interaction with magnetic, electric, mechanical, and thermal fields. The student was here introduced to applications of thermodynamics, quantum physics, and statistical mechanics to modern engineering problems. Outside the department the new curriculum included additional depth in mathematics and physics.

The old options of power, communications, and electronic applications gave way to a core curriculum as the foundation for individual programs worked out by each student in conference with his educational advisor. The undergraduate program could prepare its students to enter industry at the bachelor's-degree level or strengthen their scientific base in preparation for postgraduate studies.

Brown and his Undergraduate Policy Committee, having developed a scientifically strong core curriculum, next set out to build a special curriculum, to be called Course VI-B, Electrical Science and Engineering, for students of outstanding ability who could anticipate graduate work. It would be a five-year course leading to the S.B. and S.M. degrees, with the expectation that significant numbers of students would continue to the doctorate. In the fall of 1957, 23 juniors and 18 sophomores registered in Course VI-B. Experience during the first year revealed that these able students were probing so deeply that they deserved the special attention of the experts in their EE subjects. In the fall of 1958 Professor Lan Jen Chu conducted a special section of Energy Transmission and Radiation for the VI-B seniors, while VI-B juniors and sophomores were beginning to get specially prepared and taught 6.00, 6.02, and 6.03. In 1959–60, 38 sophomores registered in Course VI-B and were taught the introductory subject, 6.003, in two sections by Harry B. Lee, a doctoral student who developed new class notes to meet the needs of these especially sharp students who were seeking solid foundations for their circuit theory. Professor Zimmermann taught 6.023, Professor Mason 6.033 in the first term and 6.053 in the second term, Professor White 6.063, Professor Kenneth Stevens 6.073, and Professor Epstein 6.083. (The final digit 3 was added to indicate the Course VI-B subject.) The new course, which attracted many unusually able students, was guided by Professor Lan Jen Chu as his special responsibility.

Resurgence of Electric Power Systems Research
By 1954 Brown had made sure that energy conversion at MIT would no longer deal with the testing and operation of standard rotating machines. He removed the 10-ton traveling crane in the Dynamo Laboratory

and discarded most of the machines, flooring over the crane well to provide additional work space for new energy-conversion projects. He took these actions despite continued objections. His disposal of the network analyzer was another of his controversial changes in electrical power.

New textbook material was needed to express the unified viewpoint. Ever since the original 1940 publication of the Blue Series of textbooks (bound in blue cloth), a broader view of energy conversion had been envisioned. Volume 1, *Electric Circuits,* contains a chapter, ''Electromechanically Coupled Systems,'' analyzing the moving-coil telephone, moving-iron galvanometers and other mechanisms, and, as an example of electric-field coupling, the capacitor microphone. Harold Hazen, as department head from 1938 to 1952, had advocated and encouraged an expanded variety of energy-conversion processes. It was during that period that Professors Fitzgerald and Kingsley developed their more fundamental text, *Electric Machinery, Basic Theory of Electromechanical Energy Conversion.*

Now, under Brown's revitalization of the electric power field, the expansion of the Blue Series began in earnest, supported by several groups of able and progressive department members. David C. White, a Ph.D. in physics from Stanford University (1949), brought together an initial group—Herbert H. Woodson, Mahmoud Riaz, Richard H. Frazier, and visiting professors Robert M. Saunders and Herman Koenig —to develop text material on electromechanical energy conversion. This material was strongly related to the new electromagnetic field treatments of Fano, Adler, and Chu and was influenced by the matrix and tensor treatments of Gabriel Kron of the General Electric Company and of W. J. Gibbs and Bernard Adkins, whose books were sponsored by the British Thomson-Houston Company.[8] Class notes were first issued to a group of seniors in the fall of 1954 and were contin-

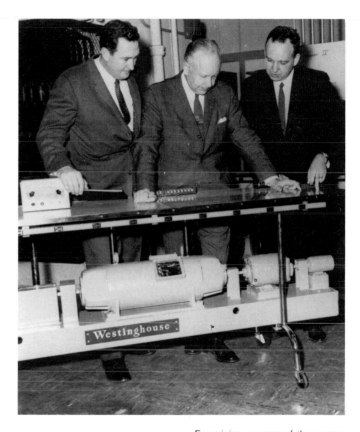

Examining a copy of the generalized rotating energy converter are (*left to right*) Paul B. Shiring of the Westinghouse Educational Foundation, MIT president Julius Stratton, and David C. White.

ually revised until the White and Woodson textbook, *Electromechanical Energy Conversion,* was published in 1959. Based on Hamilton's Principle and the Lagrangian equations of motion, the new treatment involved the concept of state functions and tied electromechanical energy conversion to basic principles of thermodynamics and mechanics.

White and Woodson conceived and built a *generalized rotating energy converter* to illustrate the action of a direct-current generator or motor, or, with a change in connections, that of a synchronous or induction alternating-current machine. When the generalized machine was demonstrated to Gabriel Kron, he placed his hand over his heart and said, "This is the realization of my dream of twenty-five years," and then, with a sigh, "but it is twenty-five years too late." These developments were presented at a meeting of AIEE in January 1956 in a paper, "A New Educational Program in Energy Conversion," by Brown, Kusko, and White.

Brown's strengths in department administration were evident in several ways. He built up loyal and enthusiastic teams of workers to develop the core curriculum subjects and kept the research and graduate education at the forefront of progress. He was an energetic salesman for his new program and got financial backing for it, especially from the General Electric Company through the interest of Clarence Linder, vice-president of GE. In September 1957, in response to the countrywide interest and with financial support from the National Science Foundation, the General Electric Company, and the Westinghouse Electric Company, the department conducted a curriculum workshop attended by professors from 104 educational institutions. Because of the interest aroused by this workshop, Dr. John A. Hutcheson, vice-president for engineering and research at Westinghouse, arranged for his company to produce enough copies of the generalized machine to give one to each of the accredited EE departments in the United States.

White and his group worked with Westinghouse engineers on the design of the machine and on the preparation of a manual for its use. A laboratory model was now available that lent itself to the study, not only of conventional rotating machines, but of unconventional ones as well.

Two new energy-conversion projects were begun in 1956–57. One, initiated by White, was a study of thermal-electric conversion, utilizing semiconductors. He was soon joined by Professor Richard B. Adler, and later by Professors Arthur C. Smith, John Blair, and Paul E. Gray. A well-equipped laboratory with crystal-growing apparatus was developed. After Brown became dean of engineering in 1959, this activity was federated with other MIT projects under the committee chairmanship of White, moving in 1965 to its new home, the Center for Materials Science and Engineering, in the Vannevar Bush Building.

The other new energy-conversion project of 1956–57 was Professor Osman K. Mawardi's study of the behavior of moving electrical conducting fluids—liquids and gases—in strong magnetic fields. The following year Professor Mawardi moved to the Department of Mechanical Engineering to join Professor Ascher H. Shapiro as that department started its program in magnetohydrodynamics (MHD). Professor Woodson took over the MHD research in 1957 and was joined in this work by Professor William D. Jackson, formerly of Manchester University in England. Graduate subjects in MHD were taught by Woodson and Jackson. In the fall of 1959 James R. Melcher, who had won an Atomic Energy Fellowship at Iowa State College, entered the MIT Graduate School to study MHD with Woodson. By 1961 Woodson and Melcher were fashioning a new energy-conversion treatment based on Maxwell's Equations; after many versions of class notes, their *Electromechanical Dynamics,* in three volumes, appeared in 1968. This is a scholarly and pedagogically well planned treatise, first reducing the field equations to circuit concepts already familiar to students, and then coupling magnetic and electric fields with mechanical systems. It includes linear and nonlinear systems and leads finally into a treatment of conducting fluids, such as are involved in the "magnetic bottle"—the container for the hot plasmas of nuclear fusion—as well as the hot gases of combustion used in magnetohydrodynamic electric generators. The book describes real-world embodiments of the various developments. The authors aimed at making this work "the mathematics, the physics, and most of all, the engineering combined into one." It has proved an effective and enduring textbook.

Melcher's doctoral thesis in 1962 was entitled "Electrohydrodynamic and Magnetohydrodynamic Surface Waves and Instabilities"; by this time he was a strong partner for Woodson in energy conversion research and teaching. He developed his own graduate subject, Continuum Electromechanics, with class notes that grew into a book by the same title. Professor Melcher made important contributions to the application of continuum electromechanics in the removal of particulates from the effluents of smokestacks and the purification of other industrial gases and liquids. There are many other practical applications. For example, charged particles play a part in the motion of fluid coolants in transformers and of fluid insulation in cables. Macroscopic charged particles on the surfaces of transmission-line insulators contribute to the objectionable 60-Hz hum. Electric forces are also used in xerographic printing and in devices for the spraying of insecticides and paints.

Melcher supervised the doctoral thesis of Alan Grodzinsky, "Electromechanics of Deformable Polyelectrolyte Membranes," in 1974. Since then Professor Grodzinsky and his students have shown that electromechanical effects play a significant role in the dynamics of connective tissue, such as cartilage.

Woodson and Jackson continued to supervise MHD research through the academic year 1964–65, including the doctoral thesis of Gerald L. Wilson, "Excitation and Detection of Magnetoacoustic Waves in a Rotating Plasma," supervised by Woodson. By this time Woodson had become interested in electric power system engineering; he arranged to spend the academic year 1965–66 in the New York City offices of the American Electric Power Service Corporation (AEP). Upon his return to MIT he initiated some new researches and subjects of instruction in power system engineering. Alexander Volk, an AEP engineer, was a visiting MIT faculty member in 1966–67 and shared in the new program while Professor Wilson was on leave at AEP. Professor Fred C. Schweppe spent part of the academic year 1967–68 investigating the control problems of AEP.

A New Electric Power Systems Laboratory
Through the exchange of personnel between AEP and MIT, the Power Systems Engineering Group was able to bring several practical problems into the laboratory for study. As one after another component of a power system was simulated by a laboratory model, the basement of Building 10 under the central dome became known as the Electric Power Systems Engineering Laboratory (EPSEL), an interdepartmental laboratory of the School of Engineering, with Woodson as its director.

Philip Sporn, the president of AEP, took a personal interest in the resurgence of activity in electric power systems at MIT; he solicited funds from several public utility and manufacturing companies, and with these and Sporn's own personal contribution MIT established the Philip Sporn Professorship of Energy Processing. The new professorship was first awarded to Woodson in 1968.

The MIT-EPSEL model power
system, the so-called transmission system simulator
(TSS), used to study many
areas of power-system dynamics and control, with (*left to
right*) Gerald Wilson, John
Kassakian, and James Kirtley.

An elective subject in power system engineering was taught in the spring of 1968 by Professors Wilson, Woodson, and Gerald I. Stillman, the last a visiting faculty member in 1967–68 from AEP. This classroom experience led to the graduate sequence Power Systems Engineering I and II, taught in the fall term of 1968–69 by Wilson and Charles Kingsley and in the spring by Wilson.

A model power system, built around four model generators designed to reproduce accurately the electrical and electromechanical properties of four specific large solid-rotor turbogenerators while scaled down in power rating by a factor of one million and operating at laboratory-compatible voltages, was constructed in EPSEL. Called the transmission system simulator (TSS), it comprised three-phase transmission lines, compensation reactors, transformers, circuit breakers, and their actuating relays, and represented a three-phase transmission line in a new way that is interesting to contrast with the Hazen network analyzer. In the analyzer a power system could be modeled at the standard 60-Hz frequency. Active and reactive power flows could be observed through readings of ammeters, voltmeters, and wattmeters. The new EPSEL transmission line model was designed for the study of lightning and switching surges, transient phenomena that required simulation over a range of frequencies. Extra-high-voltage lines, 700 kV and above, were already in operation and were not transposed, transposition of the line conductors having been predominant in the high-voltage (110–220 kV) era. The new transmission line model was designed and built under the supervision of Professor Wilson by Kenneth A. Schmidt and William E. Feero, who completed their graduate theses in 1971 and 1972. John G. Kassakian, in his doctoral dissertation of 1973, studied the effects of nontransposition and frequency-dependent neutral modeling on the simulation of power transmission systems.

Earlier work (1964–66) by Woodson and Zdenek J. Stekly (MIT Sc.D. in mechanical engineering) at the Avco-Everett Research Laboratory had shown that supercooled field windings were successful in small alternators.[9] A new research program in the development of large steam-turbine alternators with superconducting field windings was initiated in 1967–68 at MIT jointly by Woodson of the Electrical Engineering Department and Joseph L. Smith, Jr., of the Mechanical Engineering Department's Cryogenic Laboratory. In principle a rotating field winding with no iron, cooled by liquid helium to four degrees Kelvin, eliminates most of the iron losses and copper losses of conventional field windings and leads to the design of alternators having increased efficiency, lighter weight, smaller size, and higher generated voltage than conventional designs. The Edison Electric Institute gave financial support to this program. The first model was rated about 80 kVA and was tested in 1971 at 45 kVA as a synchronous condenser. The Edison Electric Institute has also encouraged the industrial development of these machines; several companies, including Westinghouse and General Electric, began their researches on supercooled field windings about 1970. With the experience gained from the first MIT model, a program was begun for the design and construction of a 3-MVA alternator. The armature for this machine was the subject of a doctoral thesis by James L. Kirtley, Jr., in 1971. Professor Kingsley was deeply involved in planning and carrying out the tests. A new research and development organization founded by the electric utility industry, the Electric Power Research Institute (EPRI), also sponsored a strong program on superconducting generators starting in the mid-1970s, based in part on this early MIT work.[10] Bringing together the industrial studies as well as the MIT experience, a new five-year program, funded by the United States Department of Energy, was begun in 1976.[11] For MIT these and other projects, rich in graduate thesis opportunities, together with a great diversity of electrical and mechanical problems, made evident the importance of interdepartmental cooperation.

Toward the Idea of Centers

Interdepartmental cooperation has, in fact, been a hallmark of the postwar era. The technical problems of the twenties were usually solvable within the walls of a single department, although even in those years there were notable examples of interdepartmental cooperation. The Second World War suddenly introduced military problems, such as the automatic control of large guns and the development of radar systems, that demanded the cooperative effort of experts in several disciplines. Following the war came the interdepartmental laboratories, such as the Research Laboratory of Electronics, the Acoustics Laboratory, and the Laboratory for Nuclear Science and Engineering. Research centers such as the Communication Sciences Center and the Center for Cancer Research were organized to investigate special areas requiring uniquely trained staff and uncommon equipment. Recognition of the need for new ways of organizing research and relating such research to engineering education began to crystallize soon after Brown became head of the EE Department, and he became a strong proponent of the idea of dynamic and interactive centers of research, an idea then beginning to take hold at MIT.

In his first report to the president in the fall of 1954, Dean Soderberg noted the changing character of engineering education under his predecessors, Professor Thomas K. Sherwood and Admiral Edward L. Cochrane, and then made some prophetic statements. Referring to the various engineering disciplines by MIT departments, he said,

In actual fact, the external world of technology never was divided in quite the precise pattern implied by the designations; a glance at the present programs of the various departments also suggests that the picture is no longer true for the Institute itself. There is a blurring of the professional pattern which has increased markedly during the recent years of intensive creative activity. As this process continues, the various departments will continue to respond, more or less, to the whole external complex of technology. They will thus tend to differ more through methods of approach and emphasis than through different syntheses of sharply defined professional fields. Most of the major subdivisions of applied science will be taught in several departments. The departments will tend to become units of administration through which the initiative of their leading personalities will find expression. . . . It is my conviction that this tendency, even though it gives rise to baffling administrative problems, is a gratifying sign of strength and healthy development. Many of the major advances in technology of recent years have been due to intermixing of fields, so that the earlier and rigid synthesis is no longer a reflection of reality. What has been called a blurring of the professional pattern, therefore, is nothing more nor less than a belated introduction into our educational system of the facts of life.

The development of Soderberg's educational ideas was greatly encouraged when he learned that the Ford Foundation was beginning to think of significant financial aid to engineering education. This prospect stimulated the engineering faculty to analyze the educational programs of the Engineering School and to propose radical advances that could be achieved with adequate funding. A Committee on Engineering Education was appointed by Soderberg in October 1956, with Professor Edwin R. Gilliland as chairman and Laurens Troost and Jerome B. Wiesner as members. In early 1957 this committee set about the preparation of a proposal requesting Ford Foundation funds for the Engineering School. An early draft of their proposal contained a review of engineering education, suggestions for new ventures for the immediate future, and specific innovations in teaching and curriculum development, stressing greater depth in science and "a longer, better-integrated educational period for engineers."

Gordon Brown set forth in the proposal the new Course VI-B, which was to be launched in the fall of 1957. It was a five-year program called Electrical Science and Engineering, leading to the S.B. and S.M. degrees and designed for students of outstanding ability. He wrote, "We expect that a large fraction of the students who elect this program will continue to the doctorate." After delineating the task of developing new engineering syntheses based upon the deeper foundation in the physical sciences, he asserted, "A new class of faculty person will emerge when this task has been accomplished, for unless the effort entails much new learning by faculties, it will have amounted to mere rearrangement of old knowledge and thereby will have fallen far short of the desirable goals." The proposal was long and needed restudy and rewriting. Brown, in a later criticism, stressed the need for a new kind of environment that would "couple more dynamically and spiritually with related disciplines to give it the needed recognition of human values. . . . I suggest we focus our attention on four human needs that are challenges to tomorrow's engineering leader: communications, energy exploitation, materials, and mobility." After describing each of these human needs, he proposed the creation of four new centers that would by their interaction "change the traditional environment and nature of engineering education and thereby equip tomorrow's new leaders with the knowledge, skills, and breadth of vision to permit them to play the role of truly professional men in our democratic society."

In a memorandum to Soderberg, President Killian said, "As our document grows, it becomes increasingly clear that we are clarifying ideas in the process and that we are seeing increasing opportunities to augment the boldness and originality of our program. . . . I propose, therefore, that we move to another general revision." He mentioned the Brown criticism and also one by Hazen (dean of the Graduate School, 1952–1967) that emphasized the importance of understanding the human aspects of engineering. Hazen had written that, in aiding developing countries, there are "many examples of highly competent technical work that prove to be social nonsense."

The ideas concerning innovations, experiments, and improvements in engineering education continued to evolve under many hands. During the spring and early summer of 1958 Professor Wiesner of the Electrical Engineering Department and Professor Shapiro of the Mechanical Engineering Department worked on the fourth draft of the proposal. Its twenty pages contained their own innovative contributions and those of the Committee on Engineering Education, Soderberg, the department heads, Edwin Land (president of the Polaroid Corporation), and Elting E. Morrison (professor of industrial history, a regular member of the committee). This draft mentioned the increasing number and importance of MIT research centers, which Brown had stressed in the spring of 1958 to the Visiting Committee of the Electrical Engineering Department.

An important participant at this point was Julius Stratton, who had become MIT's acting president as well as chancellor after Killian became advisor to President Eisenhower in 1957. Among other matters, Stratton urged that further work be done on the concept of research centers.

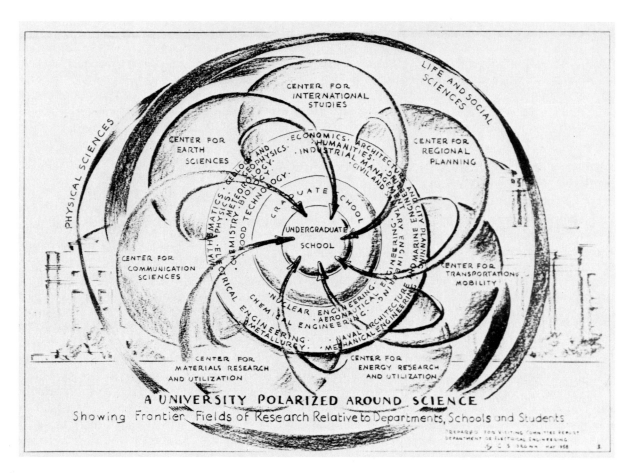

Sketch by Gordon Brown suggesting the dynamics of his concept of research centers and indicating how research could vitalize engineering education.

By March 1959 the proposal had reached its final stage. Entitled "A Proposal for Experimental Developments in Engineering Education," it was the culmination of intensive cooperative efforts and led to a Ford Foundation grant of nearly $9,275,000, the largest grant ever made to MIT, for the advancement of engineering education.

Among the arguments marshaled in favor of ambitious innovations in engineering education were the following:

• The increase in the rate of technological development is illustrated strikingly by the 130 years required to understand completely the thermodynamics of the steam engine, as compared to the postwar development of controlled nuclear fission. "Here, the application of science and engineering on a vast scale has compressed into a few years a development process which under the old conditions might have required many decades, or which might have been impossible altogether." This "new pace of technological development" calls for a new kind of engineering education. Indoctrination in current practice is no longer useful. "When an engineer sets out to build something today, he frequently builds something new, something never built before." This is difficult to teach and can be learned only "by doing it, and by doing it under real conditions in company with those who are themselves active in creative work."

• Research has always been the keystone of graduate education at MIT. "We now believe that a much improved *undergraduate* education could be given in an environment where there can be an intimate and informal association between students and faculty and where the student can experience the many benefits which come from study in the research environment." Additional research centers should be established. The Research Laboratory of Electronics is a convincing example of such a center, "bringing together the communications and computer engineers, the psychologists, the biologists, the linguists, and the mathematicians who formed the nucleus of the new Center for Communication Sciences." Thus, one of the influential outcomes of the program was that undergraduate students would become involved in real research projects.

Soon after the release of the final proposal, Gordon Brown was asked to be the new dean of engineering when Soderberg relinquished the position officially on June 30, 1959. It would be Brown's responsibility to administer the funds of the Ford Foundation grant. Brown served as dean of engineering from July 1, 1959, to November 1, 1968, longer than any predecessor. He then asked to be relieved of administrative duties and was appointed the first Dugald Caleb Jackson Professor of Electrical Engineering. In 1972 he was elected Institute Professor by his faculty colleagues.

The TX-0, first of a new family of TX computers at Lincoln Laboratory that were to help lead the way to a growing MIT involvement in computers and computer science. In the foreground is the console; the computer is on the panel behind it. On the left in the background is the first 64-K magnetic-core memory bank ever built. At center right in the background are the panels for the TX-2 computer under construction. After the TX-0 had served its purpose at Lincoln Lab, it was given to the Electrical Engineering Department and has been used extensively by students ever since.

Part V

Into the Computer Age

Introduction
A New Identity

In the years covered in this last section—roughly the last quarter of a century—there have been so many significant concurrent developments in electrical engineering that they could hardly be dealt with adequately even in a number of volumes. One of these, however, stands out in terms of its vigorous growth, its influence on many other fields, and the changes it has brought about in electrical engineering during this time. Accordingly, in these last chapters we focus on the growth of digital computation at MIT, on the inception and evolution of time sharing, and on the progressive integration of computer science and electrical engineering in the educational curriculum.

Even in limiting ourselves to developments related to computers, we must be highly selective, for the fact is that the electronic digital computer has begun to change human perceptions of the world, human thought, and human action. All really powerful tools have effected such change to some degree; but the digital computer is a tool unlike any other. The very concept of digital computation, bearing more than a resemblance to the functioning of the living nervous system, is altering the character of scientific and technological education.

In the 1966–67 report of the School of Engineering, MIT's Dean Gordon S. Brown wrote about the growing impact of this new machine:

The digital computer promises to alter the way of life of human beings within the next decade as much as, if not more than, Gutenberg's invention of printing from movable type. . . . The frame of reference is so changed that the student may ask questions that heretofore would not have made sense. Herein lies the potential for a revolution in the teaching-learning process that today can be only dimly perceived. It promises not only to change the content of what we teach to students, but also to revolutionize the methodology whereby we do it. . . .

We have concluded that rather than become preoccupied with computers and computer centers, we must realize that we are being carried along by the pressures of an open-ended, self-generating, campus-wide information-processing program. The computer per se is the small part of it.

It appears that right within our grasp is a new conceptual movement which has a potential never before available for unifying the goals and purposes of the Institute. The common interest in the computer shown by scholars in disciplines so diverse that heretofore they have never encountered one another, let alone shared knowledge and experience, is indeed amazing. The challenge we see ahead is to mobilize this common interest in ways that will convert this potential for greater unity of purpose into a reality—to achieve a different and better kind of university.

Such a possibility—that alternative education, which had long stood apart from the classical universities, might now lead to a new type of university and a new identity—began to be seriously debated for the first time during the period we now examine. Certainly, electrical engineering was ready to assume a new identity.

Chapter 20

The Computer and Changing Perceptions

In scanning, even superficially, some of the momentous events of this past quarter of a century, one discerns many forces other than scientific and technological ones coming into play. It is a period of changing perceptions, a period when the national recognition of the central role of science and technology soared, and then, as the social and environmental impacts of their uninhibited growth began to be felt, a profound troubling set in. There are certain characteristics of this period that resemble those of industrial Great Britain in the last quarter of the nineteenth century, when an antitechnology sentiment and questioning arose as a reaction to untrammeled technological progress during the great Age of Steam.

To begin with, the mid-1950s in the United States were years of high optimism, when anything seemed possible, when all the pent-up knowledge and achievements of the war and postwar periods were about to be exploited. In the sciences there was a kind of renaissance, the peak of the so-called interdisciplinary movements, when collaboration among many diverse disciplines was going on. It was a period of industrial expansion, based largely on the new technologies that had been incubating since the 1930s. By the late 1950s fields like electronics and communications had become glamorous, attracting a large number of engineering students.

In the spirit of "we can do anything," the United States embarked on the unparalleled moon-landing program in 1961. The computer and semiconductor electrical enterprises—two sister technologies that bootstrapped each other in an incredible growth process—were rapidly becoming multibillion-dollar enterprises; these in turn transformed many other industries. Simultaneous with this general economic expansion, the Cold War fostered a scientific and technological race, especially in the area of new and more powerful weapons. Moreover, the nationalist movement among former colonial holdings in the Third World was vigorous and helped to inspire the civil rights movement in the United States. This movement in turn became the model or metaphor for the freedom movements of many minority groups. These, in the 1970s, would begin to have a significant impact on American universities under the equal-opportunity laws, which would put to the test the striving for excellence, on the one hand, and the deeply founded American principle of opportunity for all, on the other.

In the 1950s and 1960s, too, because of the popularization of the fruits of science and technology, waves of students came into all the nation's graduate schools, seeking higher education in these fields. The schools, challenged to keep up with this deluge, had to resort to extensive recruiting for their growing staffs. At the same time there was a strong orientation to more science-based studies, largely a result of war experiences and the continuing pressures of Cold War research. During this period the United States not only assumed the role of a major world power, with a concomitant sense of responsibility, it also became a mecca for learning, especially in the fields of science and engineering, fields in which the Europeans had excelled in the decades before the war.

The Vietnam War years tested the nation and the universities in a new way. Even engineering students—habitually the least politically active of all students—became involved in war resistance movements, and scientists and engineers mobilized symbolic research work stoppages. One consequence of this was that some universities divested themselves of laboratories that were predominantly occupied with weapons research. On the East Coast, MIT divested itself of Draper Laboratory in 1973, the year in which the last American troops left Vietnam; in that same year, on the West Coast, Stanford University divested itself of the Stanford Research Institute, in the wake of a powerful peace movement that at its peak involved as many as 10,000 students.

In 1973, also, the major oil exporting nations organized the Oil-Producing Economic Community (OPEC) and began using oil as a political weapon against the highly industrialized countries, steadily raising their prices and precipitating the ''energy crisis.'' Thus began a period of inflation, of economic instability and retrenchment, which would have widespread effects, not least in the universities.

The environmental movement, partially triggered by the discovery that smog in cities was connected with automobile and power-plant exhausts, was internationalized virtually overnight by the 1969 moon landing: that dramatic event brought home the sense that our biosphere was indeed limited and fragile. Hand in hand with the new environmental consciousness, serious studies went on in the scientific and technical communities into the consequences of uninhibited technological activities. ''Technology assessment'' began to grow as a field in itself, and environmental impact considerations began to come into engineering design. Whereas traditionally the practice of engineering had been guided by the balancing of two major factors—engineering considerations, such as efficiency and effectiveness, and economic considerations, such as overall costs and marketability—there now arose the possibility that social and environmental impacts would become a third major factor, in the form of new standards and regulations.

In schools like MIT a new concern began to emerge about the ways in which students could be sensitized to the consequences of their future work as well as trained to do the work itself. However, contradictions that had been near the heart of alternative education from its beginnings, more than a century earlier, surfaced again. Could a school like MIT give real weight to the humanities and liberal arts, as, say, did Harvard? If it attempted to do so, would the Institute become only a second-rate Harvard and a second-rate engineering school? Could a school primarily dedicated to science and technology serve the vital role in its next century that it had in its first?

A perception shared by many people at MIT is that as an educational institution it has remained alive through its emphasis on continuing research, by staying at the cutting edge, by attracting teachers and students of excellence, by maintaining an internal debate. All of these characteristics are illustrated in the story of the development of computer science at MIT. The relation of computer science to electrical engineering is an important issue under debate; but the computer cuts across—and will continue to cut across—a range of educational issues, as well as some of the most difficult life issues of our time.

Computer Science at MIT

Computer science got its start at MIT in electrical engineering, though in some schools it had its genesis in mathematics departments; at Carnegie Mellon, perhaps a unique case, it was begun in the psychology department. In general, however, in the United States, research and teaching about computers came principally out of electrical engineering. Computer science has gone through a period of struggle to acquire academic legitimacy and a distinct identity. In some schools there are now completely separate computer science departments, but at MIT recognition came in the form of a new name—the Electrical Engineering and Computer Science Department.

There is at least one noteworthy parallel between the emergence of computer science in this recent period and the emergence of electrical engineering at the end of the last century. With the growth of electrical engineering great new industries were formed, and electrical machines took over many of the laborious

tasks performed by men and animals. Electric utilities grew and soon delivered electric power for multiple uses to nearly every business and home; they helped to foster a great dispersal of the work force—they changed *where* people did their work, as well as *how* they did their work. Similarly, with the growth of computers great new industries have been formed and laborious tasks taken over, but this time in the mental arena—which is why Norbert Wiener called this second Industrial Revolution the Information Revolution. And computer systems have also grown as did the electric utilities, allowing many users to draw upon their computational powers through local outlets. The origin of these new informational utilities—time-sharing systems—is treated in this last section.

Computers, like the earlier electrical machinery that replaced muscle power, are bringing a form of automation into the workplace. Whereas electrical machinery dislocated blue-collar workers, the new automation is affecting the white-collar work force. Computerized systems will undoubtedly cause enormous changes and dislocations over the decades to come. They may also continue the process of the dispersal of the work force.

There is a significant difference in the origins of electrical power systems and of computers, however, and it illustrates the impact of research-oriented education. The electrical machines of the nineteenth century were developed by individual inventors, who built up industries around them. From that experience electrical engineering gradually took shape in the schools. However, *all* the early digital computers were created and developed in research projects in the schools; they then moved out into industry, where they became the foundation of great new empires. This, in fact, has been the pattern in the period following World War II in a number of scientific and technical fields.

As they have come of age, computers have been put to work in an ever-widening range of applications. Starting as research tools in scientific and technical fields—as assistants in problem solving—they have come into routine use in all aspects of business, medicine, education, government—wherever data must be handled, stored, collated, retrieved, and so on. Applications include airline reservation systems, space vehicle guidance systems, medical diagnostic systems, machine-aided design, real-world modeling and simulation, total business operational systems, and individual pocket calculators. From small toys and games to giant national defense networks, computers of different sizes are at work. Their penetration in a bare quarter of a century has been extraordinary even in an extraordinarily technological age.

On the research frontier, work on computers and with computers has gone on unabated in many directions. A common feature of the research fields is that the computer researchers have long since lost interest in computation per se and instead have explored what computation can do in other realms—machine recognition and synthesis of speech, linguistics, artificial perception and manipulative systems, automatic machine translation, text handling (all the way from document creation to editing to final publication, including graphics capabilities)—in short, a broad range of heretofore human perceptual and cognitive tasks. Some scientific researchers are trying to make computers do what intelligent adults can do. Other researchers have continued to work on computer architecture, to design systems suited to tasks of ever-greater complexity.

One strand of computer research that has had a particularly long history at MIT is that of information storage and retrieval, as in libraries. That work began with Vannevar Bush in the mid-1930s, was revived by him immediately after World War II, and resulted in a project called INTREX. INTREX initially involved experiments in library automation, but these experiments touched on some deeper considerations, which Bush was beginning to worry about in 1945—the ways in which we mentally associate information, the ways in which we sometimes create startling mental collisions of old information and outlooks, and thus jump, by some inductive process, to entirely new ideas and new perceptions.

If scientists can unravel some of the mystery of how we humans squirrel away information and create new ideas, and if some of these capabilities can be incorporated in machines, then our uses of existing archives (which have long outgrown our capabilities of browsing in real time) may be made more effective. If computers, with their unique and peculiar capacity to store information, can help people follow specific trails through such information jungles, then man-machine interaction may indeed reach a new, fruitful level. It is this kind of hope that drives some of the most significant computer research in our time.

Chapter 21

The Computation Center

As early as the mid-1950s, MIT was facing something of a computational crisis. Computer use was already so heavy that Whirlwind (see chapters 15 and 17) could no longer serve the growing queues of customers. Though there were debates about whether or not computer research was appropriate in the Institute environment, there was no doubt that the scientific and technical community had come to a quick appreciation of the use and power of computers in solving special problems.

One of the people at MIT worrying about the computer overload was Professor Philip M. Morse of the Physics Department, who was then chairman of the Committee on Machine Methods of Computation. Morse, whom Compton had brought to MIT from Princeton in 1931, in part to help build a new bridge between the physicists and the electrical engineers, possessed a remarkable facility for bringing new things together and for getting new organizations going to solve pressing problems. In his memoirs, *In at the Beginnings,* Morse reflects on the state of the computer art as follows:

By 1955 the use of computers at MIT had outgrown the capacity of Whirlwind. Digital computation had won out over the older analogue computation, represented by the Bush differential analyzer. An electronic version of the differential analyzer had been built during the war, with vacuum tubes and wires replacing the gears and shafts of the original, but still having the disks and wheels. In the fifties it took up an inordinate amount of space and electric power. Although Whirlwind was overcrowded, almost no one wanted to use the much slower and less accurate differential analyzer, so it was scrapped.

Before the early fifties, nearly all digital computers had been designed and built in universities, with support from a government agency, but by 1955, commercial companies began to produce some. International Business Machines (IBM) was the biggest such firm; computer development was a natural extension of its mechanically operated, punched-card machines, although it required a completely new kind of engineer to design a digital computer. IBM began recruiting these engineers from the universities, the only place they could be found. They lured away several of the Whirlwind team; they also made arrangements to use the core memory developed by Forrester. Other experts preferred more independence than IBM offered; some of them started their own companies, and some of these became extremely successful.

Progress in programming was also made. Writing a program in machine language was a time-consuming piece of drudgery that had to be simplified. Each step in a calculation required a multiplicity of orders to the processor, each given in the proper order, all to achieve one small part of the calculation. . . .

The brighter programmers began to put together metaprograms called compilers, which would instruct the machine how to write its own machine-language program. The original commands, written out by the human programmer, could be couched in a form close to the usual mathematical formulations—sum this series, divide by the following product, and so on. These commands would involve many fewer instructions than the final, detailed machine-language program and would be less likely to be miswritten. These original instructions, written in an easily understood "compiler language," would be fed into the memory as though they were data to be used in some calculation. The compiler program would then act on these data to produce a program in machine language, all the multiplicity of details being generated with the slavish speed and accuracy that is the computer's

sole advantage over the human. The machine-constituted program in machine language could then be used to order the computer to perform the computation originally desired. Johnny von Neumann's idea of treating instructions like data was beginning to bear unforeseen fruit.

The argument over the desirability of compiler programs was intense and acrimonious. The hardware experts were horrified. It was inefficient, they protested; every computation would have to run twice through the machine, once using the compiler to prepare a program in machine language, then running this program to direct the machine to do the actual calculation. In addition, the compiler itself, the sequence of instructions to the machine to get it to translate directions in compiler language into a program in machine language, would have to be a program in machine language. And it would have to be the great-grandfather of all programs, because it would have to foresee all the different operations used in any sort of computation and would have to guard against all the logical errors that might occur. It would take dozens of man-years of the best programmers' time to write one, and, when written, it would be so big there would not be room for anything else in the machine's memory.

The software enthusiasts had answers to all these objections. They said that computers were built to perform mental drudgery. Although in the early fifties they were few enough and expensive enough that it still might be cheaper to have a person rather than a machine write a program in machine language, in ten years it would be the other way around. In the end, why should a man do what a machine can do faster and more accurately? In addition, many people who will want to use computers will not want to go through the grind of learning machine language; computers will simply have to have enough memory to ac-

commodate a compiler if they are to be useful to everyone. True, a compiler will be a tough job to put together, but, if it is done right, it needs to be done only once for each kind of machine.

The compiler proponents soon won out. In less than ten years all large computers had compilers written for them. Indeed, as time went on, different compilers were created, as people began to realize that the machine could carry out other tasks than purely numerical calculation. Machines could sort and compare, for instance, could look up a word in a stored dictionary and copy out the definition, could in the end carry out any sort of mental drudgery. It began to appear that the electronic digital computer could be to the white-collar worker what the gasoline engine had been for the blue-collar worker.[1]

By May 1954, less than a year after Whirlwind I became operational as MIT's first large-scale, reliable digital computer, Morse and his committee were anticipating the growing demand for computer use in the MIT community. In December 1954 and January 1955 Morse and Vice-President Stratton were weighing the advantages of keeping Whirlwind I in operation as against acquiring a commercial machine. The IBM 701 was available, and the 704, an advanced scientific computer with core memory, could be supplied to MIT early in 1957. In July 1955 Morse recommended the construction of a computation center to house the 704 computer, which IBM, after some prompting by Morse, was now offering to supply free of charge. In September the negotiations between IBM and MIT reached agreement, and plans for the Compton Laboratories were revised to include 18,000 square feet of floor space for the MIT Computation Center, with a special wing near the northwest corner to house the IBM 704 computer.

The MIT Computation Center was jointly announced on September 23, 1955, by President Killian and Thomas J. Watson, Jr., president of IBM. The dedica-

Philip M. Morse at the new MIT Computation Center, formed in 1955.

tion took place on Alumni Day, June 10, 1957, with Watson starting the "giant computer." The Compton Laboratories were dedicated the same day. By agreement with IBM the 704 was available free of charge to MIT seven hours a day, to 24 other New England colleges seven hours a day, and to IBM ten hours a day on the night shift. Morse was appointed director of the Computation Center and Dr. Frank M. Verzuh assistant director. Professor Dean N. Arden of the EE Department was in charge of programming research, and Dr. Fernando J. Corbató was in charge of the graduate student program. IBM also provided some financial support for research associates and assistants in computer research; the National Science Foundation provided support for computer research in the physical sciences, while the Rockefeller Foundation supported computer research in the social sciences. Thus, research in the use of computers proceeded across a broad front.

The research assistants in electrical engineering continued to be associated with a group in the Computer Components and Systems Laboratory (a research activity of the Electrical Engineering Department, and contractually administered by the Research Laboratory of Electronics). This group was an outgrowth of a teaching and research program started by Professor W. K. Linvill in 1953. His fourth-year elective—6.25, Machine-Aided Analysis—became popular; among his laboratory group was Dudley A. Buck, S.M. '52, Sc.D. '58, who in 1954 developed his "cryotron"—a superconductive, magnetically-controlled gating device. When Linvill went on leave from MIT in 1956, the undergraduate subject was taken over by Professor Arden of the Computation Center, and the research program was administered by Professor Ewan W. Fletcher. In June 1957 a Navy (Bureau of Ships) contract was activated to share with Lincoln Laboratory the funding of a program in the Computer Com-

Fernando J. Corbató at the Computation Center.

ponents and Systems Laboratory entitled High-Speed Computer System Research; this program sought to implement high-speed memories and logical elements through a basic understanding of materials at the molecular level. Professor Arthur L. Loeb led the effort in this laboratory, including a seminar on quantum mechanics attended by staff members and graduate students. An electron microscope and "clean-room" facilities were acquired, permitting an extraordinary degree of control of the geometry, purity, structure, and electric and magnetic properties of materials. Studies of thin-film magneto-optics were documented in the doctoral theses of Charles C. Robinson, S.B. '55, S.M. '57, Ph.D. '60, and F. Williams Sarles, Jr., S.M. '55, Sc.D. '61, and in the master's thesis of Joseph M. Ballantyne, S.M. '60, Ph.D. '64. The doctoral thesis of Dudley Buck, "Superconductive Electronic Components," supervised by von Hippel, carried his cryotron from its initial wire-wound model to the crossed-foil model.

In February 1961 the group in Building 10 was augmented by Professor James G. Gottling, S.M. '56, Sc.D. '60, and seven of his staff in thin-film research from the Electronic Systems Laboratory. Professor Robert E. Newnham from the Laboratory for Insulation Research moved to Building 10 in April 1961. In the period 1957–1960 this Building 10 laboratory contributed to two doctoral, nine master's, and seven bachelor's theses. Gottling's thin-film research continued until 1965, and Newnham's molecular science and engineering research until 1966.

Phasing Out Whirlwind
When in June 1957 the new Computation Center with its IBM 704 computer took over the computing from the Digital Computer Laboratory (Whirlwind I), full responsibility for Whirlwind I was transferred to Lincoln Laboratory, effective July 31, 1957. When Whirlwind I

finished its work in the air-defense program of Lincoln Laboratory, MIT recommended that the machine be scrapped. In spite of this recommendation William M. Wolf and some of his associates who had worked with the machine believed it could be successfully relocated and operated. By letter of June 30, 1959, the Office of Naval Research confirmed a "proposed lease" of Whirlwind I to the Wolf Research and Development Corporation, which was to assume responsibility for its custody and control.

In May 1960 Whirlwind I was moved to a South Boston warehouse and a new building was constructed for it in West Concord, where the computer was reassembled in late 1962. Albert V. Shortell, Jr., S.B. '50, S.M. '56, joined the Wolf Research and Development Corporation as vice-president and, with Richard F. Jenney, S.B. '52, S.M. '57, and others formerly associated with Project Whirlwind, got the machine into operation again in 1963. It was used in contract work for the Air Force and in the solution of scientific and business problems. It was dismantled in the spring of 1973, and in January 1976 the MITRE Corporation delivered an exhibit containing essential Whirlwind I components to the Smithsonian Institution in Washington, D.C. The Digital Equipment Corporation Computer Museum has on display many Whirlwind I components, including one of the two original coincident-current magnetic-core Whirlwind memories.

Early Programming
The problem of man's communication with the computer has enlarged as the computer art has expanded. Hardware—machines—and software—programming—have had a modifying influence upon each other and have enabled computers to accomplish tasks of almost unbelievable complexity. The earliest programmers had to write their programs in the 0's and 1's of the computer language. The Whirlwind I storage had registers of 16-bit words, and these

were of two kinds, *arithmetic binary numbers* and *orders*. The first bit of a number specified its sign (0 for +, 1 for −). The binary point was fixed between the first and second bits, the remaining 15 bits to the right of the point representing the magnitude of the number. The orders were stored in consecutive registers and extracted by the control of the computer in the same sequence. Thus, the arrangement of orders in storage corresponded to the program, or routine, though provision could be made in the program breaking into the sequence at any desired point. A word containing an order had the *operation* coded in the first 5 bits; the remaining 11 bits gave the *address,* or register location in storage.

Professor Maurice V. Wilkes of Cambridge University reports that the first comprehensive conversion routine for translating coded orders into machine language was developed by Charles W. Adams, S.B. '48, XVIII S.M. '49, and his Whirlwind group at MIT. As Whirlwind I became usefully operational in 1951, Adams assisted staff members and students in writing their own programs for problems to be run on the computer. He soon found himself in charge of a group called the Science and Engineering Computation Group, with John T. Gilmore, Jr., a graduate of Boston College, as his right-hand man. Forrester and Everett were by this time busy with the classified air-defense program, leaving Adams and his group with the academic applications of the computer. The interchange of ideas among the contemporary computer groups began to yield a programming methodology that was brought into focus by Maurice Wilkes, David Wheeler, and Stanley Gill, who spent time with the various computer projects and published a book, *Preparation of Programs for an Electronic Digital Computer,* in March 1951.

The Whirlwind programming capabilities of 1952 consisted of a "comprehensive system"—a library of subroutines and a postmortem routine in addition to conversion routines. The Summer Session Computer, an imaginary machine having characteristics different from those of any existing machine, was devised by Adams to teach programming principles to a group of 105 students, mostly from industry, in an intensive two-week course in the summer of 1953. Wrote Adams, "It is intended entirely for classroom use. For this, it should be easy to learn, simple yet versatile to use, and sufficiently typical of available computers to be an effective example without being enough like any one to be offensively specialized." In each of the three years in which Adams taught such a course, he was assisted by one of the scientists from the Cambridge University computer group.

In 1952 and 1953 J. Halcombe Laning, Jr., X S.B. '40, XVIII Ph.D. '47, and Neal Zierler, both of the Instrumentation Laboratory, developed on Whirlwind I an algebraic system of coding by which an algebraic problem could be programmed for the computer. Jean Sammet, in her book *Programming Languages,* describes this as an "impressive" system and the "most significant of all the early work."[2] It appears to have been the first American system to write a program in a natural algebraic language, earlier work being that of Heinz Rutishauser at the Eidgenössische Technische Hochschule in Zürich, Switzerland.

FORTRAN and Other Computer Languages
Since the early Whirlwind I days the concept of automatic programming has evolved in the direction of making the machine itself do much of the detail work. Source languages have been developed, with strict rules for writing a program. In such a source language the user can tell the computer *what* he wants done without a complete specification as to *how* the machine carries out his command. FORTRAN (originally Formula Translation) is perhaps the most widely used system for scientific and engineering problems.

Many high-level languages have been developed. The Whirlwind engineers continued to design computer languages especially useful in the Cape Cod and SAGE systems during the 1950s, demonstrating a very early kind of time sharing and the rudiments of man-machine communication. Of those early efforts, Norman Taylor wrote years later,

Whirlwind I programmers were asked to develop software which was powerful enough to handle airplane tracking in real time, run displays so that operations personnel could assist aircraft to carry out the interceptor or bomber mission, calculate closing vectors between aircraft in real time, and handle the myriad of internal computer housekeeping and filekeeping chores associated with the air-defense mission.

Thus, this initial software effort gave birth to the true air-defense software system used later in SAGE. But more basic than these contributions to SAGE was pioneering understanding of the technical difficulties of time-sharing a small core memory with multiple programs rolled in and out of a supporting larger drum system. This was the early beginning of time sharing, relative addressing, operating systems, and the executive real-time monitors now common in most on-line real-time systems.

Not only were these concepts new, but the need for real-time software in the true aircraft-movement sense made the work doubly demanding, since the proper software had to be operated in the proper part of core in synchronism with a real-time clock pacing aircraft as they moved.

Further, the software not only shared core on a real-time clock basis, but the same software was used repeatedly in sequence to control many aircraft within a few seconds, with the attendant problems of keeping old tracks and new data properly related to each other.

Under Bob Wieser's directions, Dave Israel, Jack Arnow, Herb Benington, Charles Zraket, and many others shared in these early developments. To the software world, these activities were as basic as the core memory and the Whirlwind I computer were to the hardware world.

When these concepts were later expanded to the full FSQ-7 SAGE computer in the late 1950s, it became clear that the manual tasks of core register assignments, opening and closing interactive routines, calling programs, interpreting sequences, became a mountainous undertaking . . . and thus began the basic thinking of using the computer itself to assign its own addresses for core register assignments (now known as assemblers) and later for the automatic collection and chaining of program subroutines. . . . [These basic ideas] later developed into concepts of compilers and interpreters.

Coincident with these operational software problems, Whirlwind became the testing ground for a group of people interested in software diagnostic programs to help an operator to detect and diagnose trouble first in hardware malfunction and later in software ambiguities. This work formed the basis of building real-time systems reliable enough for military use, and capable of self-diagnosis and self-switching to an alternate mode whenever reliability was in question.[3]

The APT (Automatically Programmed Tools) and AED (Automated Engineering Design) systems of Douglas T. Ross (see chapter 14) were related to the numerical control of machine tools. Examples of other early computer languages that originated at MIT are COMIT, DYNAMO, and LISP. COMIT dealt with symbol manipulation (including algebraic differentiation), the searching of lists for specific information, theorem proving, and other types of problems. The June 1959 Computation Center report stated that the COMIT compiler occupied approximately 7,000 registers, and its interpreter approximately 5,000. This language was easy to learn and to use, and it contributed important ideas to LISP and other languages, leading eventually to Bell Telephone Laboratories' SNOBOL in the 1960s.

In Jay W. Forrester's book *Industrial Dynamics* (1961), the growth and stability of economic and industrial systems were studied by means of models comprising as many as 300 to 600 variables and first-order difference equations. Noted the author, "The character of the simulation equations directly suggested the usefulness of a compiler which could read the equations and translate them into computer coding." A compiler-generator, called DYNAMO, was written in 1958 by Phyllis Fox and others in the School of Industrial Management, including Richard K. Bennett, VI-A S.B. '50, S.M. '51, E.E. '52, and Alexander L. Pugh III, S.M. '53, E.E. '59.

Professor John McCarthy reported in the December 1958 Computer Center report, "During the past three months, the [artificial-intelligence] project has developed a programming language called LISP (List Processor) for manipulating symbolic expressions and has coded and debugged the major subroutines." The original purpose of this language was to study a list of formal declarative and imperative sentences in such a way as to make deductions from them. Developed initially for the IBM 704 computer, LISP was implemented on later machines such as the IBM 709, 7090, 7094, and System 360. Its use has gone far beyond its initial purpose and includes algebraic manipulations and theorem proving. For a description of the development of computer programming, one might refer to the book *Programming Systems and Languages* by Saul Rosen of Purdue University.[4] It contains original articles by several important contributors to the field, including Fernando Corbató and Jack B. Dennis of MIT.

On July 3, 1960, the IBM 704 was shut down in preparation for its replacement by the 709, which was installed in August. In its three years of operation the 704 had been used in the solution of more than a thousand problems, most of them reported to the public in journal articles, society-meeting papers, and theses. Out of 274 theses with major contributions from the Computation Center, 47 were in electrical engineering. The new IBM 709 took care of computer demand in late 1960, but by the beginning of 1961 the capacity of the Computation Center was again saturated. Professor Morse observed, "Perhaps the logical end is only reached when there is a time-sharing console in every classroom and every laboratory." That would very nearly happen, and then a younger generation of computerists would claim, "The logical end is only reached when everyone has his or her personal computer."

Chapter 22

The Evolution of Time Sharing

"Time sharing has been proved feasible." This was the conclusion of a January 1960 report by Herbert Teager of MIT's Computation Center, based on experimental work following the suggestion of John McCarthy a year earlier. One of the founders of the field of artificial intelligence, McCarthy, who came to MIT in 1956 as a guest of the Computation Center on leave from Dartmouth College and in 1958 became an assistant professor in the MIT Electrical Engineering Department, stirred up much early enthusiasm about computing and was a staunch advocate of time sharing. It was his suggestion that led to the beginnings of a system that would allow many users simultaneous access through separate ports of entry to a large computer—in other words, the seed of a computational utility.

In 1960, too, another scientist who would have much to do with MIT's fortunes in developing time sharing, J. C. R. Licklider, published what would prove a seminal paper, "Man-Machine Symbiosis," in the *Institute of Radio Engineers Transactions on Human Factors in Electronics*. In it he heralded a larger role for computer systems beyond their number-crunching capabilities. He saw, as did others, that computing machines were an entirely new kind of entity that would lead to dramatic changes in the ways in which people interacted with machines and solved problems. His paper embodied concepts that a generation of young computer scientists would take to heart and begin to implement in a more or less programmatic fashion.

Time sharing at MIT was actually first implemented when IBM, in December 1958, supplied a "real-time package' to the 704 in the Computation Center, which facilitated communication between the programmer and the machine. John McCarthy wrote to Morse shortly thereafter, on January 1, 1959:

This memorandum is based on the assumption that MIT will be given a transistorized IBM 709 about July 1960. I want to propose an operating system for it that will substantially reduce the time required to solve a problem on the machine. . . . Suppose that the programmer has a keyboard at the computer and is equipped with a substantial improvement on the TX-0 interrogation and intervention program. (The improvements are in the direction of expressing input and output in a good programming language.) Then he can try his program, interrogate individual pieces of data or program to find an error, make a change in the source language, and try again. If he can write his program in source language directly into the computer and have it checked as he writes it, he can save additional time. The ability to check out a program immediately after writing it saves still more time by using the fresh memory of the programmer. I think a factor of five can be gained in the speed of getting programs written and working over present practice if the above-mentioned facilities are provided. . . . The only way quick response can be provided at a bearable cost is by time sharing. That is, the computer must attend to other customers while one customer is reacting to some output.

At midyear 1958–59, Herbert M. Teager, having returned from naval service, was made an assistant professor of electrical engineering and placed in charge of the time-sharing development of the Computation Center. Early in 1959 he and his group, including Marjorie L. Merwin and Joseph R. Steinberg, were testing out a new technique in which a Flexowriter was "connected to the 704 in-out package as a debugging control center for isolating and removing errors from programs. . . . The time the computer usually sits idle during such on-line debugging will be

John McCarthy, an early advocate of time sharing at MIT, was also one of the founders of the field of artificial intelligence.

used instead by a host program under a unique *time-and-machine-sharing* system." In the January 1960 Computation Center report, Teager wrote: "Each user would, of course, like to have the full capacity of a very large system at his immediate beck and call. This is clearly impossible . . . unless many users can use the machine at virtually the same time, i.e., the machine is time-shared. . . . As of today, an electric typewriter has been connected to the 704, and a programming system written that allows the 704 to be time-shared. . . . Time sharing has been proved feasible."

A pilot survey of computer needs for the new Center of Communications Sciences (established in 1957–58) indicated that these needs ought to be embedded in the context of the larger computer policy of MIT. In view of the growing demand for computer facilities throughout the Institute and the saturation of the Computation Center, President Stratton appointed, in late spring 1960, a steering committee with Professor Hill as chairman, the other members being Professors Fano, Morse, and Wiesner. By June the committee had set up a working group; its charter was to make a complete systems study and to recommend a computer policy that could meet and advance MIT's research and teaching objectives. Intensive work by this group (later called the Study Group) during the summer of 1960 yielded a survey of MIT computer facilities and current use by the campus research groups. A review of computer manufacturers showed that IBM's STRETCH computer was the most advanced machine of 1960. An alternative to the purchase or renting of a commercially built machine was the design and construction of an advanced MIT machine.

A proposal was worked out by McCarthy and Marvin Minsky, and other advanced ideas were proposed by Douglas Ross, Jack Dennis, and Daniel Goldenberg. By fall the members of the Study Group were convinced that a new MIT-designed computer could not be produced in time to meet the needs of the MIT community and that the cost would be very high. Perhaps a STRETCH-like machine could be made to contain the advanced and desirable qualities suggested by the Study Group. Teager assembled these surveys and proposals into the January 1961 report, entitled "MIT Computation—Present and Future." The report mentioned the many students involved in the study, especially Allen L. Scherr, VI-A S.B. and S.M. '62, Ph.D. '65, whose doctoral thesis, "An Analysis of Time-Shared Computer Systems," was later supervised by Teager.

In the July 1961 Computation Center report Dr. Corbató and his staff announced, "a time-sharing system has been designed and is being debugged. This system makes use of the two Flexowriters and the clocks that have been attached, under the direction of H. M. Teager, to the Direct Data Connection of the IBM 709 on Channel D. Outwardly, the computer may have three users; one user is the background FMS (Fortran Monitor System); the other two users are those at the Flexowriters. The Flexo users will direct the operation of the computer by typing in commands." Some editing could be done in the user's symbolic language.

Corbató described this system in July 1961 in a paper he read at the University Directors' Conference in Endicott, New York. It was first demonstrated at the MIT Computation Center in November 1961, just before the 709 was shut down. The transistor model 7090 was installed in the Computation Center at the beginning of 1962, the 709 having been purchased by MIT and moved to the fourth floor of Building 10 to form the Cooperative Computing Laboratory under the direction of Professor Michael P. Barnett of the Physics Department.

At the suggestion of MIT, IBM modified its design of the 7090 for time sharing, so that the machine delivered to MIT had these modifications, including program relocation and memory protection features, built into it. The Compatible Time-Sharing System (CTSS), which evolved from the experimental time-sharing system for the IBM 709, was developed largely on the 7090 by Corbató and his group and described in a paper presented at the San Francisco Spring Joint Computer Conference in May 1962. By July 1962 Teager and his staff had modified the earlier data rack to provide the 7090 with a link for the connection of three foreground Flexowriters. Users in the foreground system could develop programs in languages compatible with the background system.

In the summer of 1962 Corbató became associate professor of electrical engineering and deputy director of the Computation Center. He led the development of the CTSS over the next year. A second bank of core memory (32,768 words), an IBM 1301 disc file (9 million words), and an IBM 7750 communication channel were added to the 7090 system, enabling a considerable increase in the number of remote consoles.

Making use of their computer experiences at MIT, Kenneth H. Olsen, his brother Stanley, and Harlan E. Anderson organized the Digital Equipment Corporation in August 1957 to manufacture computer modules similar to those designed for the TX-0 and TX-2 computers at Lincoln Laboratory. By 1961 they had evolved the idea of manufacturing a reliable, simple, easy-to-operate, short-word-length computer similar to the TX-0, but with a new transistor that Philco had just produced. The first of the now famous PDP line of computers was soon on the market, and in September 1961 a PDP-1, serial number 3, was presented to the MIT Electrical Engineering Department as a gift from the Digital Equipment Corporation.

Professor Jack B. Dennis, VI-A S.B. '53, S.M. '54, Sc.D. '58, had been given charge of the TX-0 facility in the summer of 1959 upon the resignation of Pugh; in 1961 he wrote a proposal for time sharing on the TX-0. With the arrival of the new PDP-1, Dennis set one of his graduate students, John E. Yates, VI S.B. '60, S.M. '62, to work on a thesis entitled "A Time-Sharing System for the PDP-1 Computer." The Yates design was for a seven-console system, but at the time of the thesis only two consoles were under construction. The PDP-1 time-sharing system became operational in the spring of 1963 and eventually had five consoles plus the background system. Dennis was very active in the use and modification of both the TX-0 and the PDP-1.

Another PDP-1 time-sharing system was developed in 1962 at the plant of Bolt, Beranek, and Newman (BB&N) in Cambridge. The principals were J. C. R. Licklider (then vice-president of BB&N), John McCarthy, Edward Fredkin, and Sheldon Boilen, who presented a paper, "A Time-Sharing Debugging System for a Small Computer," at the Spring Joint Computer Conference of 1963.

The MIT Computation Center staff continued to develop the Compatible Time-Sharing System. A version of CTSS was demonstrated and described by Corbató in a half-hour television program on May 17, 1963, over Station WGBH in Boston. The first system to allow 21 typewriter users, CTSS Model 13, began operation in June 1963.

Project MAC

Although the 21-terminal pioneering CTSS in the MIT Computation Center helped to verify the feasibility of time sharing, the system could not evolve in that context to a really practical scale. Basically, the Computation Center was set up to provide traditional batch-

Kenneth H. Olsen of the Digital Equipment Corporation presents MIT with a PDP-1 computer, September 1961. *Left to right*: Olsen, Charles Townes, Peter Elias.

Robert M. Fano and Marvin L. Minsky (*seated*), examining a graphic display produced by the PDP-1.

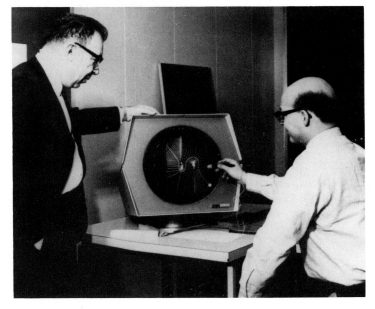

processing computational services, so that CTSS could only be used part-time. However, plans for the next big step—the building of a center dedicated to time sharing—had already crystallized in late 1962.

In 1962 J. C. R. Licklider, who had been instrumental in the PDP-1 time sharing at BB&N, was appointed by the Advanced Research Projects Agency of the Department of Defense to head up the offices of Information Processing Research and Behavioral Sciences. During his stay in Washington Licklider's conviction that the time was ripe for a massive attack on time sharing was reinforced and spurred him to look for a suitable institution and the right person to direct the task. It was Licklider's judgment that MIT was one of the best places to set up such a center, but what he needed was a scientist or engineer of stature with broad experience, willing to make a major commitment to administering and seeing through this serious enterprise. In the fall of 1962, at the First Congress on the Information System Sciences in Hot Springs, Virginia, Licklider found his person, Professor Robert Fano of MIT, a former colleague of Licklider's at RLE. In the course of a long discussion with Licklider, Fano suggested that Philip Morse, head of the MIT Computation Center, would be the "natural person"; but Morse was deeply involved in many other projects. A number of others who were already extremely competent in computational work were still very young, deeply committed to continuing research, or not temperamentally disposed to taking on a burdensome administrative role. By this time Fano himself had learned enough about computation to feel strongly that MIT should be involved in time-sharing research. Since no one else seemed ready to shoulder the responsibility for this next crucial stage of computer evolution, Fano decided he would take it on.

Fano (*foreground*) headed Project MAC, the first large-scale time-shared computer system. In the background, writing, is John Tucker, head of the Electrical Engineering and Computer Science Department's Cooperative Program.

Fano brought together such experienced people as Corbató, Dennis, Teager, Ross, Minsky, and Martin Greenberger to prepare a formal proposal to Licklider. They described the goals of the research and development program as follows:

The broad, long-term objective of the program is the evolutionary development of a computer system easily and independently accessible to a large number of people and truly flexible and responsive to individual needs. An essential part of this objective is the development of improved input, output, and display equipment, of programming aids, of public files and subroutines, and of the overall operational organization of the system. A second, concomitant objective is the fuller exploitation of computers as aids to research and education, through the promotion of closer man-machine interaction. The second objective is not only important by itself, but is also essential to the development of the computer system envisioned above, and vice versa. The third objective, which must be part of any university activity, is the long-range development of national manpower assets through education in the pertinent area: of faculty as well as of students, and outside MIT as well as within the confines of the campus. Again, this third objective is inextricably interwoven with the preceding two, because people's approach to problems will have to evolve in parallel with the computer hardware and software.

That was the beginning of Project MAC. (The acronym stood for two things—Machine-Aided Cognition, reflecting the broader research aims, and Multiple-Access Computer, the actual system.) In the world of computerists MAC was a landmark development. It proved, in the words of Fernando Corbató, "a really monumental accomplishment, the first coming of age of computer science as an organized group at MIT."[1]

Whatever others thought of the project at the time, Fano's intention from the beginning was that it would survive as a permanent MIT academic laboratory,

staffed primarily by faculty and students. He also saw it as a way of pulling together people with various computer interests who were at that time scattered around the Institute in various laboratories—RLE, the Electronic Systems Laboratory, the Communications Center, and others. Thus, one of the many tasks of Project MAC was to devise a format for the participation of these interested individuals without their abandoning existing relationships. This is one of the major reasons why MAC was called a "project."

The policy committee for Project MAC itself represented a broad range of interests at MIT. It included Professor Philip M. Morse, director of the Computation Center, as chairman; Dean Gordon S. Brown, School of Engineering; Professor Peter Elias, head of the Department of Electrical Engineering; Dean George R. Harrison, School of Science; Professor Albert G. Hill, Physics Department; Dean Howard W. Johnson, School of Industrial Management; Professor Carl F. Overhage, director of Lincoln Laboratory; and Professor Fano ex officio as director of MAC.

The initial MAC system, which was a copy of the CTSS at the Computation Center, was operational by November 1963. Within six months it was serving 200 users in ten different academic departments, who had access to 100 teletypewriter terminals, connected to MAC via the MIT telephone system. Initially, MAC could serve as many as 24 users simultaneously. Although time sharing was at the center of the MAC program, the project soon became a general laboratory for dealing with many types of computer programs, as, for example, Marvin Minsky's artificial-intelligence programs and Jack Dennis's programs for the study of the structure of computing systems.

The development of MAC and its follow-up system has been well documented; see, for instance, the succinct account by Fano in the *Encyclopedia of*

Marvin Minsky, 1968, with a product of his artificial-intelligence research: a system composed of a PDP-6 computer (not shown), a television camera, and a mechanical arm, capable of recognizing blocks of various sizes and shapes and assembling them into structures without step-by-step instructions from an operator.

Jacob Katzenelson, VI Sc.D. 1962, with the Kludge, an important contribution to the field of interactive computer graphics.

Computer Science and Technology, which points to many of the major references and milestones.[2] All through the 1960s the MAC system was steadily increasing in its functional capabilities and in its variety of users and research applications, and it was to become an important node on the ARPAnet (Advanced Research Projects Agency network), which connected many computer centers together in the United States and abroad. Among the significant advances was the development of computer graphics and of computer-aided design. For instance, a doctoral thesis that generated much interest at the time was "Sketchpad—A Man-Machine Graphical Communication System," by Ivan E. Sutherland, Ph.D. '63, supervised by Claude E. Shannon. Contemporary with Sketchpad was the Automated Engineering Design (AED) work of the Electronic Systems Laboratory under Douglas T. Ross.

The Sketchpad programs were implemented on the Lincoln Laboratory TX-2 computer and an X-Y cathode-ray tube (CRT) display that required a separate TX-2 output instruction to plot each of the hundreds of points comprising a displayed picture. Under the Air Force AED project in the Electronic Systems Laboratory, John E. Ward and Douglas Ross sought ways to reduce this heavy, repetitive computational load of real-time graphics display. Project MAC shared the cost of developing the new display system, nicknamed Kludge.

An innovative feature of the display was the use of incremental digital interpretation for line generation, a technique employed in the earlier numerical machine-tool control system developed in the Electronic Systems Laboratory. Through the use of this technique, display speed was enhanced by a factor of 15 and computational demands were reduced by a factor in excess of 100. Digital Equipment Corporation adopted the technique for its commercial display terminal, the DEC 340.

Other real-time graphic manipulations with minimal computer load were incorporated in the Kludge by adding a three-dimensional hardware picture rotation matrix multiplier, controlled by a "globe" input device. In addition, a complement of switches, buttons, and knobs, a keyboard, and a light pen provided means for user interaction in the creation and modification of graphic displays.

The Kludge was perhaps the first graphic display capable of being operated on a time-shared computer, the MIT CTSS, starting in 1963. Over a period of eight years the Kludge was used by a large number of other researchers from many academic departments at MIT. To cite a few, Professor Steven A. Coons (Mechanical Engineering) developed a surface-patch technique that was used by the Naval Architecture Department in designing ship hulls; Professor Cyrus Levinthal (Biology) experimented with three-dimensional models of molecular structures; Professor Kenneth Stevens (EE) and his group developed vocal tract displays for speech analysis; and Professor Michael Dertouzos (EE) used the display in his development of electronic circuit-design techniques.

By 1965 the MAC installation was structured as shown in the accompanying sketch. Work on the follow-up system to MAC began in 1964, culminating in the development of the so-called Multics (Multiplexed Information and Computing Service) system, which was done with Bell Telephone Laboratories and the Computer Department of the General Electric Company as collaborators. The Multics system, which took five years to complete, was to become a commercial product of Honeywell Information Systems. Multics became available to the MIT research user community in 1969; by 1971 it was serving 500 people, could service 55 simultaneous users, and was operating 24 hours a day, 7 days a week.

The emergence of Multics was to lead to a reorientation of Project MAC. Professor Fano describes it thus:

The responsibility for operation and maintenance of the Multics System was turned over to the MIT Information Processing Center in October 1969. That date, seven years after Project MAC was conceived, marks the end of its initial phase with time sharing as its primary focus. By that time, on-line use of computers was widespread and time-sharing service was being offered commercially by several companies. Also, many time-sharing systems had been developed elsewhere to fit a variety of needs and which spanned a broad range of sizes and capabilities. The original goal of Project MAC had been largely reached, and time sharing was no longer a priority research topic.

An important consequence of this change of interests and priorities in Project MAC was that its research program became more narrowly focused on computer science and engineering topics, with faculty and students being drawn largely from the Electrical Engineering and Computer Science Department. Thus the character of Project MAC changed rapidly from that of a research effort involving people from a large number of MIT academic departments and laboratories to that of a discipline-oriented research laboratory of a traditional type. The name was correspondingly changed, first to MAC Laboratory and later, in 1975, to Laboratory for Computer Science.[3]

During his tenure as director, Professor Fano had a number of times recommended a new name for Project MAC and had pushed to make the laboratory and computer science more of an integral part of the academic structure. It was actually not until February 2, 1976, that the facility officially became the Laboratory for Computer Science. Fano had served as director for Project MAC until September 1968, succeeded by J. C. R. Licklider, who served until 1971; he was

Photo taken at the Technical University of Berlin during an MIT/TUB computer conference, July 1968. The setup demonstrated the ability to engage a time-sharing computer at MIT from terminals several thousand miles away in Berlin. The blackboard diagram shows the two transatlantic cable routes used; in the center is a graphics terminal; Francis Reintjes works at the hard-copy terminal.

Structure of the MAC computer installation at the end of 1965.

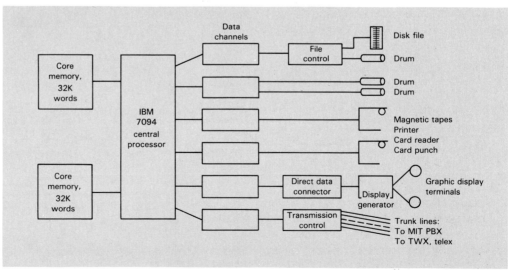

followed by Edward Fredkin, who served until 1974. Since that time Michael Dertouzos has been its director.

Computer Technology and the MIT Educational Program

The full-scale development of time sharing at MIT and the increasing involvement of faculty and students in computer research have exercised a profound effect on the MIT educational program, not only in the Electrical Engineering Department, but across all departments and interests. These recent changes have deep roots in the past. The EE Department's classes in machine computation began in the spring of 1935, when Professor Samuel H. Caldwell first offered his course 6.60, Mathematical Analysis by Mechanical Methods, a two-hour-per-week lecture subject based on problem solving with the six-integrator differential analyzer (see chapter 4). By 1947 Caldwell's courses were already offering studies of digital machines.

With the development of Whirlwind, Jay Forrester began urging the MIT Mathematics Department to expand the field of digital computation, but it was the Electrical Engineering Department that, in the fall of 1950, actually set up an enlarged program in automatic computation and numerical analysis. The subsequent progressive changes in the curriculum involved many teachers, including Caldwell, Francis B. Hildebrand of the Math Department, and William Linvill of the EE Department. In the spring of 1959 John McCarthy had 34 freshmen in his new seminar, Introduction to Automatic Computation; in 1960 the number of students in the regular course version of this subject had risen to 150. Thereafter, with the evolution of the on-line computer research community via Project MAC and its successor Multics, computer science would climb to a new plateau at MIT. The impact on the Department of Electrical Engineering is sketched in our concluding chapter.

Michael L. Dertouzos, director of the Laboratory for Computer Science, which evolved out of Project MAC.

Chapter 23

The Department of Electrical Engineering and Computer Science: Seeking a New Paradigm in Engineering

In the late 1870s, when he became director of the Department of Physics, Professor Charles Cross, the only faculty member, along with two assistants, taught all the electrical material then offered, and established within Physics the courses that would grow into the Electrical Engineering Department. A century later, out of that small nucleus, there had grown an educational program in electrical engineering and computer science involving about 110 full-time faculty members and 1,700 students. Cross's research projects in acoustics and telephony have sent out a hundred branches. How to embrace its diversity of disciplines in a coherent and manageable program has been something of a preoccupation of the department's staff during the last two decades. The emergence of computer science, which resulted in the renaming of the department, has tended to provide a dramatic focus for departmental issues, but there are other underlying factors that are equally, if not more, important.

For one thing, the growing role in modern society of engineers, who are bringing changes, for better or worse, and who are dealing with the design of ever more complex technological-social systems, is raising questions about the breadth of their education. Writing about this issue in 1973, Gordon Brown contended that engineers are really "revolutionists." "More than any other profession," he observed, "they are the primary instigators of social, economic, and political change in the society in which they live."

The realization of this revolutionary impact has come home with ever-greater force during the course of this century we have been examining, and markedly so in the last two decades. Somewhat as a consequence, in contrast to the ''we can do anything'' spirit of the mid-1950s, engineering educators have begun asking whether the time has come for a new unifying paradigm—some kind of grand synthesis—for dealing with the new level of complexity that engineering has brought into man's personal and social life. One way of seeing the issue comes to us through computer science, or, more generally, through all the sciences now involved in information processing.

Some Characteristics of Computer Science

In the decade of the 1970s the rapid growth of computer science posed fundamental questions for the Electrical Engineering Department. It was not merely a matter of adapting to computers as new tools, it was a matter of beginning to grasp their potential ramifications, first in the academic research-and-educational context and then in society at large. It was already manifest to many practitioners in the computational arts that these were tools of a different magnitude, with a potential human impact perhaps not less profound than that of atomic energy. Thus, for researchers and educators, far more than curriculum changes were at issue. In fact, as the information-processing field evolved at MIT, there emerged a strong separatist movement within the EE Department that aimed at setting up computer science as an entity independent of electrical engineering.

In order to understand how and why that movement evolved, it is important to examine some of the general characteristics of computational machines and to contrast their development with other, ''traditional'' fields. When the first computers were invented, as we have seen, it took engineers, physicists, and mathematicians to design and build them. They needed to understand the laws of physics and electronics, the behavior of materials, the organization of circuits, and so on; and they had to become versed in logic, that branch of mathematical philosophy which has seen a resurgence in this century. All of this was science and engineering in the Western experimental tradition.

The physical growth and sophistication of computers continued to depend upon the application of such science and engineering in progressive waves, as vacuum tubes gave way to transistors and magnetic cores, which gave way to small semiconductor chips, ''engraved'' with ever more numerous and richly distributed logic circuits. Finally, entire computers are constructed on silicon chips so tiny that their contents can only be seen through a microscope. This is science and engineering at the atomic level.

Together with this scaling down in hardware and scaling up in theoretical understanding, there has been a steady growth in software—in the computer languages with which the machines have been programmed. There has almost always been some relationship between the specific kind of hardware and the software, though in recent times there has arisen the perception that the software extends far beyond the machine per se and that, necessarily, it must take into account human habits, methods, and cultural mores as well as institutional processes. Thus, in the dialogue between people and information-processing systems, each party causes evolution and change in the other. To facilitate this ongoing dialogue, there has been a tendency in computer science to design languages that are more and more natural. Accordingly, many linguists and philosophers have been drawn into computer research; for in working with artificial languages they can pose ever more interesting questions about the character of our natural languages. Indeed, some computerists assert that the whole future of psychology and philosophy, as well as engineering, is encompassed in the computer field. In their evolving forms, it is argued, computers can serve as the most eloquent paradigms and mirrors of all human cognitive activities.

Computer science has shared some of the characteristics of the established physical sciences in their earlier formative periods, when a large part of the work was gathering bits of information, building up taxonomies, seeking more inclusive theories and laws, and building them on a rigorous mathematical footing. The electrical sciences, for instance, began with the exploration of seemingly different electrical phenomena—the static electricity that would arise from the rubbing together of dissimilar materials; the chemical electricity discovered in the first batteries made by Volta in 1800; the animal electricity manifested in the twitching of frogs' legs induced by Galvani in 1780; the phenomenon of lightning, which would lead to Ben Franklin's kite experiment. These were initially viewed as independent phenomena, not as instances of something more universal. The nineteenth century was especially dedicated to experiment, to the accumulation of great quantities of detail, and, finally, as in James Clerk Maxwell's work, to the unfolding of universal laws that applied to all electrical phenomena.

In other respects, however, computer science bears more resemblance to the so-called life sciences—biology and its subdisciplines—than it does to the physical sciences. Like the life sciences, computer science has been rife with submodels, and it has been difficult to arrive at good foundational ideas or concepts that provide a broadly unifying outlook. Some of the impacts of the attempts to deal with computer science in engineering education and research, especially in the Electrical Engineering and Computer Science Department, are summarized in the statements by current department leaders that appear at the conclusion of this chapter.

Restructuring a Department—Inward and Outward Pressures

Given the kinds of problems and issues that computer science has brought to the fore in the past two decades, it is not surprising, in retrospect, to see how long the separatist debate within the department went on. Being neither a simple nor a single-issue debate, but involving many different perceptions of an emerging science, it spanned three successive administrations of the EE Department—those of Peter Elias (1960–1966), Louis D. Smullin (1966–1974), and Wilbur B. Davenport, Jr. (1974–1978). It was during Davenport's tenure as department head that the name was officially changed to the Department of Electrical Engineering and Computer Science, signaling a consensus that a deepening integration of the diverse disciplines that had emerged in electrical engineering would prove the healthier and more vigorous course of action.

The great diversity of electrical engineering has been both a strength and a weakness. As we noted at the very beginning of this volume, electrical engineering has been continually expansive in what it was willing to take on, continually enlarging its identity, as well as attracting different types of students. Thus, for instance, the following fields have all become a "normal" part of EE at MIT: power, illumination, electronics, communications, servomechanisms, microwaves, electromagnetic systems, electrical materials, biomedical systems, psychoacoustics, information theory, plasma fusion and solar energy, control theory, semiconductor technology, computational systems, linguistics, and neurophysiology. Electrical engineers work at problems in all of these and many other areas. Computer science, with its peculiar and unique characteristics, has added greatly to this diversity.

A serious educational dilemma about diversity is that individuals, besides displaying different skills, talents, and dispositions, frequently have fundamentally different ways of perceiving objects, fields, people, the world; that is, they likely have different ways of mentally modeling the world in which they operate. In an educational department or school with more than 1,000 students and more than 100 faculty, and with very many specialized disciplines, profound questions must be confronted daily about the most effective ways to educate. The size and diversity of the EE Department in the 1970s—larger than several other schools combined, and larger than all MIT in the 1890s—gave rise to forces for fragmentation.

On the other hand, this large staff and student body also brought important benefits in terms of "critical mass." With more students, more sections in a given subject, more teachers, there was the enhanced stimulation of different viewpoints, always vital to the educational process. A department, for instance, with a sophomore class of 200 students could embark on projects a department of 50 students would seldom even consider.

Added to these positive and negative factors were the effects of the nationwide economic troubles of the 1970s, exacerbated by the energy crisis, and leading to tightened departmental budgets. At the same time a countervailing force—the promotion of research, which is so important to technological education—called for an optimistic and expansive allocation of new resources rather than retrenchment. Thus, the period was one in which pressures were on the rise.

In an earlier chapter we alluded to the divided feelings people had about computer work—the sense that Whirlwind might be getting out of hand in terms of costs, some "cultural" feeling in the immediate postwar RLE period that computers did not constitute "proper academic research," growing apprehensions about whether this field, especially in its software aspects, embodied sufficient fundamentals to be re-

Peter Elias. Louis Smullin.

garded as science, and a disposition (considering the enormous costs of computer development) to leave the hardware aspects to industry. At the same time the electrical engineering department, which for all the early decades of its growth had been principally engineering-oriented, had been steered through Gordon Brown's efforts in the postwar era toward a science-based orientation. That is, the engineers—those who liked to design artifacts, to build them and make them work, without necessarily acquiring a profound grasp of theoretical principles—had become in some ways, albeit unintentionally, second-class citizens. As Davenport says in retrospect, "Gordon kicked the pendulum over to the applied science side and it latched there." Davenport, in his tenure as department head, attempted to redress this imbalance by bringing up more people for tenure who were closer in disposition "to the real world of engineering." However, at the time the computer courses were being initiated, the science orientation still held sway.

When he became department head in 1960, Peter Elias saw computer science as residing largely in the research stage, as in Project MAC, with undergraduate courses just beginning to evolve. It was clear that there was an immense demand for computational services, but as yet no clear indication as to the direction the satisfaction of that demand should take, or how it would affect the EE curriculum. It was, in a sense, the beginning of a new transitional phase.

The EE curriculum was fairly well established in terms of having incorporated the World War II experience, which had now been structured for the undergraduate courses, but there was still no adequate treatment of the science of electrical materials, which had assumed a new importance, especially with the development of semiconductor devices. Though this solid-state technology was moving toward integrated circuits in semiconductor materials, there was still no indication of how fast this new field would grow.

In the mid-1950s the department was awarding a handful of doctorates a year; within the next 20 years the number of graduate students jumped enormously and the department was producing about 50 doctorates a year. In 1981 the faculty was having to deal with over 500 graduate students, as well as 1,000 undergraduates. During Elias's tenure, in the early 1960s, changes were occurring in the curriculum—tutorials in small groups, take-home laboratory kits in electronics, more emphasis on the honors students—and, except for some dissatisfaction over iterations among the computer science courses, things were continuing at a lively pace in all the EE and CS disciplines. There was still no formal curriculum in computer science, though there were a number of trial courses being offered at the undergraduate level, and the graduate courses that had been started by Caldwell years earlier had become formalized. But overall, because computer science was in such an early stage of evolution, it was probably not very clear to anyone exactly how the subject should be taught.

In fact, what seemed the big challenge for the department in the early 1960s was how to deal with the science of electrical materials—solid-state physics—and how to build the undergraduate curriculum so that this science could be assimilated more easily. During the 1960s MIT was building its large interdisciplinary Center for Materials Science and Engineering, and Lincoln Laboratory was moving heavily into materials research. Also during this period MIT ran summer courses in solid-state physics for professors from other engineering schools and, in effect, seeded laboratories in solid-state physics in EE departments around the country. Under Elias the EE Department also continued fostering the educational energy-and-information themes formulated by Gordon Brown and carried over into the materials realm.

It was only after Project MAC got going in 1963–64 that a significant momentum began to grow for a formal curriculum in computer science; for as it developed, MAC became a source of new courses and of people who could serve on a computer science faculty. MAC had become a new part of an old dynamic that had always worked at MIT—that of the faculty members coming together to work out a curriculum, then dispersing back to the laboratories where they worked at the edge of knowledge, then coming together again with new ideas taught first to graduate students and then woven into the curriculum. Building curriculum changes, and agreeing on the content and approach in core courses—which had to be taught to many sections of undergraduates—was an important form of communication among EE faculty members. By the time Elias relinquished the position of department head in 1966, the computer scientists were pushing for a separate curriculum, and the question of a separate computer science department had already been raised. Elias's own sense at the time was that such a move might be all right for the computer scientists, but that the move would not be good for the EE Department. It would also weaken the opportunity for each to reinforce the other. His view was supported by the Visiting Committee, which saw such a split as being counterproductive; American industry, for example, needed people who had an understanding of both hardware and software.

When Louis Smullin, who had grown up on the EE side of the faculty, was asked by Dean Gordon Brown to become department head, he found the central issue facing him then, and continuing to the end of his tenure in 1974, to be computers and their place in the EE curriculum. It was during Smullin's tenure that the decision was made to create a new post, Associate Department Head for Computer Science, and then, by way of organizational symmetry, to create the post of Associate Department Head for Electrical

Engineering. Professor Robert Fano, director of MAC, was to become the first associate department head for computer science.

With the strong science orientation of the department, it was a question of how to integrate the evolving computer science courses into the overall electrical engineering teaching program. According to Smullin, "the fact is, even today, computer science is not a science in the same way that Maxwell's theories underlie the electrical fields, allowing a deductive science. The teaching of computer science, however, is still a highly skilled imaginative art. Though a lot of effort goes into the teaching of it, our courses are still not great. Not because the people teaching it are not intelligent, but because it is just plain hard to know how to organize such a miscellaneous thing which has no underlying calculus, something which is universal for the rest of electrical engineering." However, Smullin foresees a time when computer science will have a solid scientific basis.

Smullin's approach during his period as department chairman was to have a number of successive committees—with members from the different electrical engineering specialties, including computerists—working on a curriculum revision that would meet the needs of everybody. During his tenure the new curriculum went through several versions.

In the meantime there was a growing sentiment among the computerists to separate from the Electrical Engineering Department. But there were real questions about the healthiest course. Smullin remembers that "there were a few of the computer people who were pushing vigorously for separation and a few electrical engineering people who essentially said, 'Let them go, and good riddance.' Then there were people like myself who felt that if we could stay together in one department, there were many advantages that would come out of it.'' For one thing, he felt that there would come a time when even the most ardent computerists would conclude that ''it wasn't as much fun to implement something like a new super-FORTRAN compiler, or whatever, but rather they would want to do something else *with* the computer.'' The computerists would need a basis, an understanding of what other things needed to be done, such as solving problems in electrical engineering, economics, the social sciences, and many other areas; and if one were a know-nothing about these other fields, one would not be equipped to take on their interesting problems. Thus, Smullin engaged in a holding operation in the faith that something else would emerge to prove cohesion the better course.

During this same period a second kind of Irredentist movement emerged, whose objective was to transform the EE Department into a *school* of electrical sciences. The rationale for the school argument lay in the great size and diversity of the department. When Smullin became head in 1966, the EE Department had about 600 undergraduates; when he retired in 1974, there were 800; by 1981 there were 1,100. In short, nearly one third of *all* undergraduates with declared majors at MIT were registered in EECS, whereas the faculty of 110 or so comprised only about one tenth of the total Institute faculty. By contrast, MIT's School of Architecture and School of Management combined were smaller than the EE Department. (It might be noted, though, that at the time Smullin was EE Department head MIT had a virtual ''electrical administration''—President Jerome Wiesner, Chancellor Paul Gray, and Provost Walter Rosenblith had all come from the Electrical Engineering Department.) Though there were numerous discussions about the school idea, apparently there was little support for it either in the rest of the Engineering School or from those concerned with setting the priorities of MIT as a whole. The school argument periodically flared up and died down again.

Meanwhile, there was a strong traditional reason for hanging on as a department. In any university, departmental crossings—people shifting from one department to another, owing to changes in their interests and inclinations—are relatively rare. Those who do so almost inevitably lose some status and must ''buck the seniority system'' already existing in any department they might switch to. On the other hand, once established within a department, a faculty member with tenure has relatively great freedom to pursue his own interests after meeting certain fundamental obligations. And, because of the extraordinary diversity of the electrical engineering field, faculty members had a very large menu of possibilities to choose from. Rather than being restricted to a relatively narrow specialty for an entire career, they could, if they wished, taste many scientific, technical, and human activities; because of this fact, too, there was a greater likelihood of EE faculty members holding joint appointments with other departments.

Given the great surge of undergraduates choosing to enter electrical engineering, a large proportion of whom were entering computer science—a reflection of the highly visible prosperity of these fields in the late 1960s—the EE faculty had a growing workload. Yet despite this surge there was no strong push for expansion of the EE faculty at the expense of other departments, even though there was some feeling that the EE Department was not getting its proper share of the budget. For one thing, the increase in EE undergraduates might be temporary; moreover, had such an expanded recruitment program been initiated, it would have been limited by the dearth of competent people willing to take on a university position. At this period the electronics and computer industries were in a strong hiring mode, with pay far outpacing academic salaries; furthermore, the research environment in some industries had become at least equivalent, if not superior, to that in the universities, with the added bonus of having much greater resources to back up the researchers. Companies like Xerox (which set up an entirely new computer research division in 1972), IBM, and Bell Telephone Labs offered prestigious and exciting positions to people who might otherwise have stayed in university environments. And these people could do pioneering research without the burden of "publish or perish" that prevails in school research settings.

We shall not attempt to follow all the twists and turns of the separate department issue, except to note that sentiment finally leaned toward the idea that it would be best for electrical engineering and computer science if the groups remained together. A number of the leading younger computer scientists, like Joel Moses, who became department head in 1981, came gradually around to this view. Also, in 1969–70, there were significant changes in Project MAC, and Marvin Minsky's artificial-intelligence group, which was part of MAC, did become a separate AI laboratory.

The period during which Smullin was department head was a tumultuous one in many respects—an era of national political ferment, student unrest, constant growth in the department despite shrinking budgets, and ceaseless arguments over the separation of computer science from electrical engineering. But Smullin's hopes for an integrated curriculum were realized as he saw a new consensus in the department gaining strength.

All during this time schools around the nation were watching each other to see how they would handle the computer science issue. Stanford, which unlike MIT had a very small undergraduate body in EE, but a large graduate school, established a separate computer science department; the University of California at Berkeley, which was similar in complexion in many respects to MIT, set up an electrical engineering and computer science department, later combining it with a separate CS department that had grown out of the mathematics department. In American industry similar battles went on between traditional science and engineering groups and computer science groups.

The specific struggles reflected many kinds of motives, from genuine concerns about the nature and vitality of science to "empire building" by opportunists. The conflict was, as much as anything, a clash of cultures—between engineering that was outward-oriented, concerned with a world describable by mathematical equations, and a science that was inward-oriented, concerned with the ways in which people perceived things and organized their knowledge, the realm of human cognition and intelligence. To some of the participants this was as distinct a shift as, say, that in the world of painting from realism to cubism.

When Smullin decided to step down in 1974, the choice of a new department head centered upon Wilbur Davenport, the person all factions trusted and wanted. It was during his headship that the department officially became Electrical Engineering and Computer Science. It was a time for unity and a thrust toward new frontiers, and Davenport seemed to be the person who could make this happen.

A New Paradigm in Electrical Engineering?
During Davenport's tenure a different kind of educational issue began to take shape. It was one that would acquire progressive definition and growing significance during the three most recent administrations of the department—those of Davenport (1974–1976), Gerald Wilson (1978–1981), and Joel Moses (appointed in 1981), Moses being the first computer scientist to serve in this post.

The problem, the tackling of which could lead to the emergence of a new paradigm in electrical engineering and computer science, has to do with "the handling of complexity." It is a problem not yet wholly and clearly defined, which makes it right for the educational context. Though at this stage most clearly perceived in EE and CS, it may become central to the progress of all modern engineering over the next decade or two. At least two major engineering schools—MIT on the East Coast and Stanford University on the West Coast—are bringing growing attention to the handling of complexity; some people speculate that the theoretical and design disciplines inherent in computer science and artificial intelligence could play a key role in defining some of the underlying issues.

Why has the handling of complexity become one of the foreground problems in engineering in the latter part of the twentieth century? Whereas a century ago engineers were pushing to master (in the words of an earlier day) an "untamed universe"—a universe not well understood, but whose secrets seemed passive and waiting—the engineers have come to discover (like everyone else) that nothing is simple and straightforward anymore. To push a new railroad across a virgin continent, to build electric power systems reaching ever further, to set up radio and television systems operating on a global scale—these were indeed accomplishments of a high order. They took minds and muscles stretched to their utmost. Yet with respect to the world of today they were simple.

What has been learned, painfully and slowly, is that the world is indeed one of incredibly rich interrelationships. There is nothing that is done, from the micro-level to the macro-level, that does not make a difference somewhere else. The examples have multiplied many times over in the industrialized societies as man has increasingly created a "man-made" world, in which the dangers, as well as the benefits, of each innovation have become overwhelmingly manifest—environmental disasters; pollution of the earth, the water, and the air; chemical and radiation hazards to human health; disruption of the biosphere; social dislocations; human psychological stress. Thus, it is evident that the impact of each innovation must be taken into account by scientists and engineers. New generations of scientists and engineers must somehow be educated in the handling of these new complexities.

The Very Large Scale Integration (VLSI) Project, now officially the MIT Microsystems Program, which began about 1980, is an example of an effort to deal with complexity in the micro-domain, presenting problems that call for a wide range of disciplines across the entire EECS Department. It also brings the department back into an area of research that it had neglected for a relatively long time.

In the late 1950s, with the maturation of transistor technology, the semiconductor industry began its steady march in putting more and more components on silicon chips, and gradually evolved circuit design methods in which classic components like resistors and capacitors were no longer viewed as discrete elements, but were "distributed." Circuits were now hand drawn in very large scale, then photographed and reduced, and used as the screens governing the deposition of the circuits on the semiconductive base. These large master circuit drawings became ever more elaborate, intricate, and beautiful, very much like Eastern carpet designs. This technology—still more an art than a science—led to the deposition of what once was an entire computer on a single silicon chip. Computer circuitry was ideal as an application, since the same basic computational circuits needed to be replicated hundreds of times. Thus, the photo "masks" or screens did not require customized design for each circuit. Also, with close packing of active components, communication time between elements was reduced and computational speeds were accordingly increased. Power needs were reduced, there was less heat to be dissipated, and so on. The semiconductor industry thus fed the computer industry and both grew together in a near-perfect symbiosis.

The investments in capital equipment required for making these new integrated circuits were large, however; by the time the industry began putting a thousand and more elements on a single chip (called Large-Scale Integration or LSI), capital costs were even higher. The Electrical Engineering Department at MIT came to the decision that it should pursue only a small fraction of the technology of the silicon integrated-circuit art. To do more was both too expensive and inappropriate—the technology seemed clearly to belong to the industry, where it was being well handled. MIT, in both its EE and Physics Departments, continued to teach the fundamental physics underlying the semiconductor field, but limited its research primarily to semiconductor devices and materials other than silicon, to complement its educational program. At the same time, the emphasis in the computer science area tended to be on software rather than hardware. Also, the semiconductor industry, which had started in the East, migrated to the West Coast (William Shockley, VIII Ph.D. '36, co-inventor of the transistor, had selected Palo Alto as the place to start a company that then spawned many other leading companies); the University of California at Berkeley and Stanford University, not the geographically remote MIT, took leading roles in silicon integrated-circuit developments.

Wilbur Davenport reports that when he became department head in 1974, ''we were a poor third in the semiconductor business.'' But Davenport began to question the premises that had kept the Institute out of this field and to realize that there were new areas in which the Institute might make contributions, particularly with the development of submicron devices. ''You could put so many of these devices on a chip,'' notes Davenport, ''that the problems were no longer *process* problems—that is, the physics and technology of converting the raw materials to a finished chip—but *what* to put on the chip!'' About that time, the president of Fairchild Semiconductor, Wilfred Corrigan, and the chairman of Intel, Robert Noyce, who was also to serve on the MIT Visiting Committee, began to urge MIT to tackle some of the serious complexity problems being confronted in the design of the next generation of semiconductor circuits, those that have come to be called VLSI circuits. As it happened, the United States Air Force sponsored a new VLSI design project at MIT; in addition, some of the work at Lincoln Laboratories on submicron devices was brought to the MIT campus in a parallel program. Davenport brought Lynn Conway of the Xerox Corporation to introduce VLSI design to the campus, using the subsequently famous Mead and Conway text, *Introduction to VLSI Systems,* in manuscript form. Thus, the department entered the silicon integrated-circuit field in an increasingly strong way.

When he became department head in 1978, Gerald Wilson (now dean of engineering) continued to push hard at the VLSI project. As an example of the kind of complexity problem the VLSI project was trying to do something about, Wilson cites the experience of one of the large semiconductor companies in developing one of its most advanced chip designs. ''An entire bank of engineers,'' says Wilson, ''spent six to eight months together in a large room doing nothing but laying out the circuit for the chip [the Motorola 68,000], making huge drawings using colored crayons. At the end of the project, most of them quit. That is not the kind of life an engineer wants to lead, and the industry now recognizes it. The design process has, in short, reached the ultimate human limits.''

Wilbur B. Davenport, Jr.

Gerald Wilson.

Ways need to be found to automate the entire design and fabrication process, for it no longer makes sense for people to do it by hand. "This is an area involving complexity," asserts Wilson, "and that really is the essence of what computer science is all about. How do you handle very complex situations, be they human situations, social situations, or technical engineering-design situations? Here we saw that we could make fundamental contributions in the development of the automated systems that would design these VLSI chips—either to supplant the engineer with a machine or, at the very least, to provide fantastic aids for the engineers."

What is required in such an immense project is the collaboration of many specialized disciplines—materials people, computer software designers, circuit designers, physicists (both theoretical and experimental), semiconductor specialists, computer system design specialists, artificial-intelligence specialists, and so on—in short, the range of skills now sheltered under the umbrella of the Electrical Engineering and Computer Science Department. Those who worked hard to keep the department together during the separatist strife see the emergence of this new kind of research and development program, and its educational implications, as almost the raison d'etre for their struggle—their faith that something like this might be coming along, and that it would be a lot easier to do it together under a single aegis than as separate specialties.

In his tenure as department head and then as dean of engineering, Wilson, who has been described as having an "activist impatience, a great energy, and a forceful pursuit," pushed along many important ongoing projects in the department and initiated some new ones that have grown to monumental proportions. For instance, he helped bring about the realization of the EG&G Education Center, the MIT Microsystems Program (which includes the Technology Research Laboratories, where VLSI chip fabrication and other integrated-circuit research facilities have been set up), and the EECS Department Computer Facilities; he also gave strong support to the robotics program and initiated an S.M. program for students in industry, a forerunner of the lifelong education idea that became a central theme of the department's centennial celebration in 1982.

At MIT the complexity issue has come into focus during a relatively short period when many diverse streams of development were invigorating electrical engineering (see table 23.1, which lists graduate curriculum offerings in EECS for 1982, assembled under six broad program areas). At the same time the EECS Department's faculty have become participants in research going on in many different laboratories at MIT (see table 23.2). The very breadth of these programs has led to a compacted but fluid "common core" that all students in the department must master before moving into their specialties. The question arises, how can an approach to complexity be folded into the already packed common core? If the complexity issue is as fundamental as appears likely, then dealing with it must also become a fundamental part of the modern engineer's equipage, if he or she is not to run the danger of being too narrowly specialized.

Here again computation has shown its capacity for facilitating change in the entire academic arena. In 1982 Gerald Wilson, as dean of engineering, enlarged his drive to integrate computers and communications into all aspects of education to encompass the entire School of Engineering. Very soon it became apparent that the effort should be expanded to include the entire Institute, and Project Athena was established for this purpose in mid-1983, with Wilson chairing the committee setting it in motion. Project Athena, which has already received a commitment of $50 million from the Digital Equipment Corporation

and IBM, with an additional $20 million being raised by MIT, aims at nothing less than a grand experiment of interlinked networks of computer work stations through which the students and staff can explore ways of enhancing the quality and effectiveness of MIT's educational programs.

Periodically over the past century, as we have seen, it has been necessary to hammer out new paradigms in electrical engineering to provide a unification in teaching and research. Whether such concepts will emerge out of the complexity issue or the ambitious Athena program—both of which go far beyond engineering as it has been understood in the past, now embracing human, social, and moral questions of the profoundest kind—remains to be seen. But one thing is clear in this endeavor once called "alternative education": just as science and engineering have become bonded in the period we have been examining, so research and education have become inextricably entwined.

Postscript

We hope that this historical account, in which we have tried to exemplify how the past has shaped the present, will help the young generation be better prepared for the unknown researches and learning that lie ahead. The perceptions of the department's current leadership, expressed in the statements that follow, may also give some guidance on the shape of things to come.

• Joel Moses, Professor of Computer Science and Engineering, Head of the Department (1981–):

Research and teaching in the area of Very Large Scale Integration (VLSI) has been a major focus of the EECS Department for the past half-dozen years. VLSI has acted as a coalescing force in the department and has proved the wisdom of keeping the department a single entity. VLSI has involved department faculty working on such diverse areas as materials,

Table 23.1
EECS graduate curriculum offerings, 1982

Area I: Systems, Communications, and Control
Professors A. W. Drake and R. G. Gallager, Co-Chairmen

Acoustic Communications
Aerospace Systems
Communication, Command, and Control Systems
Data Communication Networks
Decentralized and Hierarchical Control
Delivery of Urban Services
Digital Control Systems
Estimation/Identification Theory and Algorithms
Human Information Processing and Decision Making
Information Storage and Retrieval
Information Theory
Mathematical Optimization and System Theory
Multivariable Control System Design
Nonlinear Systems and Filtering
Operations Research
Optical and Quantum Communications
Optimal Control
Power Systems Modeling and Control
Stochastic and Adaptive Systems

Area II: Computer Science
Professors P. Elias and P. Solovits, Co-Chairmen

Programming Languages
Software Engineering
Operating Systems
Computer Architecture
Distributed Computation
Computer Networks
Office Automation
Artificial Intelligence
Natural Language Processing
Robots, Vision, and Industrial Automation
Symbolic Computation
Machine-Aided Medical Diagnosis
Theory of Computation
VLSI Design Aids and Algorithms
Formal Semantics of Computer Languages and Systems
Algorithms and Computation Complexity
Computers in Education
Social Issues

Area V: Materials and Devices
Professor D. J. Epstein, Chairman

Semiconductor Materials and Devices
Microelectronics and Integrated Circuits
VLSI Processing and Device Technology
Submicron Structures and Devices
Amorphous Semiconductors
Dielectric, Optical, and Magnetic Materials and Devices
Solid-State Lasers and Laser Materials
Quantum Electronics
Adaptive Optics
Microwave and Acoustic Wave Materials and Devices
Solid-State Physics and Applications

Area VII: Bioelectrical Engineering
Professor L. S. Frishkopf, Chairman

Sensory Physiology and Psychophysics
Speech and Hearing
Computer Synthesis and Recognition of Speech
Sensory Aids
Biological Signal Processing
Biological Image Processing and Pattern Recognition
Biomedical Electronics and Transducers
Microprocessor-Based Clinical Instrumentation
Cardiac Electrophysiology and Rhythm Analysis
Medical Information Systems
Biophysics and Physiology
Remodeling and Regeneration of Connective Tissues

Area III: Electronics, Computers, and Systems
Professor D. H. Staelin, Chairman

Digital Signal Processing
Audio Signal Processing
Image Processing
Mini- and Microcomputer Applications
Analog and Digital System Design
Computer-Aided Circuit Design
Software/Hardware System Design and Computer
  Architecture
Integrated Circuit Design and Design Aids
VLSI Circuit Design and Analysis
Electronic Instrumentation
Power-Control Electronics
Array Processing
Radio and Optical Systems
Stroboscopic Photography
Theory of Nonlinear Circuits and Systems

Area IV: Energy and Electromagnetic Systems
Professor J. R. Melcher, Chairman

Electromagnetic Fields and Waves
Microwave Devices and Systems
Plasmas and Controlled Thermonuclear Fusion
Continuum Electromechanics
Environmental Control
Power Systems and Control
Energy Conversion Components and Systems
High-Voltage Research
Quantum Electronics and Nonlinear Optics
Picosecond Optics
Optical Wave Guides
Radio Astronomy and Remote Sensing
Solar Energy Systems
Fields in Biological Tissue and Cells

Table 23.2
Laboratories involving EECS faculty, 1982

Laboratory for Electromagnetic and Electronic Systems
Director: Professor T. H. Lee

Research Laboratory of Electronics
Director: Professor J. Allen

Laboratory for Computer Science
Director: Professor M. L. Dertouzos

Artificial Intelligence Laboratory
Director: Professor P. H. Winston

Energy Laboratory
Director: Professor D. C. White

Laboratory for Information and Decision Systems
Director: Professor S. K. Mitter

Operations Research Center
Co-Directors: Professors R. C. Larson (EECS) and J. F.
Shapiro (Management)

Center for Materials Science and Engineering
Director: Professor J. D. Litster (Physics)

Plasma Fusion Center
Director: Professor R. C. Davidson (Physics)

MIT Stroboscopic Light Laboratory
Director: Professor H. E. Edgerton

MIT Lincoln Laboratory
Director: Professor W. E. Morrow

circuit analysis and design, algorithm design, and artificial intelligence. VLSI chips having a million components are almost available.

This presents a challenge to teachers of electrical engineering as well as computer science. One outcome that I hope for is a new conceptual synthesis of computer science and the systems side of EE, which includes circuits, signal processing, communication, and control. Such a synthesis would have to deal with what has long been recognized as a key issue in computer science, namely, how one copes with the complexity inherent in large systems. Progress on the complexity issue would have impact not only in EECS, but also in many other areas of human endeavor. In particular, progress on the complexity issue will yield new methodologies for designing large systems.
Many have long felt that one cannot teach design in a systematic fashion. While this view may be true for many problems in engineering and the arts, we simply must learn to systematize most aspects of the design of systems containing millions of similar components.

VLSI is not the only area in EE and CS that is growing fast at this time; in fact, most areas in EE and CS are undergoing major changes and are having a profound impact on our society. There is a structural change going on in our society in which electrical engineers and computer scientists are playing a dominant role. We can expect this structural change to a knowledge-based society to go on for many years to come. This will mitigate against the usual ebbs and flows of engineering employment for our students. As a result of this trend, we anticipate an increasing interest in our educational and research programs in the coming years.

The past century has witnessed the evolution of the department's mission from purely undergraduate education, to include research and some graduate education early in this century, to the current coequal emphasis on undergraduate education and graduate education and research. At the centennial of the department, celebrated in October 1982, we presented the conclusions of a study entitled "Lifelong Cooperative Education." The study, chaired by Professors Fano and Smullin, argues that the department, in cooperation with industry, should also undertake the mission of providing education throughout the working lifetime of engineers. The need for lifelong education is paramount in fields as fast-changing as EE and CS. One of the first examples of such lifelong education the department has undertaken has been the very popular offering of our first common core subject in programming to most of our EE faculty, proving that lifelong education begins at home.

Considering the great stresses on the department at this time from large undergraduate enrollments as well as large graduate enrollments and research volume, we may not be able to devote many resources to lifelong education for some time. Nevertheless, we are witnessing once again in the second century of the department its willingness to accept the challenge of changes in goals and organizational mechanisms.

• Richard B. Adler, Professor of Electrical Engineering, Associate Head of the Department for Electrical Science and Engineering (1978–):

In the last two decades, miniaturization of electronics, especially of computers, has propelled the technical world with a vengeance into the age of complex systems. Not only the ability to make things small, but also the consequent ability to process signals digitally at extremely rapid rates, has already produced a proliferation of "smart" systems into nearly every niche of daily life. Surely we can look forward to the extrapolation of this trend until the prostheses for the handicapped perform better than the original organ they replace, and the "real" robots of science-fiction stumble onto the scene.

Joel Moses.

Richard B. Adler.

The attendant massive accumulation of data, and its circulation worldwide at unprecedented rates, demands extensive rapid communication at almost unimaginably high traffic densities. We can already witness the start of the optical domination of communication technology, and the emergence of extremely wide-bandwidth communication channels over modest distances.

Whether we will continue to build all of our systems with electronics as we now know it, or whether for some of them the line between electronics and genetic engineering will become blurred, it is not possible to discern now; but the control of ''bodies'' on the scale of man's own, as well as on the scale of the international scene, will undoubtedly be the order of the future day.

To achieve the complex, intelligent systems of tomorrow, mastery of at least molecular-scale technology and intrusions into the quantum world are necessary. And as man reaches out further into space, which he seems destined to do, the relativistic science of the massive and the faraway must move closer to the engineering purview.

So there seems little that we can stop teaching, and much to be added; but we can hope that the world of complex systems will force us further to merge the principles of computer science and those of systems design, with a chance (however slim it may now appear) of thereby achieving some economy of curriculum.

• Peter Elias, Professor of Electrical Engineering, Acting Associate Head of the Department for Computer Science and Engineering (1981–1983):

For most of the period from the late nineteenth century until World War II, electrical engineering was largely the exploitation of three kinds of devices: the rotating machine, the telephone, and the incandes-

cent lamp. A fourth device, the vacuum tube, became important toward the end of that period. A few researchers were needed who had a deep physical understanding of each device in order to find the appropriate mathematical models to use in analysis and design. The bulk of engineering students throughout the country, however, learned the models rather than the physics. It was hardly necessary for each electrical engineer to learn the physics required to generate useful models of new devices, since there were so few new devices.

All of this changed dramatically during and immediately after World War II. Radar, a whole new domain of electrical engineering, was developed largely by physicists, because few electrical engineers knew enough about electromagnetics to deal with microwaves. New devices appeared at an increasing rate. It soon became clear that teaching students only device models was not enough, since current devices might be of historical rather than applicable interest by graduation. The new electrical engineering curricula adopted widely in the 1950s taught classical physics with engineering applications and modeling techniques, and added some quantum and solid-state physics to deal with semiconductor devices as they appeared.

The rate of development of devices has continued to increase most dramatically in the digital domain. The complexity of a typical system composed of those devices has grown even more rapidly. Earlier problems of complexity had to do largely with complex and deep physics—the nonlinear electromechanics of a rotating machine or a network of such machines, the physical mechanisms and mathematical characterization of atmospheric noise or ionospheric propagation. Current devices like logic gates are operated in modes in which their physics is easily modeled: the complexity of such systems lies in the complexity of their interrelations, not in the complexity of their components.

Nowadays, a small system may have millions of components and do hundreds of interdependent but distinguishable jobs. There is a group of electrical engineers who work with applied physicists and materials scientists in making faster, smaller components and new types of devices. But the central problem faced by most electrical engineers and computer scientists is coping with organizational, not physical complexity. To some extent this problem is an extension of earlier system problems in electrical engineering: Guillemin's RLC network theory and Bush's network analyzer also dealt with systems of many interconnected components. But those systems had only a few kinds of nodes and branches, with structure at only one or two levels. The variety of possible interrelations is much greater in systems with many kinds of nodes and branches, each with a hierarchy of levels of internal structure, exhibiting memory and conditional and recursive behavior.

It is an interesting question whether there is a curriculum that deals with these issues and survives the rapid changes in technology or whether we remain in our current state in computer science, teaching mostly topics that we do not understand because by the time we understand them they are technologically obsolete and not worth teaching. It seems likely that indeed it will never again be possible to learn enough in four years—or eight—to deal with all of the new technologies and fields of physics and mathematics (and perhaps of biophysics) that will become professionally important to an electrical engineer or computer scientist, and that the departmental program just started on the recommendation of the Centennial Study Committee to deal with lifelong cooperative education will become more and more central to the department's educational mission with time.

• Fernando Corbató, Professor of Computer Science and Engineering, Acting Associate Head of the Department for Computer Science and Engineering (1983–):

In a few short decades the modern digital computer has signaled the beginning of a second industrial revolution, involving the manipulation of information instead of the harnessing of energy sources. Already the pace of change has been explosive, with several order-of-magnitude improvements in the size, reliability, speed, energy consumption, and cost of the underlying technology. And it would appear that changes will continue unabated for at least several more decades as fabrication techniques improve, parallelism of operation becomes exploited, and fruitful applications multiply. These are heady times indeed for all those involved in the flourishing growth and societal importance of the new technology.

But it is because of this very increased importance, both economic and social, that there are moderating forces too. Public awareness of potential uses of digital logic is increasing rapidly, especially with the proliferation of personal computers. And with this awareness comes increasing sensitivity to potential computer abuses, to the need to control both openness and privacy of information, and to the dangers of information tyrannies, whether in private or government hands.

Universities face several challenges arising from the changes produced by the computer revolution. Foremost, of course, is supplying leadership in identifying the shifting frontiers of relevant research and in helping formulate the long-term framework for technological change. Moreover, such leadership is needed not only in computer science and electrical engineering departments, but, because of the increasing diversity of computer applications, in nearly every department and field. A second major challenge facing electrical

Fernando Corbató.

engineering and computer science departments in particular is organizing and digesting for each successive wave of students the ever-increasing amount of technological lore, such that they acquire a certain technical nimbleness based on fundamental concepts rather than contemporary artifacts.

If there is a single theme behind these challenges, it is the problem of complexity. Whether it is the organization of million-gate arrays on VLSI wafers, the reliable development of software for large-scale systems such as the Social Security System, or developing architectures that allow straightforward use of parallel processing, the fundamental goal is developing systematic techniques that allow predictable and successful engineering of new ventures. Moreover, even when successfully constructed, large computer systems continue to exhibit evolutionary complexity as they develop networks and loosely coupled nodes such that all changes must be made incrementally without either shutdown or replacement.

Finally, the teaching of computer science offers special challenges. Today it is a rare student who does not acquire experience and expertise by some form of apprenticeship, for practical details and engineering wisdom are difficult to acquire otherwise. Moreover, without some practical experience it is a rare student who fully appreciates generalization and abstraction. But the difficulty is that as the body of computer technology becomes ever larger, it is clear that the curricula must contain an ever-changing sample of illuminating practical experiences. Thus the computer science curriculum must remain in constant flux to reflect the changing computational artifacts required for concreteness.

At the 1982 EECS centennial celebration: (*left to right*) Richard Adler, Gerald Wilson, Mildred Dresselhaus, Louis Smullin, Peter Elias, Gordon Brown, Joel Moses, Robert Fano, Jerome Wiesner, Fernando Corbató. (Courtesy Calvin Campbell, MIT News Office)

Appendix

The following chronology was compiled in 1980 by members of the MIT Museum staff. MIT events are listed above the lines; national and world developments—scientific and technological, social and political—below.

1846
William Barton Rogers sets down ideas for a polytechnic institution

Oregon Treaty divides U.S. and British territories at 49th parallel

Robert W. Thompson patents pneumatic tire

Smithsonian Institution founded

Elias Howe patents sewing machine

1847
Harvard and Yale establish scientific schools

American Medical Association founded

Smith Brothers develop cough drops

Ascanio Sobrero discovers nitroglycerin

Hanson C. Gregory introduces ring doughnuts

1848

Gold discovered at Sutter's Mill, California

Marx and Engels publish *The Communist Manifesto*

American Association for the Advancement of Science founded

First Women's Rights Convention held at Seneca Falls, New York

1849
William Barton Rogers marries Emma Savage, daughter of prominent Bostonian James Savage

Harriet Tubman escapes to North

Walter Hunt patents safety pin

Thomas Bowler introduces derby hat

Joseph Monier patents reinforced concrete

1850

Levi Strauss introduces "bibless overalls"

Robert W. Bunsen introduces gas burner

Sparrows imported to U.S. for experimental purposes

Second Law of Thermodynamics promulgated by Rudolph Clausius

1851

The New York Times begins publication

Scott Archer develops collodion photographic process

Harriet Beecher Stowe publishes *Uncle Tom's Cabin*; Herman Melville, *Moby Dick*

Joseph Paxton erects Crystal Palace

1852

"Uncle Sam" first appears as cartoon

Boston Public Library opens

American Society of Civil Engineers founded

Elisha G. Otis invents safety elevator

Gadsden Purchase Treaty signed

1853

William Barton Rogers resigns as professor of natural philosophy at University of Virginia and moves to Boston

New York Central becomes first major U.S. rail combine

George Crum invents potato chip

First U.S. World's Fair held in New York

Pocket watches mass-produced in Waltham, Massachusetts

1854

Henry David Thoreau publishes *Walden*

John Beardsley makes paper from wood pulp

U.S. Mint opens branch in San Francisco; buys gold at $16 an ounce

Benjamin Silliman achieves fractional distillation of crude petroleum

1855

Boston Society of Natural History and other local institutions begin search for solutions to pressing space problems

Alexander Parkes patents celluloid

Joshua C. Stoddard invents calliope

David Edward Hughes patents teleprinter

Gail Borden patents condensed milk

1856

Indenture adopted between Commonwealth, City of Boston, and various private proprietors relating to Back Bay—later confirmation by legislature opens way for filling of Back Bay

Western Union organized

Treaty of Yedo opens Japan to outside world

Hamilton L. Smith patents tintype camera

National Republican Party formed

1857

William Barton Rogers lectures at Lowell Institute on elementary laws of physics

Dred Scott decision enrages abolitionists

Atlantic Monthly begins publication in Boston

Louis Pasteur proves living organisms spoil milk

Currier and Ives, lithographers, start printing

1858

Cyrus W. Field completes first Atlantic cable

Boston begins filling in Back Bay

New York's Central Park opens

Ferdinand P. A. Carré devises refrigerator

John Landis Mason patents Mason jar

1859

Boston Society of Natural History and group of associated institutions file petitions with legislature for grant of land in Back Bay for use of Society and for what "might be known as the Massachusetts Conservatory of Art and Science"–rejected

R. L. G. Plante develops practical storage battery

George M. Pullman designs railroad sleeping car

Charles Darwin publishes *Origin of Species*

Work begins on Suez Canal

Comstock Lode discovered

1860

Committee representing associated institutions again petitions state for land, this time presenting a more definite plan that mentions possibility of a polytechnic college

William Barton Rogers assumes leadership role and prepares a plan entitled "Objects and Plan for an Institute of Technology"

Pony Express established

Frederic Walton invents linoleum

John Etienne Lenoir patents internal combustion engine

First aerial photographs taken from balloon tethered over Boston

1861

Boston Society of Natural History renews petition with legislature for grant of Back Bay land

Committee of Associated Institutions, William Barton Rogers, chairman, petitions legislature for charter for Massachusetts Institute of Technology

April 10–Governor Andrew signs act chartering the Massachusetts Institute of Technology and granting block of land in Back Bay, to include "a Society of Arts, a Museum of Arts, and a School of Industrial Science"

Fort Sumter bombarded– Civil War begins April 12

Congress levies first income tax; money used to support Union forces

Vassar Female College chartered

Oliver Wendell Holmes invents stereoscope

1862

Fifty-four incorporators elect William Barton Rogers first president of MIT–Institute organized and begins to function as a Society of Arts

Stimulated by Morrill Act for public land grants and funds for Massachusetts colleges, Governor Andrew suggests that there be a merger of all educational facilities in Boston area, which would make Harvard predominant. Rogers opposes this first attempt at thwarting independence and wins.

Lincoln issues Emancipation Proclamation

Merrimac battles *Monitor* at Hampton Roads

Léon Foucault measures speed of light

Gambling begins in Monte Carlo

Pasteur develops "germ theory"

U.S. begins printing paper money

1863

Institute secures portion of state's Morrill Act land grant funds

William J. Walker gives $60,000 to complete guaranty fund and save Institute charter

Construction of Back Bay building, designed by William G. Preston, begins

---

National Academy of Sciences established

Thanksgiving Day becomes national holiday

Definitive rules for soccer drawn up in England

Perrier water introduced commercially

1864

William Barton Rogers's "Scope and Plan of the School of Industrial Science," outlining proposed organization and plan of instruction, adopted by Corporation of Institute

William Barton Rogers makes trip to Europe and visits scientific schools, laboratories, and workshops

Official seal of Institute approved by Corporation

"In God We Trust" appears on U.S. coins

Theta Xi, professional engineering and science fraternity, founded

James Slater patents drive chain

International Red Cross established

1865

Eli Forbes enrolls as first student

February 20–First classes held in rented space in Mercantile Building in downtown Boston (corner of Summer and Hawley Streets), with studies "suited to the various professions of the Mechanician, the Civil Engineer, the Builder and Architect, the Mining Engineer, and the practical chemist"

Lowell Free Courses begin with evening classes open to both men and women taught by MIT faculty under auspices of Lowell Institute

---

April 9–Lee surrenders to Grant at Appomattox

April 15–Abraham Lincoln assassinated

Gregor Mendel enunciates his "Law of Heredity"

Chicago Union stockyards open

First train holdup in North Bend, Ohio

1866

MIT Building in Back Bay completed and occupied—later named in honor of William Barton Rogers

Tuition:
$100 for first-year students
$125 for second-year students
$150 for upperclassmen

"L" Street baths open in Boston

Alfred Nobel invents dynamite

Charles E. Hires concocts root beer

Mount Washington cog railroad construction begun

James H. Mason patents percolator

1867

Charles W. Eliot and Francis H. Storer publish *Manual of Inorganic Chemistry*, which revolutionizes teaching of subject

Robert H. Richards, a member of the first class, writes, "We learned to observe, record, collate, and draw conclusions from our experiments. . . . It was all one continuous question—What is Truth?"

Alaska purchased from Russia for $7.2 million

Georges Leclanche invents practical dry-cell battery

William Underwood and Company can deviled ham

Marquess of Queensberry boxing rules formulated

1868

Department of Architecture organized by William Robert Ware

William Barton Rogers suffers stroke at faculty meeting—John D. Runkle appointed acting president

MIT given power by Massachusetts legislature to grant degrees

First graduating class of 14 receives Graduate of Institute degrees with course designation

---

Christopher L. Scholes patents practical typewriter

George Westinghouse, Jr., invents air brakes

Navajo Nation forced to sign treaty with U.S.; removal to reservation follows

First U.S. open-hearth furnace built in Trenton, New Jersey

1869

Edward C. Pickering develops physics teaching laboratory

Charles W. Eliot becomes president of Harvard and soon urges merger with MIT

Suez Canal opens

Transcontinental railroad completed

William F. Semple patents chewing gum

Ives W. McGaffey invents vacuum cleaner

Memorial Day observance begun

1870

William Barton Rogers resigns presidency because of ill health

John D. Runkle becomes second president and remains steadfast in belief that MIT must remain independent from Harvard

---

British Post Office issues first postcards

Atlantic City boardwalk erected

DNA is discovered, but not fully understood

Four million buffalo roam American plains

1871

Ellen H. Swallow admitted as special student in chemistry

Alexander Graham Bell moves to Boston, attends lectures by Lewis B. Monroe, and works in physics laboratory with Edward C. Pickering and Charles R. Cross '70 on sound and acoustics problems

---

Portland cement invented

P. T. Barnum opens Greatest Show on Earth

Great Chicago Fire levels city

American Public Health Association founded

1872

Student battalion patrols after Boston fire

MIT degree title changed to Bachelor of Science with course designation

Advanced courses established

Montgomery Ward catalogue published

Eadweard Muybridge catches motion in photographs

Great Boston Fire destroys business district

Yellowstone National Park authorized

1873

First student publication, *The Spectrum*, begins monthly publication

Ellen H. Swallow receives S.B. degree in chemistry

Nationwide economic depression opens period of desperate efforts by MIT to survive

---

Joseph F. Glidden patents barbed wire

U.S. adopts gold standard

Cable streetcars operate in San Francisco

Preakness has first running

Economic depression of 1873

1874

Practical experience in laboratory, shop, and field espoused by John D. Runkle leads to student summer visits to engineering and mining operations

First formal graduation exercises held

French Impressionists show paintings in Paris

Margarine introduced in U.S.

Madison Square Garden opens under name of Barnum's Hippodrome

Robert Green invents ice-cream soda

1875

Robert H. Richards '68 founds Alumni Association and is elected first president

Gym classes begin in new gym and drill building

Ellen H. Swallow '73 marries Robert H. Richards '68

Matthew Webb swims English Channel

Hollis Hunnewell introduces court tennis

George F. Green invents electric dental drill

First roller skating rink opens

Wellesley Female Seminary founded

1876

Women's Laboratory opened in cooperation with Women's Education Association of Boston, provides for special instruction in chemistry, mineralogy, and related subjects

Alexander Graham Bell gives first demonstration of telephone to Society of Arts

Cardinal red and silver gray chosen as MIT colors

Two hundred faculty, students, and graduates travel via boat and train to Philadelphia for Centennial Exposition

Alexander Graham Bell patents telephone

Centennial Exposition opens in Philadelphia

Custer dies at Little Big Horn

American Chemical Society organized

Thomas Alva Edison patents mimeograph

1877

John D. Runkle establishes School of Mechanic Arts based on Russian system of training through shopwork

Swan boats appear in Boston Public Garden

Nathanael G. Herreshoff '70 patents catamaran

Edison invents phonograph

A. Downs and T. P. Blunt discover germicidal properties of ultraviolet rays

1878

John D. Runkle resigns because of poor health–William Barton Rogers again accepts presidency with conditions that $100,000 be raised to ensure Institute's stability and that a search for a successor begin immediately

Howard A. Carson '69 becomes first alumnus elected to Corporation

Second attempt by Harvard to absorb MIT as its engineering school

---

Commercial telephone exchange opens in New Haven

Goldfish imported from Japan

Louis Comfort Tiffany begins manufacture of Tiffany glass

President Hayes initiates annual Easter egg roll

1879

Grade reporting system changed from numerical system to

H = passed with honor
C = passed creditably
P = passed
F = failed

F. W. Woolworth opens first store

Edison invents practical incandescent lamp

James J. Ritty invents cash register

Women obtain right to practice law before U.S. Supreme Court

Ira Remsen discovers saccharin

1880

American Society of Mechanical Engineers founded

New York City population exceeds one million

Halftone photographic illustrations appear in newspapers

Samuel Bath Thomas introduces English muffins

1881

Francis Amasa Walker, economist, becomes third president

Laboratory of Applied Mechanics established

Student newspaper, *The Tech*, begins publication

Ellen Swallow Richards '73, Marion Talbot '88, and others meet at MIT to form Association of Collegiate Alumnae–now American Association of University Women

---

President Garfield assassinated

Clara Barton founds American Red Cross

Albert Abraham Michelson invents interferometer

Henry Lee Higginson founds Boston Symphony Orchestra

1882
William Barton Rogers dies while speaking at commencement exercises

Charles R. Cross '70 establishes first electrical engineering course in Department of Physics

Sigma Chi, first social fraternity at Institute, established

Charles H. Kip proposes Harvard Cooperative Society

William Horlich invents malted milk

Elihu Thomson's dynamo powers street lights

U.S. population exceeds 50 million

1883
"New Building" is completed—later named in honor of Francis A. Walker

---

Brooklyn Bridge opens

Orient Express, Europe's first transcontinental train, makes first run

Buffalo Bill's Wild West Show opens

Alexander Graham Bell founds *Science* magazine

1884
Biology Laboratory established

Association formed to play intercollegiate football with Williams, Amherst, and Tufts

---

Washington Monument completed

Charles Parsons tests first practical steam turbine

Lewis Waterman invents fountain pen

1885
First student yearbook, *Technique*, published by Class of '89

Francis A. Walker is elected first president of newly formed American Economic Association

---

Pasteur successfully vaccinates against rabies

George Eastman produces coated photographic paper

William Stanley and George Westinghouse perfect transformer for electrical networks

Leland Stanford, Jr., University founded

1886

First advanced degree given to Frederick Fox, Jr., '85—Master of Science in chemistry

---

Statue of Liberty unveiled

Geronimo captured; Indian Wars end

Johnson and Johnson Company produces ready-to-use bandages

Jacob's Pharmacy, Atlanta, Georgia, introduces Coca-Cola

Thomson invents electric welding process

1887

Technology Quarterly begins publication

Alfred E. Burton establishes summer school for Civil Engineering

Thomas A. Edison gives Institute dynamo and 150 lamps to use in teaching electrical engineering

---

Florida legislature passes first "Jim Crow" law

Heinrich Hertz discovers radio waves

Edison invents motor-driven phonograph

Sears, Roebuck Company founded

1888

First classes in chemical engineering offered by Lewis M. Norton

Engineering Building built to house Mechanical and Civil Engineering Departments

---

First commercial aluminum produced

Jack the Ripper murders six women in London

Gregg shorthand system introduced

Alexander Graham Bell publishes *National Geographic*

1889

Department of Natural History becomes Department of Biology, with William T. Sedgwick as head

School of Mechanic Arts discontinued

Clement W. Andrews appointed librarian of Institute

Uncoached Tech football team wins intercollegiate pennant

Herman Hollerith develops punch-card system

Eiffel Tower completed

Thomson invents recording wattmeter

Flexible Flyer sled introduced

1890
Geology splits from Mining Engineering to become separate course

---

Charles Dana Gibson creates ''Gibson Girl''

Daughters of the American Revolution organize

First steel-framed building erected in Chicago

Alphonse D. Rockwell invents electric chair

1891
Lunchroom opened in basement of Rogers Building by Ellen A. King

---

Immigrant depot at Ellis Island opens

Wireless telegraphy begins

Carnegie Hall opens

Walter Camp writes football rule book

1892
Enrollment reaches 1,000

Rudolph Diesel patents internal combustion engine

Sierra Club founded

U.S. flour sent to relieve Russian famine

Chicago's Loop becomes operational

1893
Department of Naval Architecture founded by Cecil H. Peabody '77

Cataloguing of library holdings accomplished

First open-heart surgery performed

Henry Ford builds first car

Lizzie Borden acquitted of murdering parents

Hershey Chocolate Company founded

Economic depression of 1893

1894

Oil discovered in Corsicana, Texas

Radcliffe College for Women opens

U.S. produces 1.5 million tons of machine-made ice

Labor Day established as national holiday

1895

William Lyman Underwood '96 and Samuel C. Prescott '94 begin studies that will provide scientific foundation for modern canning

W. K. Roentgen discovers X rays

King C. Gillette invents safety razor

Charles W. Post introduces Grape Nuts

First U.S. pizzeria opens in New York

Niagara Falls AC power system completed

1896

Emma Savage Rogers, assisted by William T. Sedgwick, edits *Life and Letters of William Barton Rogers*

Small wind tunnel with speed of 15 mph built by students under guidance of Gaetano Lanza using air current from Institute's ventilating system

Henri Becquerel discovers radioactivity

Chop suey concocted in New York

Klondike Gold Rush begins

Olympic Games revived in Athens

1897

Francis A. Walker dies suddenly of apoplexy

James Mason Crafts, world-renowned for Friedel-Crafts Reaction, takes charge of Institute as chairman of faculty—later elected fourth president—resigns after three years

J. J. Thomson discovers electron

Boston Marathon first run

First portion of Boston subway opens

Library of Congress completed

1898

Department of Architecture moves to new Pierce Building

Membership in Alumni Association opened to those who had not received a degree

Students form "Tech Battalion" and offer themselves for Spanish-American War duty

---

U.S. annexes Hawaiian Islands

Maine sunk off Havana; U.S. declares war on Spain

Marie and Pierre Curie isolate radium

Zeppelin builds airship

1899

First Tech Show held

Technology Review begins publication by Alumni Association

Boll weevil crosses Rio Grande

American Physical Society formed

Stanley Steamer automobile climbs Mt. Washington

Felix Hoffman and Hermann Dresser perfect aspirin

1900

Henry Smith Pritchett, astronomer and mathematician, inaugurated fifth president

Cane Rush ends in death of student—event replaced by Field Day following year

---

Max Planck formulates quantum theory

General Electric founds Industrial Research Laboratory, supervised by Willis R. Whitney '90

R. A. Fessenden transmits speech via radio waves

E. R. Thomas Motor Company manufactures practical motorcycle

1901

Intercollegiate football team disbanded by student vote of 119 to 117

Guglielmo Marconi sends wireless messages across Atlantic

Peter C. Hewitt invents mercury-vapor lamp

National Bureau of Standards established; Samuel W. Stratton first director

U.S. College Entrance Examination Board conducts first tests

1902

Henry S. Pritchett introduces beer and music at new Tech Union to dismay of local community

Department of Electrical Engineering formed

Alfred E. Burton named first dean of Institute

Rayon, "artificial silk," patented by Arthur D. Little '85 and Harry S. Monic

Morris Michtonn introduces Teddy Bear

American Automobile Association founded

Binney and Smith introduce Crayons

1903

Lydia G. Weld '03 becomes first woman to receive engineering degree (in naval architecture)

Arthur A. Noyes '86 organizes Research Laboratory of Physical Chemistry

Lowell Institute School begins evening classes with MIT faculty serving as instructors and Charles F. Park '92 as director

---

Automobile crosses U.S. in 65 days

Edwin S. Porter produces *Great Train Robbery* film

Wright brothers fly at Kitty Hawk

Massachusetts issues first automobile license plates

1904

Harvard-MIT merger again proposed, faculty and alumni object, but Corporation votes for merger. Massachusetts Supreme Court holds that Institute cannot sell its land, killing plan

Harvard-Tech student riot injures 50

Alumni set up five-year plan to raise funds to meet Institute needs

First All-Tech Alumni Reunion held

Register of Graduates first published

Rhodes Scholars program begins

Construction of Panama Canal started

Harvard Stadium completed

Helen Keller graduates from Radcliffe

1905

Plywood first commercially produced

Einstein introduces equation $E = mc^2$ in "The Electrodynamics of Moving Bodies"

Theodore Roosevelt wins Nobel Prize for Peace—first U.S. winner

Fessenden broadcasts voice

1906
Research Laboratory of Applied Chemistry formed by William H. Walker

Term "suffragette" coined

Earthquake destroys San Francisco

Lee DeForest invents three-element radio vacuum tube

Charles S. Rolls and Frederick H. Royce incorporate Rolls-Royce Ltd.

1907
First Ph.D.'s granted to three students in physical chemistry

Henry S. Pritchett resigns to become head of Carnegie Foundation for Advancement of Teaching

Arthur A. Noyes '86 becomes acting president

Dugald C. Jackson begins 28-year tenure as head of Department of Electrical Engineering

Lee DeForest invents high-frequency "radio" surgery

United Press founded

Hurley Machine Company introduces electric washing machine

Wireless telegraphy service begins between U.S. and Ireland

1908

William Coolidge '96 produces ductile tungsten for incandescent lamp

Ford produces Model T

National Interfraternity Conference first meets

Hans Geiger and Ernest Rutherford develop Geiger counter

1909
Richard Cockburn Maclaurin, physicist and mathematician, inaugurated sixth president during Second All-Tech Reunion

MIT Aero Club formed

Richard C. Maclaurin chooses site in Cambridge as future home of "The New Technology"

Robert Peary and Matthew Henson reach North Pole

Karl Landsteiner establishes existence of different blood types

NAACP organized

Synthetic rubber produced by Karl Hoffman

Rose O'Neill patents Kewpie Doll

1910
Civil Engineering Summer Camp at East Machias, Maine, established

Barney Oldfield drives Benz 133 mph

Charles River Dam completed

Camp Fire Girls of America founded by Luther Halsey Gulick

Edouard Benedictus patents safety glass

1911
Congress of Technology, celebrating 50th anniversary of MIT charter, draws national attention to quality of Institute graduates and convinces legislature state aid for education important

T. Coleman du Pont '84 gives $500,000 to purchase Cambridge land for new campus

Emma Savage Rogers and Ellen Swallow Richards '73 die

Calbraith P. Rogers flies across America in Burgess-Wright biplane

Boy Scouts of America incorporated

Irving Berlin's "Alexander's Ragtime Band" popularizes ragtime

C. F. Kettering invents electric self-starter for motorcar engines

1912
"Mr. Smith" gives $2.5 million for construction of "The New Technology" on condition he remain anonymous

Titanic sinks

First self-service grocery opens

SOS in Morse code adopted as universal distress signal

Jacques E. Brandenberger and David H. McAlpin perfect cellophane

1913
Jerome C. Hunsaker '12 initiates aeronautical engineering course

William Welles Bosworth '89 selected architect and Charles A. Stone '88 and Edwin S. Webster '88 engineers of "The New Technology"

U.S. federal income tax introduced

Lincoln Highway opens; first coast-to-coast paved road

Neils Bohr formulates theory of atomic structure

Arthur Wynne devises crossword puzzle

Albert Schweitzer opens hospital in Lambaréné, French Congo

1914

Davis R. Dewey organizes course in engineering administration

Beaver becomes official MIT mascot: "Of all the animals of the world, the beaver is noted for his engineering and mechanical skill and habits of industry. His habits are nocturnal, he does his best work in the dark."

Archduke Ferdinand assassinated at Sarajevo

Electric traffic lights first installed in Cleveland

Panama Canal opens

Robert H. Goddard begins experiments in rocketry

1915

Telephone call made from New York to San Francisco

Boston Red Sox win World Series

Eugene Sullivan and William Taylor develop Pyrex

D. W. Griffith produces *Birth of a Nation*

1916

"The New Technology" dedicated—4,000 alumni, and Franklin Delano Roosevelt, attend Third All-Tech Reunion

Ceremonial crossing of Charles River made in *Bucentaur*

"Masque of Power" pageant written and staged by Ralph Adams Cram

Cornerstone of Francis A. Walker Memorial laid—funds raised by alumni subscription

Department of Architecture remains in Rogers Building in Boston

Vannevar Bush receives Doctor of Engineering degree, fifth to be awarded

Einstein announces his General Theory of Relativity

Blood is refrigerated for transfusions

Margaret Sanger opens birth-control clinic

Federal child labor law passes

1917

Specialized schools for Army and Navy aviators, aviation engineers, radio engineers organized on campus in support of World War I effort

Arthur D. Little '85 initiates Chemical Engineering Practice School

U.S. enters World War I

Sigmund Freud writes *Introduction to Psychoanalysis*

Fashionable women bob hair

James Montgomery Flagg designs Uncle Sam "I Want You" poster

1918

Enima Rogers Organization of Technology Women formed by Margaret Alice Maclaurin—later MIT Matrons, and now Women's League

Harlow Shapley measures Milky Way

Daylight Saving Time introduced in U.S.

Regular Air Mail service established—Earle Ovington '04 first pilot

World War I ends

1919

VooDoo, student humor magazine, first published

Division of Industrial Co-operation and Research formed—later known as Division of Sponsored Research, and now as Office of Sponsored Programs

Boston Police strike

NC-4, designed by Jerome C. Hunsaker '12, makes first transatlantic crossing by heavier-than-air machine

Dial telephones introduced

George B. Hansburg patents pogo stick

1920

"Mr. Smith" revealed to be George Eastman

Richard C. Maclaurin dies of pneumonia

Elihu Thomson named acting president—Institute run by administrative committee

Tech Engineering News begins publication

Department of Chemical Engineering established with Warren K. Lewis '05 as head

Varsity crew organized

League of Nations organized—U.S. fails to join

Bradley Dewey '09 develops water-based latex sealing compound for canning

American women get vote

Prohibition becomes law

Regularly scheduled radio broadcasting begins (KDKA, Pittsburgh)

1921

Ernest Fox Nichols, astrophysicist, inaugurated seventh president—resigns after five months because of poor health

Pratt School of Naval Architecture and Marine Engineering Building (5) completed

Warren K. Lewis '05, William H. Walker, and William H. McAdams '17 publish *Principles of Chemical Engineering*

Monument to Unknown Soldier built in Arlington National Cemetery

Herbert T. Kalmus '03 develops technicolor process

Berlin Autobahn is first highway designed exclusively for motor traffic

Iodized salt introduced

1922

Tent collapses during commencement exercises in Great Court

Edward L. Bowles '22 and Carlton E. Tucker '18 develop concepts for communications curriculum

---

Tomb of King Tutankhamen discovered by Howard Carter and others

Lincoln Memorial dedicated—statue by Daniel Chester French '73

E. V. McCollum isolates Vitamin D

Bacteriologist Gaston Ramon develops tetanus toxoid

1923

Class of '23 introduces caps and gowns at commencement

Samuel W. Stratton, director of U.S. Bureau of Standards, inaugurated eighth president

Edward H. R. Green invites Institute to use his Round Hill estate at Dartmouth, Massachusetts, for research purposes—Vannevar Bush '16, Edward L. Bowles '22, and others conduct research in fog dispersal, navigation, radio and light studies, and microwave communications in Institute's first interdisciplinary laboratory

---

Jacob Schick patents electric razor

Adolf Hitler stages "Beer Hall Putsch"

Benito Mussolini secures fascist Italian dictatorship

New York's Yankee Stadium opens

1924

Edwin H. Blashfield '69 begins murals in Walker Memorial

Patent for iconoscope (TV) issued to Vladimir K. Zworykin

Lenin dies; Stalin begins power struggle with Trotsky

"Little Orphan Annie" comic strip introduced

First Macy's Thanksgiving Day Parade

Frank Epperson patents popsicle

1925

George Eastman gives $4.5 million as unrestricted gift to MIT

Institute returns to semester system

Vannevar Bush '16, Herbert R. Stewart '24, and Frank D. Gage '22 build product integraph, Institute's first analog computer

---

Hitler publishes *Mein Kampf*

Robert A. Millikan discovers cosmic rays in upper atmosphere

Scopes "Monkey Trial" held in Tennessee

Westinghouse demonstrates photoelectric cell

1926

George J. Leness '26 wins New England Intercollegiate half-mile championship

James R. Killian, Jr., '26 joins *Technology Review* staff

Goddard launches first liquid-fuel rocket

Electrolux Servel Corporation introduces gas refrigerator

Kodak produces 16mm movie film

Book-of-the-Month Club established

1927

Vannevar Bush '16 and Harold L. Hazen '24 build second product integraph with two integrators

Ralph T. Jope '28 introduces Glove Fight to Field Day

Telecast of image and sound over distance demonstrated

Sacco and Vanzetti executed

Charles A. Lindbergh flies solo to Paris in *Spirit of St. Louis*

The Jazz Singer, first talkie, opens

Academy of Motion Picture Arts and Sciences presents first "Oscars"

1928

Carl-Gustaf A. Rossby establishes meteorology course

War outlawed; 65 nations sign Kellog-Briand Pact

Alexander Fleming discovers penicillin

First scheduled television broadcast

Color motion pictures introduced

1929

Committee chaired by Theodore A. Riehl '30 selects Standard Technology Ring featuring beaver—Brass Rat

Robert E. "Tubby" Rogers advises students on how to succeed—"Be a snob and marry the boss's daughter."

Nationwide highway numbering system adopted

Black Friday; stock market crash

William Green develops automatic pilot

Richard E. Byrd flies over South Pole

1930

Karl Taylor Compton, physicist, inaugurated ninth president

Samuel W. Stratton elected first chairman of Corporation

Erwin H. Schell '12 named head of new Department of Business and Engineering Administration

John C. Slater comes from Harvard to head and reorganize Department of Physics

George R. Harrison, from Stanford, selected director of Research Laboratory of Experimental Physics

---

Clyde W. Tombaugh discovers planet Pluto

Pinball machines manufactured

William Chalmers invents plexiglass

Grant Wood paints *American Gothic*

1931

Schools organized: Vannevar Bush '16 chosen dean of engineering; Samuel C. Prescott '94, dean of science; and William Emerson, dean of architecture

Differential analyzer based on design by Vannevar Bush '16 put into operation

---

Harold C. Urey identifies heavy water, D_2O

E. O. Lawrence invents cyclotron

"Star Spangled Banner" designated national anthem

Empire State Building opens

1932

Eastman Laboratories (Bldg. 6) completed to house Departments of Physics and Chemistry

Vannevar Bush '16 becomes first vice-president

Robert J. Van de Graaff builds large high-voltage generator at Round Hill

Institute patent policy established

Tuition raised from $400 to $500 per year

Amelia Earhart flies solo across Atlantic

James Chadwick discovers neutron

Federal gasoline tax enacted

Lindbergh baby kidnapped

Edwin Land invents synthetic light polarizer

1933

Karl T. Compton appointed chairman of National Science Advisory Board by Franklin Delano Roosevelt

---

National Science Advisory Board authorized; headed by Karl T. Compton

Adolf Hitler named chancellor of Germany

Prohibition repealed

Tennessee Valley Authority established

Gold standard abandoned

1934

Harold E. Edgerton '27 and Kenneth J. Germeshausen '31 devise electrical circuitry making possible high-speed photography

---

Rudolph Kuhnold tests first practical radar

P. Travers publishes *Mary Poppins*

Shirley Temple becomes star

Dionne quintuplets born

1935

Tech Dinghy, designed by George Owen '94, launched

115-foot Cape Cod Canal model built by Civil Engineering for environmental study

---

Robert Watson Watt builds aircraft detection radar equipment

Krueger Brewing Company cans beer

John L. Lewis organizes C.I.O

Social Security Act passed

Will Rogers and Wiley Post killed in plane crash

1936

John B. Wilbur '26 develops simultaneous calculator

MIT Nautical Association formed under guidance of Walter C. "Jack" Wood '17

Jesse Owens wins four gold medals at Berlin Olympics

Howard Johnson opens first restaurant

Boulder (Hoover) Dam completed

Isador Rabi develops magnetic resonance methods for observing spectra of atoms and molecules

1937

Department of Mining Engineering becomes Department of Metallurgy

Nathaniel M. Sage '13 becomes director of Division of Industrial Cooperation

Warren K. Lewis '05 and Edwin R. Gilliland '33 invent fluid-bed method of catalytic cracking

---

Amelia Earhart lost in Pacific

Insulin controls diabetes

Hindenburg crashes at Lakehurst, New Jersey

Pablo Picasso paints *Guernica*

1938

New Rogers Building (7) completed–Department of Architecture moves to Cambridge

Gordon S. Brown '31 develops cinema integraph

Charles S. Draper '26 begins inertial guidance studies

Alfred P. Sloan ('95) Fellowship Program begun

Vannevar Bush '16 resigns to become president of Carnegie Institution in Washington

---

Lajos Biro invents ballpoint pen

Walt Disney's *Snow White and the Seven Dwarfs*, first full-length animated cartoon feature, shown

1939

James R. Killian, Jr., '26 becomes executive assistant to President Compton

Vannevar Bush '16 appointed chairman of National Defense Research Committee by Franklin Delano Roosevelt

Hoyt C. Hottel '24 and Albert G. H. Dietz '32 design first MIT solar house with funds from Godfrey L. Cabot ('81) Foundation

Old Rogers Building in Boston torn down

Igor Sikorsky builds first helicopter

Germany invades Poland; World War II begins

General Electric introduces fluorescent lighting

DDT, low-cost pesticide, introduced

1940

Master's thesis by Claude E. Shannon '40 lays groundwork for design of electronic digital computers

Alumni Pool, designed by Lawrence B. Anderson '30 and Herbert L. Beckwith '26, opens

Radiation Laboratory established to develop microwave radar–150 radar systems developed in five years–staff numbers over 4,000 by end of war

---

Zworykin invents electron microscope

German forces occupy Paris

Howard Florey develops penicillin as practical antibiotic

Maurice and Richard McDonald open drive-in hamburger stand

1941

Radar on roof of Building 6 used to "see" buildings in Boston

Charles S. Draper '26 devises "shoebox" gun sight

Harold E. Edgerton '27 commissioned to adapt ultra-highspeed stroboscopic photography for night aerial reconnaissance

---

Japan attacks Pearl Harbor; U.S. enters World War II

"Manhattan Project" begins

Eleanor Roosevelt endorses birth control

1942

Development of molecular biology begins under Francis O. Schmitt

---

Bell Aircraft tests first U.S. jet

Fire in Boston's Coconut Grove nightclub kills 492

Enrico Fermi demonstrates self-sustaining nuclear chain reaction

Sugar, gasoline, and coffee rationed in U.S.

1943

First MIT Personnel Office opens

"Big Inch" oil pipeline completed

Jefferson Memorial dedicated

Albert Hoffman discovers hallucinogenic properties of LSD

The Pentagon, world's largest office building, constructed

1944

School of Architecture becomes School of Architecture and Planning under William W. Wurster '17

Jay W. Forrester '45 and Robert R. Everett '43 of Servomechanisms Laboratory begin research that will lead to development of Whirlwind computer

---

Mark I computer developed at Harvard

Franklin D. Roosevelt elected to fourth term

Selman A. Waksman isolates streptomycin

Allied troops land on Normandy beaches; D day

1945

July 16–5:30 A.M.– Vannevar Bush '16, James B. Conant, Philip Morrison, Leslie R. Groves '17, J. R. Oppenheimer, and others witness explosion of first atomic bomb

Department of Food Technology organized

Residence for women opens at 120 Bay State Road

Franklin Roosevelt dies; Harry S. Truman becomes president

V-E Day–May 8

Hiroshima and Nagasaki bombed, V-J Day–August 14

CARE packages sent to Europe

Aerosol sprays introduced

1946

Research Laboratory of Electronics, peacetime sequel to Radiation Laboratory, established with Julius A. Stratton '23 as director

Laboratory for Nuclear Science and Engineering established with Jerrold R. Zacharias as director

---

Carbon 13 discovered

United Nations General Assembly holds first session

ENIAC (electronic numerical integrator and computer) is first automatic electronic digital computer

Vinylite phonograph records sold

1947

Klaus Liepmann establishes music curriculum

Harold E. Lobdell '17 named executive vice-president of Alumni Association

---

Chuck Yeager breaks sound barrier

Jackie Robinson becomes first black in major league baseball

Edwin H. Land demonstrates Polaroid camera

More than one million veterans enroll in college under G.I. Bill of Rights

1948

Norbert Wiener publishes *Cybernetics*

Goal of $20 million announced for Mid-Century Development Fund Drive

---

United Nations establishes World Health Organization

Bernard Baruch coins "Cold War"

Wire Recording Corporation of America announces magnetic recorder

Peter Goldmark invents long-playing record

1949

Winston Churchill speaks at Mid-Century Convocation on social implications of scientific progress

James R. Killian, Jr., '26 inaugurated tenth president—first alumnus to hold office

Karl T. Compton becomes second chairman of Corporation

Julius A. Stratton '23 becomes provost

Alvar Aalto designs new dormitory, later named for Everett M. Baker, dean of students

Servomechanisms Laboratory begins research on numerical control systems

---

U.S.S.R. tests its first atomic bomb

Apartheid becomes law in South Africa

John Glenn flies Air Force jet across U.S. in 3 hours, 46 minutes

Radio Free Europe begins broadcasting to countries behind Iron Curtain

1950

Committee on Educational Survey, Warren K. Lewis '05 chairman, presents landmark report on education at MIT

Alfred P. Sloan ('95) Foundation grants $5,250,000 to establish School of Industrial Management, with Edward P. Brooks '17 as dean, and to purchase building at 50 Memorial Drive

John E. Burchard '23 named dean of new School of Humanities and Social Science

Charles Hayden ('90) Memorial Library dedicated

McCarthy era begins

Korean War starts, General MacArthur serves as commander of UN forces

"Peanuts" by Charles Schulz distributed by UPS

Smokey the Bear becomes symbol for forest-fire prevention

1951

Course in nuclear engineering organized by Manson Benedict '32

Industrial Liaison Office established

Center for International Studies established with Max F. Millikan as director

$25,668,532 raised during Mid-Century Development Fund Drive

Julius A. Stratton '23 becomes vice-president and provost

Transcontinental television inaugurated by President Truman

Coast-to-coast direct-dial phone service established

Color television introduced in U.S.

Atomic reactor in Arcon, Idaho, produces electric power

1952

Lincoln Laboratory develops plans for Distant Early Warning (DEW) line—later SAGE (Semi-Automatic Ground Environment) digital computers used to assimilate information received by DEW line

Jerome B. Wiesner and William H. Radford '32 develop new idea for over-the-horizon communication: tropospheric scatter

Elspeth D. Rostow becomes first woman to hold professorial rank at Institute—assistant professor of history in Economics and Social Science Department

Hydrogen bomb exploded at Eniwetok

Polio epidemic spreads across U.S.

Sony introduces pocket-size transistor radios

John Mullin and Wayne Johnson demonstrate videotape

1953

Whirlwind I comes on-line as high-capacity, high-speed, reliable digital computer

Charles S. Draper's ('26) inertial guidance system monitors unpiloted flight from Massachusetts to California

Hillary and Tenzing climb Mount Everest

Julius and Ethel Rosenberg executed for treason

Lung cancer linked to cigarette smoking

Arthur Miller's *The Crucible* parallels 1692 Salem witch trials with McCarthyism and Red Scare

1954

Samuel C. Prescott '94 publishes *When MIT Was "Boston Tech," 1861–1916*

Lightweight crew wins Thames Challenge Cup at Henley

Atomic-powered submarine, USS *Nautilus*, launched

Supreme Court rules segregated schools unconstitutional

Jonas Salk develops polio vaccine serum

Gregory C. Pincus and Hudson Hoagland develop oral contraceptive pill

1955

Sebastian S. Kresge Auditorium and Chapel, designed by Eero Saarinen, dedicated

Julius A. Stratton '23 named chancellor

The Bronze Beaver established as highest Alumni Association award

Tappan Stove Company develops microwave oven

Disneyland opens

The Guinness Book of World Records published

Ann Landers advice column appears in Chicago *Sun-Times*

Automatic toll collectors introduced

1956

William Shockley '36 shares Nobel Prize for development of transistor—first MIT alumnus to win Nobel Prize

"In God We Trust" becomes official U.S. motto

Don Larsen pitches perfect no-hit, no-run, no-walk World Series game

U.S.S.R. suppresses Hungarian revolution

Elvis Presley records "Blue Suede Shoes"

1957

Tech Talk, Institute's newspaper, begins publication

Computation Center opens

James R. Killian, Jr., '26 becomes science advisor to President Eisenhower

Julius A. Stratton '23 appointed acting president

Vannevar Bush '16 elected third chairman of Corporation

John C. Sheehan achieves chemical synthesis of penicillin

Karl Taylor Compton Laboratories dedicated

European Economic Community (Common Market) established

U.S.S.R. launches *Sputnik 1*

President Eisenhower sends troops to forestall violence in Little Rock, Arkansas, desegregation crisis

Sperry Rand Corporation develops UNIVAC, solid-state computer

1958

On-campus nuclear reactor goes critical

Application of radar astronomy at Lincoln Laboratory begins with study of Venus's atmosphere and mapping of Mars's surface

---

U.S. satellite, *Explorer 1*, put into orbit

Transatlantic jet service inaugurated by BOAC in Boeing 707

USS *Nautilus* passes under North Pole

Beatnik movement spreads throughout America and Europe

1959

Joint Center for Urban Studies established with Harvard

Julius A. Stratton '23 becomes eleventh president

James R. Killian, Jr., '26 elected fourth chairman of Corporation

Vannevar Bush '16 made first honorary chairman of Corporation

Gordon Brown becomes dean of engineering

Fidel Castro becomes premier of Cuba

Project Mercury astronauts begin training

U.S.S.R. launches rocket with two monkeys–*Lunik I* reaches moon

Saint Lawrence Seaway opens

Lunik III photographs moon

1960

Second Century Fund campaign launched to raise $66 million for new buildings, endowed chairs, and research development

National Magnet Laboratory formed under Francis Bitter

Echo I communications satellite launched

Theodore Maiman develops laser

Organization of Petroleum Exporting Countries (OPEC) formed

Computerized typesetting introduced

1961

Mechanical Engineering Steam Laboratory dismantled

Charles H. Townes becomes provost

Engineering Projects Laboratory established under Robert W. Mann '50 and Kenneth R. Wadleigh '43

MIT Centennial celebrated by week-long gala chaired by John E. Burchard '23–speakers include Prime Minister Harold Macmillan of England, J. Robert Oppenheimer, and Aldous Huxley

---

Peace Corps created

Yuri Gagarin first human launched into space

Berlin Wall built

Alan B. Shepard becomes first American in space

U.S. advisors sent to Vietnam

Invasion of Cuba fails at Bay of Pigs

1962

Neurosciences Research Program for brain research organized by Francis O. Schmitt

Bruno Rossi and Riccardo Giacconi spot first observed X-ray source other than sun, ScoX-1

John F. Enders produces successful measles vaccine

Rachel Carson writes *Silent Spring*

John Glenn orbits earth

Ranger IV satellite reaches moon

Cuban missile crisis

1963

Time-sharing computer systems developed by Robert M. Fano '41, Fernando J. Corbató '56, and others in Project MAC (Machine-Aided Cognition and Multiple-Access Computer)

Center for Space Research organized

Second Century Fund Campaign raises $98 million

First tower of Stanley R. McCormick Hall, women's dormitory, completed—gift of Katharine Dexter McCormick '04

Beatles recording "I Want to Hold Your Hand" becomes hit

Winston Churchill made first honorary U.S. citizen

President Kennedy assassinated—November 22

Quasars discovered

Roche Laboratories introduce Valium

1964

Charles H. Townes shares Nobel Prize for invention of laser

Department of Psychology established under Hans-Lukas Teuber

School of Industrial Management renamed Alfred P. Sloan ('95) School of Management

Cecil H. ('23) and Ida M. Green Center for Earth Sciences, designed by I. M. Pei '40, completed

Department of Political Science formed with Robert C. Wood as head

---

U.S. Post Office adopts ZIP Codes

Ranger VII returns close-up photos of moon's surface

Civil Rights Act passed

World's longest suspension bridge, Verrazano Narrows, opens

1965

New Student Center, designed by Eduardo Catalano, named in honor of Julius A. Stratton '23

First Lincoln Experimental Satellite (*LES-1*) launched to study satellite communications

Graduate House renamed in honor of Avery A. Ashdown '24

Richard P. Feynmann '39 shares Nobel Prize for contribution in quantum electrodynamics

Robert B. Woodward '36 receives Nobel Prize for synthesizing quinine and other compounds

---

Astronaut Edward White walks in space

Congress authorizes use of U.S. ground troops in Vietnam

President Johnson outlines program for a "Great Society"

Power failure blacks out seven states in northeast U.S.

1966

Alexander Calder's stabile *La Grande Voile* becomes centerpiece of McDermott Court

Julius A. Stratton '23 retires as president—becomes chairman of Ford Foundation

Howard W. Johnson becomes twelfth president

Jerome B. Wiesner becomes provost

Robert S. Mulliken '17 awarded Nobel Prize in Chemistry

Apollo guidance system developed by Instrumentation Lab

Surveyor 1 makes soft landing on moon

Seventy-eight species declared rare and endangered

Medicare program goes into effect

"Star Trek" becomes popular television series

1967

Center for Advanced Visual Studies organized by Gyorgy Kepes

Jerome Y. Lettvin '47 and Timothy Leary debate individual's right to "turn on, tune in, drop out" with LSD

MIT-Wellesley cross-registration program announced

Dow Chemical Company recruiting protested by antiwar demonstrators

---

Dr. Christiaan Barnard performs heart transplant

Synthetic DNA produced

Arab-Israeli War lasts six days

Race riots rock 127 U.S. cities

1968

Sophomores boycott Field Day, ending 64-year tradition

Libraries add millionth volume—first edition of Walt Whitman's *Leaves of Grass*, given by I. Austin Kelly '26

Pass/Fail grading introduced for freshmen

Joint program with Woods Hole Oceanographic Institute initiated

Har G. Khorana receives Nobel Prize for work with DNA

MIT and California Institute of Technology compete in Great Electric Car Race

---

Manned *Apollo 8* orbits moon

Martin Luther King, Jr., assassinated

James D. Watson's *The Double Helix* published

Robert F. Kennedy assassinated

Peggy Fleming wins only U.S. gold medal in Winter Olympics

1969

Murray Gell-Mann '51 wins Nobel Prize in Physics

Salvador E. Luria shares Nobel Prize for discovery of mutations in viruses

Forty-eight MIT professors lead "research stoppage" as protest to "government misuse of scientific knowledge"

Science Action Coordinating Committee organizes protests of defense research, ROTC, and Vietnam War

MIT Community Service Fund established

Coed dormitories gain official sanction

Antiwar protests spread across U.S.

U.S. removes cyclamates from market, limits use of MSG, bans DDT

Neil Armstrong and Edwin Aldrin '63 land on moon

Mariner photographs reveal Mars's surface

1970

Offices of president and chairman occupied by 60 people

Joint Program in Health Sciences and Technology developed with Harvard

Instrumentation Laboratory renamed Charles Stark Draper ('26) Laboratory

Paul A. Samuelson wins Nobel Prize in Economic Science

Forty-three colleges compete in Clean Air Car Race from MIT to California

Undergraduate Research Opportunities Program (UROP) begins under leadership of Margaret L. A. MacVicar '65

Unmanned Soviet *Venera 7* lands on Venus

448 U.S. universities and colleges closed following Kent State slayings

Price of gold on free market falls below $35 per ounce

150-inch reflecting telescope at Kitt Peak Observatory in Arizona completed

1971

Independent Activities Period initiated

Jerome B. Wiesner inaugurated thirteenth president

Paul E. Gray '54 becomes chancellor

Howard W. Johnson becomes fifth chairman of Corporation

James R. Killian, Jr., '26 becomes second honorary chairman of Corporation

Walter A. Rosenblith named provost

MIT Council for Arts formed

MIT Historical Collections established

National Women's Political Caucus founded

Daniel Ellsberg leaks "Pentagon Papers"

American astronomers discover two "new" galaxies adjacent to Milky Way

Attica Prison riot leaves 43 dead

1968 My Lai massacre exposed

1972

Energy Laboratory established with David C. White as director

Carola B. Eisenberg named first woman dean for student affairs

MIT team wins Tiddlywink Championship in England

Cambridge police tear-gas antiwar protesters on Massachusetts Avenue

---

Okinawa returned to Japan after 27 years of U.S. occupation

Eleven Israeli athletes killed at Summer Olympics in Munich

World Trade Center, designed by Minoru Yamasaki, opens

Equal Rights Amendment goes to states for ratification

1973

Center for Cancer Research, headed by Salvador E. Luria, established to study fundamentals of cancer biology

Draper Laboratory divested

College of Health Science, Technology, and Management established

One Hundred Years of the New Woman convocation commemorates centennial of Ellen Swallow Richards's graduation

MIT Symphony Orchestra makes first national concert tour under leadership of David Epstein

---

Spiro T. Agnew resigns vice-presidency

Pioneer 10 transmits television pictures from within 81,000 miles of Jupiter

Last U.S. troops leave South Vietnam

Energy crisis grips world following Arab oil embargo

1974

Great Court renamed Killian Court in honor of James R. Killian, Jr., '26

MIT Housing Program in Cambridge provides 684 new homes for elderly

World's largest yo-yo has tryout off side of Green Building

Quarter Century Club and Silver Club merge to include all 25-year MIT employees

Institute athletic teams compete in 22 intercollegiate sports

Harvard-MIT Rehabilitation Engineering Center established

Watergate cover-up hearings begin

Hank Aaron betters Babe Ruth's 714 career home run record

"Streaking" becomes U.S. fad

Richard M. Nixon resigns

1975

Five-year "MIT Leadership Campaign" to raise $250 million announced

Chemical Engineering Building, designed by I. M. Pei '40, named in honor of Ralph Landau '41

Seeley G. Mudd Building, housing Center for Cancer Research, dedicated

SAS-3 (Small Astronomy Satellite 3) launched by Center for Space Research to search for X-ray sources

David Baltimore '61 shares Nobel Prize for Medicine and Physiology

Haldeman, Mitchell, and Erlichman convicted in Watergate case

Surrender of South Vietnamese ends Vietnam War

Apollo-Soyuz rendezvous in space

Boston begins city-wide busing to achieve integration

1976

Francis E. Wylie publishes *MIT in Perspective*

Project MAC renamed Laboratory for Computer Science

MIT scientists led by Har G. Khorana complete synthesis of first man-made gene that is fully functional in a living cell

Samuel C. C. Ting wins Nobel Prize for "J" particle

Convocation celebrating centennial of telephone held at MIT

MIT becomes first private institution designated as Sea Grant College

Concorde begins transatlantic passenger service

Viking 1 lands on Mars

Queen Elizabeth II is first British monarch to visit U.S.

Mao Tse-tung dies

Patricia Hearst convicted of armed robbery

1977

Alumni Center established in Building 10

Frank Press becomes science advisor to Jimmy Carter

College of Health Science, Technology, and Management named in honor of Helen F. and U. A. Whitaker '23

Margaret Hutchinson Compton Gallery opens—gift of Class of '38

Center for Materials Research in Archaeology and Ethnology established under direction of Heather N. Lechtman

Alex Haley's *Roots* seen by 130 million viewers

Alaskan pipeline opens

Anwar Sadat of Egypt visits Israel

President Carter grants conditional pardon for Vietnam draft evaders

1978

Worst blizzard of century closes Institute for week

Henry G. Steinbrenner ('27) Stadium dedicated

Gymnast Leslie Harris '81 becomes first MIT woman to win All-American status

Research Laboratory of Physical Chemistry celebrates 75th anniversary

---

Pope John Paul II elected first non-Italian pope in four centuries

Egypt and Israel sign peace agreement

Panama Canal Treaty ratified

Louise Brown, "test-tube baby," born in England

1979

Jay W. Forrester '45 inducted into National Inventors' Hall of Fame for core memory invention

Sheila E. Widnall '60 becomes first woman elected chairman of MIT faculty

Tuition: $5,300

His Highness the Aga Khan gives $11.5 million to endow Islamic architecture program at MIT and Harvard

World Conference on Faith, Science, and the Future held in Kresge Auditorium by World Council of Churches

---

Idi Amin Dada of Uganda overthrown

Three Mile Island nuclear power plant malfunctions

John F. Kennedy Library opens in Boston

During anti-Shah demonstrations 50 Americans taken hostage in American Embassy in Teheran

1980

Ellen Swallow Richards '73 Lobby dedicated

MIT College Bowl team, Brian E. Clouse '80 captain, wins 14 games in regional and national competition

MIT Leadership Campaign raises $250,232,000

Francis E. Low appointed provost

Paul E. Gray '54 becomes fourteenth president of MIT

U.S.S.R. invades Afghanistan

U.S. speed skater Eric Heiden wins five gold medals at Winter Olympics

U.S. boycotts Summer Olympics

Notes

Abbreviations Used in Journal Titles

AIEE
American Institute
of Electrical Engineering

BSTJ
Bell System
Technical Journal

EPRI
Electric Power
Research Institute

IEE
Institution of
Electrical Engineers
(British)

IEEE
Institute of Electrical and
Electronics Engineers

IRE
Institute of Radio
Engineers

SPEE
Society for the Promotion
of Engineering Education

Chapter 1

1. Samuel C. Prescott, *When MIT Was "Boston Tech," 1861–1916* (Cambridge, Massachusetts: Technology Press of MIT, 1954).

Chapter 2

1. F. E. Terman, "Electrical Engineering Education," *Student Quarterly, IRE* (May 1962): 44–47; F. E. Terman, "A Brief History of Electrical

Engineering," *IEEE Proceedings* 64 (1976): 1399–1407; Mendell P. Weinbach, "The College of Engineering," in *The University of Missouri: A Centennial History*, ed. Jonas Viles (1939), 384–385 (contains the story of early electrical engineering instruction); *EPRI Journal*, special issue commemorating Edison's invention of the electric lighting system in 1879, guest editor, Nilo A. Lindgren (March 1979): 36.

2. D. F. Berth and H. G. Smith, "When the Sparks Began to Fly, A Century of Electrical Engineering at Cornell," *Engineering Cornell Quarterly* 11, no. 2 (Summer 1976): 2–24.

3. C. S. Derganc, "Thomas Edison and His Electric Lighting System," *IEEE Spectrum* 16, (February 1979): 50–59; Robert Conot, *A Streak of Luck* (New York: Seaview Books, 1979), a detailed story of Edison's life, including his invention of the carbon filament lamp; *EPRI Journal*, special issue commemorating Edison's invention of electric lighting (March 1979): 22–23.

4. Weinbach, ''College of Engineering.''

5. Robert V. Bruce, *Bell: Alexander Graham Bell and the Conquest of Solitude* (Boston: Little, Brown and Company, 1973).

6. Here and in the following paragraphs we quote from Bruce, *Bell.*

7. Bruce, *Bell*; John Brooks, *Telephone: The First Hundred Years* (New York: Harper and Row, 1976); *A History of Engineering and Science in the Bell System. The Early Years (1875–1925)*, prepared by members of the technical staff, Bell Telephone Laboratories, M. D. Fagen, ed. (1976).

8. *EPRI Journal*, special issue commemorating Edison's invention of electric lighting (March 1979).

9. Silas W. Holman, *Matter, Energy, Force, and Work* (New York: Macmillan, 1898).

Chapter 3

1. Dugald C. Jackson, ''The Typical College Course Dealing with the Professional and Theoretical Phases of Electrical Engineering,'' *Science* 18, no. 466 (1903): 710–716.

2. Dugald C. Jackson, ''Preliminary Report on the Joint Committee on Engineering Education,'' *SPEE Proceedings* 16 (1908): 47–60.

3. Harold Pender, *Principles of Electrical Engineering* (New York: McGraw-Hill, 1910).

4. Harold Pender, *American Handbook for Electrical Engineers* (New York: Wiley, 1914), 2,023 pages.

5. Samuel C. Prescott, *When MIT was ''Boston Tech,'' 1861–1916* (Cambridge, Massachusetts: Technology Press of MIT, 1954).

6. Henry S. Pritchett, ''A Tale of Two Presidents,'' *Technology Review* 25 (1923): 199–200.

7. Arthur E. Kennelly and Uhachi Nabeshima, ''The Transient Process of Establishing a Steadily Alternating Current on a Long Line, from Laboratory Measurements of an Artificial Line,'' *Proceedings of the American Philosophical Society* 59 (1920): 325–370; Louis F. Woodruff, *Electric Power Transmission and Distribution* (New York: Wiley, 1925).

8. Arthur E. Kennelly, ''Impedance,'' *AIEE Transactions* 10 (1893): 175–232; Charles P. Steinmetz, ''Complex Quantities and Their Use in Electrical Engineering,'' *Proceedings of the International Electrical Congress* (New York: AIEE, 1894), 33–75.

9. Vannevar Bush, ''Arthur Edwin Kennelly,'' *National Academy of Sciences, Biographical Memoirs* 22 (1940): 83–119.

10. Waldo V. Lyon, ''Transient Conditions in Electrical Machinery,'' *AIEE Journal* 42 (1923): 388–407; Charles Kingsley, Jr., ''Saturated Synchronous Reactance,'' *Electrical Engineering* 54 (1935): 300–305; Yu Hsiu Ku, ''Transient Analysis of A-C Machinery,'' *AIEE Transactions* 48 (1929): 707–715; Waldo V. Lyon, *Transient Analysis of Alternating-Current Machinery* (Cambridge, Massachusetts, and New York: Technology Press–Wiley, 1954); Vannevar Bush and Parry Moon, ''A Precision Measurement of Puncture Voltage,'' *AIEE Transactions* 46 (1927): 1025–1038.

11. Parry Moon, *The Scientific Basis of Illuminating Engineering* (New York: McGraw-Hill, 1936).

12. Parry Moon, ''The New Approach to Room Lighting,'' *Illuminating Engineering* 44 (1949): 221; Parry Moon and Domina Eberle Spencer, ''How to Design Your Luminous Ceilings,'' *Illuminating Engineering* 46 (1951): 295; Parry Moon and Domina Eberle Spencer, *Field Theory for Engineers* (New York: Van Nostrand, 1961).

13. Dugald C. Jackson, ''Japanese Higher Education and Research in the Physical Sciences,'' *Science* 24 (1936): 189–192; ''Industrial and Cultural Japan,'' *Electrical Engineering* 56 (1937): 208–215; ''Engineering's Part in the Development of Civilization,'' *Mechanical Engineering* (July–December 1938), reprinted in book form by the American Society of Mechanical Engineers (February 1939), 114 pp.; *Present Status and Trends of Engineering Education in the United States*, Report to the Committee on Engineering Schools, Engineers' Council for Professional Development (1939), 177 pp.; ''Trends in Engineering Education,'' *Science* 92 (1940): 183–189; ''Engineering Education,'' *Journal of Engineering Education* 29 (June 1939).

Part II Introduction

1. Arthur Amos Noyes, Class Dinner at Tech Union, November 1, 1908.

Chapter 4

1. Vannevar Bush, *Pieces of the Action* (New York: Morrow, 1970).

2. Oliver Heaviside, *Electrical Papers* (New York and London: Macmillan, 1892).

3. Bush, *Pieces of the Action.*

4. Vannevar Bush, U.S. Patent No. 1,048,649, *Profile Tracer*, issued December 31, 1912.

5. Bush, *Pieces of the Action.*

6. Vannevar Bush and Ralph D. Booth, ''Power System Transients,'' *AIEE Transactions* 44 (1925): 80–103.

7. John R. Carson, ''Theory of Transient Oscillations in Electric Networks,'' *AIEE Transactions* 38 (1919): 345.

8. James and Sir William Thomson, ''An Integrating Machine'' and ''An Instrument for Calculating the Integral of a Product,'' *Proceedings of the Royal Society of London* 24 (1876): 262–268.

9. Vannevar Bush, Frank D. Gage, and Herbert R. Stewart, ''A Continuous Integraph,'' *Journal of the Franklin Institute* 208 (1927): 63–84.

10. Vannevar Bush and Harold L. Hazen, ''Integraph Solutions of Differential Equations,'' *Journal of the Franklin Institute* 208 (1927): 575–615.

11. Vannevar Bush, ''The Differential Analyzer: A New Machine for Solving Differential Equations,'' *Journal of the Franklin Institute* 212 (1931): 447–488.

12. Truman S. Gray, ''A Photoelectric Integraph,'' *Journal of the Franklin Institute* 212 (1931): 77–102.

Chapter 5

1. W. S. Murray, *Superpower: Its Genesis and Future* (New York: McGraw-Hill, 1925).

2. *AIEE Transactions* 43 (1924): 1–103. Six papers on power systems with 32 pages of discussion presented at the Midwinter Convention, February 1924, in Philadelphia.

3. W. W. Lewis, ''A New Short-Circuit Calculating Table,'' *General Electric Review* (1920): 669.

4. O. R. Schurig, ''A Miniature AC Transmission System for the Practical Solution of Network and Transmission System Problems,'' *AIEE Transactions* 42 (1923): 831.

5. H. H. Spencer and H. L. Hazen, ''Artificial Representation of Power Systems,'' *AIEE Transactions* 44 (1925): 72.

6. G. S. Brown, ''Eloge: Harold Locke Hazen, 1901–1980,'' *Annals of the History of Computing* 3, no. 1 (January 1981).

7. Eric T. B. Gross, ''Network Analyzer Facilities Keep Pace with Industry Growth,'' *Electric Light and Power*, March 15, 1956.

Chapter 6

1. Julius A. Stratton, ''The M.I.T. Radio Society,'' *Tech Engineering News* 3 (1922): 176.

2. Julius A. Stratton, ''Manuel Sandoval Vallarta,'' *Year Book of the American Philosophical Society* 1978: 108–113.

3. Timothy E. Shea, *Transmission Networks and Wave Filters* (New York: Van Nostrand, 1929).

4. Arthur H. Lewis, *The Day They Shook the Plum Tree* (New York: Harcourt, Brace & World, 1963).

5. Gordon G. Macintosh, ''IXV-IXAN, An Experimental Station with the World as Its Laboratory,'' *QST*, July 1928, 19–22.

6. Julius A. Stratton, ''Streuungs Koeffizient von Wasserstoff nach der Wellenmechanik,'' *Helvetica Physica Acta* 1, no. 1 (1928): 47–74.

7. Julius A. Stratton, ''The Effects of Rain and Fog on the Propagation of Very Short Waves,'' *IRE Proceedings* 18 (1930): 1064.

8. Julius A. Stratton and Henry E. Houghton, ''A Theoretical Investigation of the Transmission of Light Through Fog,'' *Physical Review* 38 (1931): 159–165.

9. Henry G. Houghton, ''Transmission of Visible Light through Fog,'' *Physical Review* 38 (1931): 152–158.

10. Henry G. Houghton, ''The Size and Size Distribution of Fog Particles,'' *Physics* 2 (1932): 467–471.

11. Henry G. Houghton, ''A Study of the Evaporation of Small Water Drops,'' *Physics* 4 (1933): 419–424.

12. Henry G. Houghton and William H. Radford, ''On the Local Dissipation of Natural Fog,'' in *Papers in Physical Oceanography and Meteorology*, MIT and Woods Hole Oceanographic Institution, vol. 6, no. 3 (October 1938), in four parts, with prefatory note by Edward L. Bowles, 3–63; vol. 6, no. 4 (November 1938), in two parts, 2–31.

13. Julius A. Stratton and Howard A. Chinn, ''The Radiation Characteristics of a Vertical Half-Wave Antenna,'' *IRE Proceedings* 20 (1932): 1892–1913.

14. Julius A. Stratton, *Electromagnetic Theory*, (New York: McGraw-Hill, 1941).

15. Lord Rayleigh, ''On the Passage of Electric Waves through Tubes, or The Vibrations of Dielectric Cylinders,'' *Philosophical Magazine* 43 (1897): 125–132.

16. Wilmer L. Barrow, ''Transmission of Electromagnetic Waves in Hollow Tubes of Metal,'' *IRE Proceedings* 24 (October 1936): 1298–1328.

17. George C. Southworth, ''Hyper-Frequency Wave Guides—General Considerations and Experimental Results,'' *BSTJ* 15 (April 1936): 284–309; John R. Carson, Sallie P. Mead, and Sergei A. Schelkunoff, ''Hyper-Frequency Wave Guides—Mathematical Theory,'' *BSTJ* 15 (April 1936): 310–333.

18. Edward L. Bowles, Wilmer L. Barrow, William M. Hall, Frank D. Lewis, and Donald E. Kerr, ''The CAA-MIT Microwave Instrument Landing System,'' *AIEE Transactions* 59 (1940): 859–865.

19. Edward L. Bowles, ''Science at Round Hill,'' *Technology Review* 37 (1934): 18.

Chapter 7

1. Clyde W. Park, ''The Cooperative System of Education,'' *University of Cincinnati Studies*, vol. 11, pt. 1 (1925).

2. Magnus W. Alexander, "A New Method of Training Engineers," *AIEE Transactions* 27 (1908): 1459–1498.

3. David F. Noble, *America by Design: Science, Technology, and the Rise of Corporate Capitalism* (New York: Knopf, 1977).

4. Karl L. Wildes, "Cooperative Courses—Their Development and Operating Principles," *AIEE Transactions* 49 (1930): 1086–1094.

5. Magnus W. Alexander and Dugald C. Jackson, "Requirements of the Engineering Industries and the Education of Engineers," *Mechanical Engineering* 43 (1921): 391–395.

6. M. F. Gardner and J. L. Barnes, *Transients in Linear Systems* (New York: Wiley, 1942).

7. *Hand in Hand* (Medford, Massachusetts: Gordon, 1958).

Chapter 8

1. H. E. Edgerton and J. R. Killian, *Moments of Vision* (Cambridge, Massachusetts: MIT Press, 1979).

2. W. V. Lyon and H. E. Edgerton, "Transient Angular Oscillation of Synchronous Machines," *AIEE Transactions* 49 (1930): 686–699.

3. I. H. Summers and J. B. McClure, "Progress in the Study of System Stability," *AIEE Transactions* 49 (1930): 132.

4. H. E. Edgerton and F. J. Zak, "The Pulling into Step of a Synchronous Induction Motor (An Application of the MIT Integraph)," *IEE Journal* 68 (1930): 1205–1210.

5. H. E. Edgerton and Paul Fourmarier, "The Pulling into Step of a Salient-Pole Synchronous Motor," *AIEE Transactions* 50 (1931): 769–781.

6. Gjon Mili, *Photographs and Recollections* (Boston: Little, Brown [A New York Graphic Society Book], 1980).

7. H. E. Edgerton and F. E. Barstow, "Further Studies of Glass Fracture with High-Speed Photography," *Journal of the American Ceramic Society* 24 (April 1941).

8. J. E. Burchard, *MIT in World War II* (New York: Tech Press–Wiley, 1948).

Chapter 9

1. Ernst A. Guillemin, *Communication Networks*, vol. 1, *The Classical Theory of Lumped Constant Networks* (New York: Wiley, 1931).

2. Ernst A. Guillemin, *Communication Networks*, vol. 2 *The Classical Theory of Long Lines, Filters, and Related Networks* (New York: Wiley, 1935).

3. Ronald M. Foster, "A Reactance Theorem," *BSTJ* 3 (1924): 259.

4. Wilhelm Cauer, "Die Verwirklichung von Wechselstromwiderständen vorgeschriebener Frequenzabhängigkeit," *Archiv fur Elektrotechnik* 17 (1927): 355.

5. George A. Campbell, *Collected Papers* (New York: AT&T, 1937).

6. M. D. Fagen, ed., *A History of Engineering and Science in the Bell System: The Early Years, 1875–1925* (Murray Hill, New Jersey: Bell Laboratories, 1975).

7. Ernst A. Guillemin, *Synthesis of Passive Networks* (New York: Wiley, 1957).

8. David F. Tuttle, Jr., *Network Synthesis* (New York: Wiley, 1958).

9. David F. Tuttle, Jr., *Electric Networks* (New York: McGraw-Hill, 1965).

10. M. E. Van Valkenburg, *Introduction to Modern Network Synthesis* (New York: Wiley, 1960).

11. Louis Weinberg, *Network Analysis and Synthesis* (New York: McGraw-Hill, 1962).

12. Ernst A. Guillemin, *Introductory Circuit Theory* (New York: Wiley, 1953).

13. Ernst A. Guillemin, *The Mathematics of Circuit Theory* (New York: Technology Press–Wiley, 1949).

14. Ernst A. Guillemin, *Theory of Linear Physical Systems, from the Viewpoint of Classical Dynamics, Including Fourier Methods* (New York: Wiley, 1963).

15. Rudolf E. Kalman and Nicholas DeClaris, eds., *Aspects of Network and System Theory* (New York: Holt, Rinehart and Winston, 1971).

Chapter 10

1. D. J. Kevles, *The Physicists* (New York: Vintage Books, 1979), chap. 14.

2. J. G. Trump, H. F. Hare, et al., "Two-Million-Volt Roentgen-Ray Therapy Using Rotation," *American Journal of Roentgenology and Radium Therapy* 66 (1951): 613; J. G. Trump, "High-Energy Electron Treatment of Wastewater Liquid Residuals," *Final Report to the National Science Foundation* (1980); J. G. Trump, "Energized Electrons Tackle Municipal Sludge," *American Scientist* 69 (1981): 276–284.

Chapter 11

1. A. R. von Hippel, *Dielectrics and Waves* (New York: Wiley, 1954).

2. A. R. von Hippel, ed., *Dielectric Materials and Applications* (New York: Technology Press–Wiley, 1954).

3. A. R. von Hippel, ed., *Molecular Science and Molecular Engineering* (New York: Technology Press–Wiley, 1959).

4. A. R. von Hippel, ed., *The Molecular Designing of Materials and Devices* (Cambridge, Massachusetts: MIT Press, 1965).

Chapter 12

1. Vannevar Bush, *Pieces of the Action* (New York: Morrow, 1970), 45.

2. A. Rosenblueth, N. Wiener, and J. Bigelow, "Behavior, Purpose, and Teleology," *Philosophy of Science* 10 (1943): 18–24.

Chapter 14

1. J. C. Maxwell, "On Governors," *Proceedings of the Royal Society* 16 (1868): 270–283; E. J. Routh, *Stability of a Given State of Motion* (London: Macmillan, 1877).

2. N. Minorsky, "Directional Stability of Automatically Steered Bodies," *Journal of the American Society of Naval Engineers* (May 1922).

3. H. L. Hazen, "Theory of Servo-Mechanisms," *Journal of the Franklin Institute* (September 1934); "Design and Test of a High-Performance Servo-Mechanism," *Journal of the Franklin Institute* (November 1934).

4. G. S. Brown and D. P. Campbell, "Instrument Engineering, Its Growth and Promise in Process Control," *Mechanical Engineering* (February 1950): 124–127.

5. Later books on numerical control include *Programming for Numerical Control*, by Roberts and Prentice, and *Numerical Control User's Handbook*, with text material written by a group of experts under the editorship of W. H. P. Leslie, head of the Numerical Control Division of the National Engineering Laboratory, East Kilbride, Scotland.

Chapter 15

1. Kent C. Redmond and Thomas M. Smith, *Project Whirlwind: The History of a Pioneer Computer* (Bedford, Massachusetts: Digital Press, 1980).

2. MIT files on this research have been lost, but Edward L. Bowles supplied a copy of the Radford report (88 pages plus appendices).

3. Robert S. Oelman, chairman of the National Cash Register Company, supplied copies of the Overbeck reports (1939–1942).

Chapter 17

1. Steve J. Heims, *John von Neumann and Norbert Wiener: From Mathematics to the Technologies of Life and Death* (Cambridge, Massachusetts: MIT Press, 1980), 187–188; confirmed in interview with James Killian.

2. Heims, *John von Neumann and Norbert Wiener*, 189 and 470.

3. Interview with James Killian.

4. Interview with Killian.

5. Kent C. Redmond and Thomas M. Smith, *Project Whirlwind: The History of a Pioneer Computer* (Bedford, Massachusetts: Digital Press, 1980).

Chapter 18

1. Herman Feshbach and Uno Ingard, eds., *In Honor of Philip M. Morse* (Cambridge, Massachusetts: MIT Press, 1969).

2. Philip M. Morse, *In at the Beginnings* (Cambridge, Massachusetts: MIT Press, 1977).

3. Philip M. Morse, *Vibration and Sound* (New York: McGraw-Hill, 1936).

4. Charles S. Draper and Philip M. Morse, "Acoustical Analysis of the Pressure Waves Accompanying Detonation in the Internal-Combustion Engine," *Proceedings of the Fifth International Congress for Applied Mechanics* (1939): 727.

5. Morse, *In at the Beginnings*, 156–212.

6. James Phinney Baxter, III, *Scientists against Time* (Cambridge, Massachusetts: MIT Press paperback, 1968 [Little, Brown, 1946]), 188–191.

7. Leo L. Beranek, "The Notebooks of Wallace C. Sabine," *Journal of the Acoustical Society of America* 61 (March 1977): 629–639. This volume also contains a valuable review of American acoustics up to 1940.

Chapter 19

1. G. S. Brown, "Educating Electrical Engineers to Exploit Science," *Electrical Engineering* 74 (1947): 110–115.

2. Gordon S. Brown, "The Modern Engineer Should Be Educated as a Scientist," *Journal of Engineering Education* 43, no. 4 (December 1952).

3. Robert M. Fano, Lan Jen Chu, and Richard B. Adler, *Electromagnetic Fields, Energy, and Forces* (New York: Wiley, 1960).

4. Henry J. Zimmermann and Samuel J. Mason, *Electronic Circuit Theory* (New York: Wiley, 1959).

5. Truman S. Gray, *Applied Electronics*, 2nd ed. (New York: Technology Press–Wiley, 1954).

6. Richard B. Adler, Lan Jen Chu, and Robert M. Fano, *Electromagnetic Energy Transmission and Radiation* (New York: Wiley, 1960).

7. David C. White and Herbert H. Woodson, *Electromechanical Energy Conversion* (New York: Wiley, 1959).

8. Gabriel Kron, "The Application of Tensors to the Analysis of Rotating Electric Machinery," parts I-XVIII, *General Electric Review* (1935); W. J. Gibbs, *Tensors in Electrical Machine Theory* (London: Chapman and Hall, 1952); Bernard Adkins, *The General Theory of Electrical Machines* (New York: Wiley, 1957).

9. Zdenek J. Stekly and H. H. Woodson, "Rotating Machinery Using Superconductors," *IEEE Transactions on Aerospace and Electronic Systems* AES-2, no. 2 (April 1964): 826–842.

10. Charles Kingsley, Jr., Gerald L. Wilson, James L. Kirtley, Jr., Thomas A. Keim, Joseph L. Smith, Jr., and Philip Thullen, "Steady-State Electrical Tests on the MIT-EPRI 3-MVA Superconducting Generator," paper presented at the IEEE Joint Power Generation Conference, September 1975.

11. "Demonstration of an Advanced Superconducting Generator," interim report prepared by the MIT Cryogenic Engineering Laboratory and Electric Power Systems Engineering Laboratory for the United States Department of Energy, April 11, 1979.

Chapter 21

1. Philip M. Morse, *In at the Beginnings* (Cambridge, Massachusetts: MIT Press, 1977), 299–303.

2. Jean E. Sammet, *Programming Languages: History and Fundamentals*, ed. George Forsythe (Englewood Cliffs, New Jersey: Prentice-Hall, 1969).

3. Private communication, Norman Taylor to Karl Wildes, 1975.

4. Saul Rosen, ed., *Programming Systems and Languages* (New York: McGraw-Hill, 1967).

Chapter 22

1. Author's interview with Fernando Corbató.

2. R. M. Fano, "Project MAC," in *Encyclopedia of Computer Science and Technology*, vol. 12 (New York: Marcel Dekker, 1979), 339–360.

3. Fano, "Project MAC," 352.

Index

Page numbers in italics indicate photographs or other illustrations.

Acoustics Laboratory, *22*, 257, 267, 302–308
AC synchronous motor, 162, *163*
Adams, Charles W., 290, 339
Adams, Comfort A., 52, 63
Adams, Roger, 185
Adler, Richard B., 137, 173–175, *174*, 248, 257–258, *258*, 269, 272, 316, 318, 369–370, *369*, *373*
AED (Automated Engineering Design), 225
Aigrain, Pierre, 173
Alba, C. J., 121
Alexander, Magnus W., 128
Alexander, Robert M., 198
Alger, John R. M., 132
Alger, Philip L., 52
Allen, A. J., 197
Allen, C. Francis, 28
Allen, Jonathan, *270*, 271, 275, 278
Allis, W. P., 247, 271, 303
Alternating current, 30
Alvarez, Luis, 198
American Bell Telephone Company, 25
American Telephone and Telegraph Company, 30, 48, 54, 75, 156
Analog machines, 90–92
Anderson, Harlan E., 345
Anderson, John W., 213, 215, 216
Anderson, Lawrence B., 304
Angell, James B., 63, 248
Anthony, William A., 23
APT (Automatically Programmed Tools), 224–225
Archambault, Bennett, 201
Arden, Dean N., 134, 337
Arguimbau, Lawrence B., 248
Arnow, Jack, 340
Athans, Michael, 226
Athena, Project, 365
Atomic Energy Commission, 289

Augustus Lowell Laboratories, 38, 41
Automated Engineering Design (AED), 225
Automatically Programmed Tools (APT), 224–225

Babbage, Charles, 90
Bailey, George W., 185
Bainbridge, Kenneth T., 196–197
Ballantyne, Joseph M., 338
Bangratz, Ernst G., *71*
Barker, George F., 23
Barker, Joseph W., *71*
Barnes, John A., 85
Barnett, Michael P., 344
Barrett, Alan H., 272
Barrow, Wilmer L., 71, 106, 119–121, *122*, 196, 209
Barss, William R., 303
Barstow, Frederick E., 150
Barta Building, *286*
Bartlett, Dana P., 29–30
Baruch, Jordan J., 305, *306*, 307
Basic Research Division, 207
Basore, Bennett L., 268
Bassett, Preston R., 186
Battin, Richard H., 265
Baumann, F. W., 121
Bavelas, Alex, 262, 267
Beal, R. R., 196
Beardsley, Kenneth D., 145
Bekefi, George, 271
Bell, Alexander Graham, 16, 19, 23–25, 54
Bell System, 156
Bell Telephone Laboratories, 113, 121, 174, 184, 186, 187, 195, 229, 244, 255, 257, 284, 287, 300, 307, 341, 351
Benington, Herbert D., 133, 340
Bennett, Ralph D., *183*, 188
Bennett, Richard K., 341
Beranek, Leo L., 303, 304–305, *306*, 308
Bergvall, R. C., 99
Berker, Nihat, 273
Bers, Abraham, 271

Beverage, Harold H., 199
Bibber, Harold C., *64*
Bicknell, Joseph, 284
Bigelow, Julian, 188
Bilter, F., 247
Binary digit (bit), 287
Bingham, Lloyd A., *71*
Birgeneau, Robert, 273
Bit (binary digit), 287
Bittenbender, Robert, 218
Blair, John, 318
Blanchard, Ednah, 72, *73*, 74
Blodgett, George W., 29
"Blue books," 74
Boehne, Eugene W., 128, 132–133, *133*
Bohr, Niels, 166, *168*
Boilen, Sheldon, 345
Bolt, Richard H., 303, 304, *306*, 308
Boot, A. H., 195
Booth, Ralph D., 87, 99, 289
Bose, Amar G., *254*, *258*, 262
Boston University, 24
"Boston Tech," 12, 14
Bosworth, Welles, 51
Bourland, Hardy M., *219*
Bowen, Edward G., *191*, 193, 195
Bowen, Harold G., 184
Bowles, Edward, 45, 69, 70, *71*, 74, 75, 81, 100, 106–123, *107*, 148, 156, 195, 196, 199–209, *204*, 255
Brainerd, J. G., 284
Braida, Louis, 278
Briggs, Richard, 271
Brockett, Roger W., 226
Brown, Gordon S., 70, 74, 77, 95, 103–104, 117, 128, 134, 147, 166, 173, *183*, 192, 196, 210–224, *212*, *217*, 280–281, 289, 307, 308, 310–325, *311*, *312*, *313*, 348, 358
Brown, Sanborn C., 247, 271
Bruce, Robert, 24
Brune, Otto, 157
Buck, Dudley A., 175, 337, 338

Buckley, Page S., 217–218
Building 6, *179*
Bunker, John W. M., 77
Burke, Bernard, 273
Bursch, William O., 130
Bush, Vannevar, 32, 48, *64*, 68, *70*, *71*, 74–75, *79*, 81, 82–95, *83*, *93*, *95*, 99–100, 103, 109–110, 147, 182–188, 229–230, 247, 289, 290
Bush-Hazen integraph, *91*, 145
Bush-Stewart-Gage integraph, 87, *89*
Byrne, John F., 72

Caldwell, Samuel H., *71*, 74, 90, *93*, *95*, *102*, 184, 188, 230, 294, 353
Caldwell, Sidney E., *102*
Camera, high-speed drum, 148
Camp, Thomas B., 75
Campbell, D. P., 215, 216
Campbell, George A., 156
Cape Cod Canal, hydraulic model of, 75
Cape Cod Model Air Defense System, 299
Carnegie Institution, 74–75
Carr, James A., *64*
Carty, J. J., 54
Cauer, Wilhelm, 156
Cavendish Laboratory, 161
Cavity magnetron, *191*
Center of Analysis, 220, 232–233, 234, 293
Center of Communications Sciences, 344
Center for Materials Science and Engineering, 174, 175–177, *176*, 318
Cerrillo, Manuel V., *158*, 252
Ceyer, Sylvia, 274
Chadwick, James, 161
Chance, Britten, 198
Charles, Project, 268, 294–295
Chatfield, Charles H., 117
Cheatham, Thomas P., 260
Chinn, Howard A., 117–119
Chomsky, Noam, 275

Chu, Lan Jen, 121, 204, *206*, 248, *252*, 316
Cinema integraph, 95, 211, *212*
Clapp, James K., 112, 115, 117
Clements, Donald F., *135*
Clifford, Harry E., 29–30, *29*, 32, 35, 41–42, *41*, 46, 52, 63
Coast Artillery, 57
Coax instrument, 170
Cochrane, Edward L., 307, 322
Cockcroft, John D., 161, 195
Coe, Conway P., 184
Cohen, Robert, 260
Colburn, Steve, 278
Coleman, John B., *71*
COMIT, 340
Communications, expansion of, 113–123
Compatible Time-Sharing System (CTSS), 345
Compton, Arthur H., 231
Compton, Karl T., 32, 69, *70*, 74, 119, 182, 184, 195–196
Compton Laboratories, 336
Computation Center, 274, 334–341, 343
Computer-Aided Design Project, 225
Computer Applications Group, 224–225
Computer Components and Systems Group, 175
Computer Components and Systems Laboratory, 337–338
Computer Science, Laboratory for, 351–353
Conant, James B., 75, 182, 184
Continuous integraph (product integraph), *83*, 87
Conway, Lynn, 278, 363
Cook, Jackson H., 196
Cookson, Albert E., 248
Coolidge, William D., 30, 44, 128

Coons, Steven A., 351
Cooperative Computing Laboratory, 344
Cooperative program, 124–137
Coppi, Bruno, 271
Corbató, Fernando J., 294, 337, 341, 344, 345, 348, 371–372, *372*, *373*
Cornell University, 23
Correlator, digital, 260, 262
Correlator, electronic, 260
Cotton, H., 145
Council of the Graduate School of Engineering, 44
Cousteau, Jacques, 138
Cowley, Percy E. A., 257
Cox, Jerome R., Jr., 305
Crary, Selden B., 214
Crawford, Perry, 281, 284
Cross, Charles R., 2, 16, *18*, 19–21, 23–25, 28–30, 35, 38
Cryogenic Laboratory, 321
Cryotron, 337
CTSS (Compatible Time-Sharing System), 345
Cyclotron, 161

Dahl, Gustav, 69, *71*, 72, 74, 103, 130, 155
Dart, Harry E., 37
Davenport, Wilbur B., Jr., 257, 262, 268, 295, 356, 361–364, *364*
David, Edward E., Jr., 252
Davy, Sir Humphry, 2
Dawes, Lyman M., *71*
Dawson, Edward, 215
DC Short-Circuit Calculating Table, 99
de Bettencourt, J. T., 295
Dellenbaugh, Frederick S., *64*, 100
Dennis, Jack B., 341, 344, 345
Derr, Louis, 30, 38
Dertouzos, Michael, 226, 351, *353*
Dewey, Davis R., 48
Dielectric Measurement Group, 170

Differential analyzer, 84, 90, *91*, 92
Digital Computer Laboratory, 210, 224, 294, 298
Digital Equipment Corporation, 338, 345, 350, 365
Dillon, Theodore H., *64*, 65
Division of Elecrical Engineering Research, 48
Dixon, Marvin H., *71*
Doctoral degrees, first awarding of, 43–44
Dodd, Stephen H., Jr., *283*, 285–286, *292*
Doherty, Robert E., 100
Dolbear, Amos E., 28–29
Dowden, Alfred L., 133
Dozier, Leonard C., 227
Draper, C. Stark, 122, 192, 213–215, 293, 303
Dresselhaus, Mildred S., 177, *177*, *373*
Dresser, Richard, 162, 165
DuBridge, Lee A., 190, *191*, 197, 314
Duncan, Louis, 32, 35–40, *36*
du Pont, T. Coleman, 51, 66
Dupree, Tom, 271
Durlach, Nathaniel, 278
Dwight, Herbert B., *71*
Dynamic Analysis and Control Laboratory, 210, 227, 293
DYNAMO, 340
Dynamo Electric Machinery Laboratory, 54, *56*, 145

Eastman, George ("Mr. Smith"), 49, 51, 63, 66
Eastman Kodak Company, 150
Eaton-Peabody Laboratory for Auditory Physiology, 263, 275, 278
Eckert, J. Presper, Jr., 284
Economos, George, 177
Eddington, Donald, 278
Eden, Murray, 275
Edgar, Charles L., 43
Edgerton, Germeshausen, and Grier, Incorporated (EG&G), 151, 153
Edgerton, Harold, 70, *71*, 74, 81, 130, 138–153, *146*, *149*, *152*, 204, 206

Edison, Thomas A., 2, 23, 24–25
EDVAC (Electronic Discrete Variable Calculator), 232
Edwards, Charles M., 227
EG&G Education Center, 153
E (Kennelly-Heaviside) layer, 109
Electrical engineering, curriculum changes in, 54, 57
Electrical Engineering and Computer Science, Department of, 354–372
Electrical Machinery Laboratory, *27*
Electrical Measurements, Laboratory of, *27*
Electric Power Systems Engineering Laboratory, 319
Electronic Discrete Variable Calculator (EDVAC), 232
Electronic Numerical Integrator and Computer (ENIAC), 231
Electronic Systems Laboratory, 134, 175, 210, 218, 225–227, 350
Electrostatic generator, 162, *164*
Electrostatic storage tube, *283*
Elias, Peter, 258, *312*, *316*, 348, 356, *357*, 358–359, 370–371, *373*
Emerson, Merton L., 63
Energy Conversion Laboratory, 173, 175
Engels, Friedrich, 9
Engineering Corps, 57
Engineering Electronics Laboratory, *73*
Engineering Internship Program, 136
Engineering Research, Graduate School of, 40
ENIAC (Electronic Numerical Integrator and Computer), 231
Entwistle, James L., *71*
Epstein, David J., *172*, 173, 175, 177, 316
Evans, R. D., 99
Everett, Robert R., 221, 282, *285*, *292*, 295

Fahnestock, Harris, 285, 295
Fairfield, John, Jr., 257
Fan, H. Y., 167
Fano, Robert M., 207, 247, 255–257, 262, 268, 295, 316, 344, *346*, 347–353, *347*, 359, *373*
Faraday, Michael, 2
Farcot, Joseph, 213
Farmer, Moses G., 21
Fay, Richard D., *71*, 302, 304
Feero, William E., 321
Feinstein, Amiel, 256, 258
Felman, Clarence G., 225
Ferenz, Ramona, *292*
Ferguson, Louis A., 43
Feshbach, Herman, 293
Filter theory, 156
Fink, Donald G., 198–199
Fiore, John, 218
Fisch, Nathaniel, 271–272
Fitzgerald, Thomas J., 74, 132, 248, 317
Five-digit multiplier, *286*, 287
Flanagan, James L., 305, 307
Fletcher, Ewan W., 175, 337
Flight simulator, 227
Florez, Luis de, 284
Floyd, George F., 217
Ford, Hannibal C., 90
Ford, Horace, 192
Forrester, Jay W., *183*, 216, 221, 228, *229*, 233, 235, 281–301, *281*, *283*, *292*, 341, 353
FORTRAN, development of, 339–341
Fossett, R. L., Jr., 121
Foster, R. M., 157
Fourmarier, Paul, 145
Fox, Phyllis, 104, 290, 341
Fraenckel, Victor H., 201
Franck, James, 166
Frank, Nathaniel H., 307
Franklin, Benjamin, 8
Franklin, Philip, 285
Frazier, Richard H., *71*, 72, 132, 317
Fredkin, Edward, 345, 353
Fry, Thornton C., 184
Fundingsland, O. T., 246

Gager, F. Malcolm, *71*, 121
Gale, Horace B., 29
Gardner, Hartley B., *64*
Gardner, Murray, *71*, 74, 77, *79*, 85, 100, 103, 132, *183*, *312*
GCA ("ground control of approach"), 198
General Electric Corporation, 3, 29, 30, 37, 48, 75, 99, 100, 128, 130, 132, 145, 160, 173, 186, 194, 300, 318, 321, 351
Generalized Harmonic Analysis, 253
General Radio Company, 148
Generator, electrostatic, 162, *164*
Generator, Van de Graaff, 123
Germeshausen, Kenneth J., 147–153, *152*, 206
Getting, Ivan A., 77, 186, 199, 246
Gewertz, Charles E. M., 157
Gill, Stanley, 339
Gilliland, Edwin R., 322
Gilmore, John T., Jr., 339
Glaser, Raymond A., 248
Goff, Kenneth W., 307
Goldenberg, Daniel, 344
Goldman, Stanford, 257
Goldsmith, Lloyd T., 117
Goodrich, Betty, *265*
Gordon McKay Professorship of Electrical Engineering, 42
Gottling, James G., 175, 338
Gould, King E., 90, 92
Gould, Lawrence, 226
Gould, Leonard A., 134, 217–218
Granlund, John, 248
Grant, Nicholas J., 175, 177
Grass, Albert M., 103, 198
Grass Instrument Company, 103
Gray, Paul E., 173, 174, 318, 360
Gray, Truman S., *94*, 95, 134, *183*, 188, 215, 218, *219*, 247, *312*, 316
Green, Cecil H., 136
Green, Edward H. R., 69, 114–115, *114*
Green, Paul, 268–269

Grier, Herbert E., 148–153, *152*
Griggs, David T., 199, 201
Grim, William, 218
Grodzinsky, Alan, 319
Gross, Eric T. B., 104
"Ground control of approach" (GCA), 198
Guillemin, Ernst A., *71*, 77, 81, 106, 114, 130, 154–159, *155*, *158*, *159*, 193, 206, 252, 269, 315

Hada, Tsuenzo, 48
Hall, Albert C., 214–217, 227
Hall, F. T., Jr., 121
Hall, William H., 197, 302, 304
Halle, Morris, 269, 275
Hamel, G., 157
Hannibal Ford integrator, *89*
Hansen, William W., 195
Harris, Lawrence A., 248
Harrison, George R., 348
Hartley, R. V. L., 255
Hartree, Douglas, 92
Hartwell, Project, 268
Harvard Psycho-Acoustics Laboratory, 256, 258, 262
Harvard University, 8–9, 42, 49, 51–52, 62–63, 65, 303
Harvey, George G., 246, 295
Haus, Hermann A., 272–273
Hayes, Hammond V., 43, 156
Hazeltine, Allen V., 199
Hazen, Harold L., 32, 34, 70, *71*, 72, 75–77, *76*, *79*, 81, 84, 90, *93*, 95, 96–105, *97*, *102*, *105*, 130, 132, *183*, 185–186, *186*, 212–213, 245–246, 289, 304, 323
Heaviside, Oliver, 28, 84–85
Hedeman, Walter R., Jr., 95
Heisenberg, Werner, 166
Henck, John B., Jr., 25, 28
Hentz, Clifford E., *64*, *71*
Hewins, E. H., 29
Hewlett, William, 136
Heymans, Paul A., 118
High-field tokamak, 271
High-voltage research laboratory, 160–165

High-voltage X rays, 162, 165
Hildebrand, Francis B., 353
Hill, Albert G., 207, *245*, 247–248, 269, 295, 344, 348
Hitchcock, H. W., 202
Hite, George, 198
Hoadley, George B., 103
Hollerith, Herman, 90
Holman, Silas W., 24, *26*, 29
Hooper, Edwin B., 213–215
Horn antenna, *122*
Horton, J. Warren, 148
Hostetter, John C., 150
Houghton, Henry G., 115, 118–119, *120*
Hovde, Frederic L., 201
Hrones, John A., 227
Hudson, Ralph G., *64*, *71*, 100
Huffman, David A., 269
Hull, Albert W., 194
Hunsaker, Jerome C., 57
Hunt, Frederick V., 303
Hunt, Robert, *172*
Hunton, James K., 248
Hurtig, C. R., 173
Hutcheson, John A., 196, 313
Hutner, R. A., 197
Huxley, Roy D., 48
Hydrogen thyratron, 206

Illuminating Engineering Laboratory, 150
Information and Decision Systems, Laboratory for, 227
Institute for Advanced Study (Princeton), 282
Instrumentation Laboratory, 283, 192
Insulation Research, Laboratory for, 166–177, 204, 298
International Business Machines, 335–338, 341, 344, 345, 365
International Telephone and Telegraph Company, 122, 169, 196, 204
Ippen, Erich, 273
Israel, David, 340
Ivaska, Joe, 218

Jackman, J. Ralph, 150
Jackson, Dugald C., 23, 32, 34, 42–48, *43*, 60, 62, *64*, 69–70, *71*, 72, 81, 84, 106–110, 147, 150
Jackson, Dugald C., Jr., 129
Jackson, William D., 318, 319
Jacques, William W., 25, 28
Jansen, J. J., 121
Javan, Ali, 115
Jenney, Richard F., 338
Jewett, Frank B., 112, 182, 184
Joannopoulos, John, 273
Joffe, A., 85
Johns Hopkins University, 25, 38
Johnson, Howard, 49, 348
Joint Services Electronics Program, 274
Jones, Reginald L., 47, 48
Jones, Thomas F., *219*

Kaiser, James F., 226
Kassakian, John G., *320*, 321
Kastner, Mark, 273
Katzenelson, Jacob, *350*
Kear, Frank G., *79*
Keithley, Joseph F., 150
Kennard, John W. B., *64*
Kennedy, Robert S., 272
Kennelly, Arthur E., 52, 63, *64*, 68, 84, 85, 109, 111
Kennelly-Heaviside (E) layer, 109
Kenrick, Gleason W., 253
Kerr, Donald E., 121, 196, 197, 198
Kershaw, Walter F., *79*, 90
Kessler, John A., 308, *309*
Kiang, Nelson, 278
Killian, James R., *192*, 269, 282
Killian, T. J., 130
Kilpatrick, Lester L., 248
Kingsbury, Stanley, *172*
Kingsley, Charles, *71*, 317, 321
Kinnard, Isaac F., 100
Kirkpatrick, H. A., 198
Kirtley, James, *320*, 321
Kittler, Erasmus, 23
Klasens, Hendrik A., 175
Klatt, Dennis, 275

Kludge, 350–351, *350*
Klystron, 122, 195
Koenig, Herman, 317
Kopal, Zdenek, 293
Koppen, Otto C., 284
Kraft, Leon G., Jr., 262
Kremer, Waldemar R., 40
Kretzmer, Ernest R., 248
Krohn, Earl H., 198
Kron, Gabriel, 317
Krylov, Alexei N., 90
Kusko, Alexander, 316
Kyhl, R. L., 134

Labate, Samuel, 305
Laboratory for Computer Science, 351–353
Laboratory of Electrical Measurements, *27*
Laboratory for Information and Decision Systems, 227
Laboratory Instrument Computer (LINK), 301
Laboratory for Insulation Research, 166–177, 204, 298
Land, Edwin, 323
Lane, Henry M., *71*
Laning, J. Halcomb, Jr., 265, 339
Lansil, Clifford E., *64*, *71*
Laser, free-electron, 271
Lawrence, Ernest O., 161, 195, 196
Lawrence, Harry, 214
Lawrence, Ralph R., 37, 41, 47, *64*, *71*, 138
Lawry, C. C., Jr., 214
Laws, Frank A., 30, 37, 41, *64*, *71*
Lee, Everett, 314
Lee, Francis, 275
Lee, Harry B., 158, 316
Lee, Patrick, 273
Lee, Yuk Wing, 157, 187, 226, 242–243, *254*, 258–262, *259*, *261*, 295
Lennard-Jones, John E., 92
Leonard, H. Ward, 28
Lettvin, Jerome, 265, *265*
Levinthal, Cyrus, 351
Lewis, Frank M., 121, 196, 197, 199, 201
Lewis, G. N., 44

Lexington, Project, 268
Licklider, Joseph C. R., 256, 258, *259*, 262, 267, 295, 307, 342, 345, 347, 351
Lin, Chia-Chiao, 293–294
Lincoln Laboratory, 115, 173, 247, 268, 283, 295, 300–301
Linder, Clarence, 318
LINK (Laboratory Instrument Computer), 301
Linvill, John G., 132, *133*, 252
Linvill, William, 218, 290, 337, 353
LISP (List Processor), 340–341
Litchfield, Isaac W., 28
Litchfield, Paul W., 119
Litster, J. David, 177, 273
Little, Arthur D., 28
Liu, Y. S., 213, 214, 226
Livingston, M. Stanley, 161, 247
Lodge, Henry Cabot, 52, 54
Loeb, Arthur L., 338
Löf, J. C. L., 213
Long-range navigation (LORAN), 197–198, 199, 202
Loomis, Alfred, 195, 196
Loomis, F. Wheeler, 295
Loomis Laboratories, 196
LORAN (long-range navigation), 197–198, 199, 202
Lowell, Lawrence, 51, 52
Lowell Building Electrical Engineering Laboratory, *36*, *39*
Luckhardt, Stanley, 272
Lumped-constant artificial line, 68
Lyon, Dean A., 167
Lyon, Waldo V., 47, *64*, 90, 130, 138

MAC, Project, 345–353
MAC computer, *352*
McCall, Samuel W., 52
McCarthy, John S., 63, 341, 342–345, *343*, 353
McClure, Alfred M., 147
McCulloch, Warren S., 243, 263, *265*
McDonough, James O., 218, 220, 221, 224
McIlroy, Malcolm S., 209

Macintosh, Gordon G., *117*
McKay, Gordon, 49, 51
Maclaurin, Richard C., 49, 51, 52, *53*, 57, 63, 65–66
MacMillan, Archie, *172*
Macnee, Alan B., 291
Macy Meetings, 243
Magnetic-core memory, 282, 296–298
Magnetic-core plane, *281*
Magnetic memory unit, *296–297*
Magnetron, *191*, 194, 195
Marcy, H. Tyler, 216
Mark 14 gun sight, 214
Markham, John R., 284
Marshall, Lauriston C., 201
Martin, D. K., 202
Martin, Stuart T., Jr., 246
Marx generator, 160
Maon, Samuel J., 173, 246, *252*, 272, 275, 316
Massachusetts Institute of Technology, move to Cambridge, 49–54, *56*
Massey, Harrie S. W., 92
Matthias, Berndt, 172
Mawardi, Osman, 307, 318
Maxwell, A. R., 213
Maxwell, James Clerk, 108, 213
Mayflower, nonrigid dirigible, *120*
Meade, John E., 198
Megaw, Helen D., 175
Melcher, James R., 165, 318–319
Mercantile Library, *10–11*
Merriam, Charles W., III, 226
Merwin, Marjorie L., 343
Merz, Walter, 172
Metcalf, George F., 196
Meteor, Project, 248
Microsystem Program, 362–363
Microwave Committee, 196–197
Microwave Early Warning, 202
Microwave instrument landing system, 121–123
Mieher, Walter M., 196
Miles, Perry, *172*, 177
Mili, Gjon, 138, 150

Milling machine, numerically controlled, 220–224, *222*
Minsky, Marvin, 344, *346*, 348, *349*, 361
MITalk system, 275
Mitchell, E. C., 213
MITRE Corporation, 300
Mitter, Sanjoy K., 226, 227
Moler, George S., 23
Molnar, Julius P., 169
Monroe, James P., 28
Moon, Parry, 68, *71*, 132
Morash, Arthur F., *71*
Moreland, Edward L., 32, 62, 69, 72–75, *73*, *183*, 185, 207
Morey, Charles A., 24
Morgenthaler, F. R., 133, 272
Morley, Herbert M., 40
Morrison, Etting E., 323
Morrow, W. E., 173
Morse, Philip M., 111, *112*, 293, 302–308, 334, *336*, 337, 344, 347, 348
Morss, Henry A., 52
Moses, Joel, 361, 365, 368–369, *369*, *373*
Mueller, Hans, 118
Mulligan, James E., *71*
Multics system, 351
Munroe, James P., 52
Murphy, Vincent K., 132
Murray, William S., 98
Museum of Technology, 7
Mustin, Lloyd M., 213–215

National Defense Research Committee (NDRC), 75, 184–188
National Magnet Laboratory, 271
Navy Aviation School, 60, *61*
Nelson, Arthur L., *64*
Nelson, Keith, 274
Network analysis, 154–159
Network analyzer, 96–105, *102*
Network synthesis, 154–159
Network synthesizer, 291
Newman, Robert B., 305
Newnham, Robert E., 338
Newton, George C., Jr., 213, 215, 216, 220
Nichols, Ernest Fox, 66

Nims, Paul T., 95
Noise Modulation and Correlation (NOMAC), 268
Noiseux, Dennis, 307
NOMAC (Noise Modulation and Correlation), 268
Norcross, Austin S., *71*
Norinder, Harold, 132
Noyes, Arthur A., 43, 80, 128

Office of Naval Research, 289
Office of Scientific Research and Development (OSRD), 75, 185–188
Office of Statistical Services, 293
Oliphant, M. L., 195
Olsen, Kenneth H., 298, 345, *346*
Olsen, Stanley, 345
Opitz, Herwart, 225
Oppenheim, Alan, 274–275
Ordnance Corps, 57
Osborne, Harold S., 47, 48
Overbeck, Wilcox P., 231–232
Overhage, Carl F., 348

Pankowski, Bernard J., 268
Papian, William N., 175, 282, 296
Parker, Ronald, 271
Parsons, John T., 220–221
Pauli, Wolfgang, 166
Peake, William, 275
Pease, William M., 216, 218, 220, 221, 224
Pender, Harold, 46–48, 68
Penfield, Paul L., Jr., 272, 278
Pensyl, Daniel S., 195
Peterson, Arnold P. G., 196
Peterson, Chester, *71*
Phase-shifting transformer, 100, *101*
Philbrick, Frederick B., *64*
Philbrick, George A., 188
Phonautograph, 24
Photoelectric integraph, *94*, 95
Pickering, Edward C., 16, *18*, 21
Pickering, William H., 23, 24, 29
Pickernell, Frank A., 30
Pierce, George W., 62
Pierce, John A., 197

Pierson, James E., 290
Pinchot, Gifford, 98
Pitts, Walter, 262
"Plan for a Polytechnic School in Boston," 7, 12
Plan Position Indicator (PPI), 194
Plasma Fusion Center, 271, 272
Poitras, Edward J., 184
"Polytechnic Institute," 6
Porkolab, Miklos, 272
Porter, Charles H., 42
Porter, Murrice O., 71
PPI (Plan Position Indicator), 194
Prescott, Samuel C., 12, 17
Price, Robert, 268–269
Product integraph (continuous integraph), 83, 87
Profile tracer, 86
Puffer, William L., 28, 29, 35, 41
Pugh, Alexander L., III, 341
Pulse-generating network, 159
Pupin, Michael I., 54

Quinlan, Robert E., 71

Rabi, I. I., 191, 197
Radar, development of, 193–207, 203
Radar antenna, 252
Radar meteorology, development of, 198
Radar School, 121, 207, 209
Radford, William H., 115, 119, 120, 188, 209, 229–232, 248, 249, 295
Radiation integraphs, 92, 95
Radiation Laboratory, 34, 165, 190–209
Radio direction-finding technique (RDF), 194
Raether, Heinz, 175
Rafuse, Robert P., 272
Randall, J. T., 195
Rapid Arithmetical Machine, 229
Rathenau, Gert W., 175
Ratliff, William H., Jr., 196
Raytheon Manufacturing Company, 75, 199, 255, 300

RDF (radio direction-finding technique), 194
Redmond, William Guy, Jr., 307
Rees, Mina, 185, 289
Reintjes, J. Francis, 128, 133–134, 135, 221, 225–227, 312, 352
Reissner, Eric, 293–294
Research Laboratory of Electronics, 173, 207, 220, 232, 234, 242–279
Research Laboratory of Physical Chemistry, 43
Resonant-cavity magnetron, 195
Reynolds, Kenneth C., 75
Riaz, Mahmoud, 317
Rice, S. O., 253
Richards, Robert H., 28
Richmond, Harold B., 315
Ricker, Claire W., 64
Ridenour, Louis N., 199
Rines, David, 148
Rivero, Horacio, 213–215
Roberts, Shepard, 196, 204
Robertson, H. P., 202
Robinson, Charles C., 338
Robinson, Denis M., 194
Rockefeller Differential Analyzer, 92, 93
Rodriguez, Jorge E., 225
Rogers, Robert E., 66
Rogers, William B., 4, 6–12, 7, 43, 66
Rogers Building, 12, 26, 52
Rogers Laboratory of Physics, 13
Rollefson, Ragnar, 295
Roosevelt, Franklin D., 33, 52, 53, 74, 184
Rosenblith, Walter, 263, 264, 275, 307, 360
Rosenblueth, Arturo, 263
Ross, Douglas T., 135, 224–225, 344, 350
Rossby, Carl Gustav, 119
Rosseland, Svein, 92
Rotating energy converter, 318
Round Hill Research Facility, 69, 106, 114–122, 116, 117, 120
Routh, E. J., 213
Ruiz, Albert L., 186

Runyon, John N., 224
Russell, Arthur L., 71
Russell, John B., 71
Russell, S. C., 214
Rutherford, Ernest, 161
Ryan, Harris J., 160

Sabine, Wallace C., 305
Sage, Nathaniel, 192, 289
SAGE (Semi-Automatic Ground Environment), 283, 299–300, 340
St. Clair, Harry P., 103
St. George, Emory, Jr., 227
Saltzer, John M., 290
Samuel, Arthur, 111, 195
Sandvold, H., 218
Santelmann, William F., Jr., 248
Sarles, F. Williams, Jr., 338
SATC (Students' Army Training Corps), 60
Saunders, Robert M., 317
Schenectady Research Laboratory, 30
Scherr, Allen L., 344
Schmidt, Kenneth A., 321
Schneider, Herman, 128
Scholtz, Robert A., 267
School of Industrial Science, 8
Schreiber, William, 275
Schrödinger, Erwin, 166
Schurig, Otto R., 52, 99, 103
Schweppe, Fred C., 226, 319
Scott, Charles F., 43
Scott, Herman H., 148
Scott, Ronald E., 291
"Scram" equipment, 218
Searle, Campbell L., 174, 248, 278
Sedgwick, William T., 44, 66
Seifert, William W., 227
Semi-Automatic Ground Environment (SAGE), 283, 299–300, 340
Sensiper, Samuel A., 248
Servomechanisms Laboratory, 184, 192, 210–227, 284
Shaad, George C., 42, 46
Shannon, Claude E., 133, 255–256, 256, 258, 266–267, 350
Shapiro, Ascher H., 318
Shapiro, Jeffrey H., 272

Shea, Timothy E., 113–114
Sherwood, Thomas K., 291, 322
Shiring, Paul B., 317
Shockley, William, 363
Shortell, Albert V., Jr., 338
Siddall, Walter D., 115
Siebert, W. M., 134, 264, 278
Siegel, Arnold, 224
Signal Corps, 57
Silvey, J. O., 214, 215
Simulation, defined, 99
Singleton, Henry E., 260, 262
Sisson, Roger L., 290
Sketchpad, 350
Skolnikoff, Eugene B., 132–133
Slater, John C., 119, 175, 192–193, 244, 246, 304
Sloan, Alfred P., 30
Sloan Foundation, 30
Sloat, J. R., 121
Smakula, Alexander, 172, 175, 177
Smith, Arthur C., 174, 318
Smith, D. B., 121
Smith, Harold B., 40
Smith, Harrison W., 37, 60
Smith, Henry I., 274
Smith, Joseph L., Jr., 321
"Smith, Mr." (George Eastman), 49, 51, 63, 66
Smith, Robert A., 177
Smullin, Louis D., 134, 269, 271, 295, 356, 357, 359–361, 373
SNOBOL, 341
Soderberg, C. Richard, 308, 322
Sommerfeld, Arnold, 121, 155, 302
Southworth, George C., 121
Spencer, Domina E., 68
Spencer, Hugh H., 100
Spencer, Richard H., 219
Sperry Gyroscope Company, 214–216
Sporn, Philip, 103, 319
Sprague, Frank J., 2
Spuhler, Harold A., 248
Squire, C. F., 247
Staelin, David H., 272

Stanford University, 63, 122, 160, 195
Stauder, Lawrence F., 147
Steinberg, Joseph R., 343
Steinmetz, Charles P., 46, 47
Stekly, Zdenek J., 321
Stevens, Kenneth N., 275, 305, 307, 316, 351
Stevens, S. Smith, 305
Stever, H. G., 248
Steward, Duncan J., 184
Steward, Irvin, 184, 185
Stewart, Herbert R., 87
Stillman, Gerald, 321
Stockham, Thomas A., Jr., *174*
Stone, Charles A., 30, 51, 52, *53*
Stone, John, 156
Stone and Webster Engineering Corporation, 30, 48, 51
Storage tube, electrostatic, *283*
Stott, H. G., 46
Strandberg, Malcolm W. P., 271
Stratton, Julius A., 70, *71*, 106, *108*, 112, 118, 119, 196–200, 204, 207, 244–251, *245*, 269, 291, 293, 303, 304, 307, *317*, 323
Stratton, Samuel W., 66, 69
Statton-Houghton formula, 118
Strobe Lab, 148, 150–151, 153
Stroboscope, 138–153, *146*, *149*
Stroboscopic light, photographs with, *139, 140, 141, 142, 143, 144, 205*
Strong, George W., 184
Stubbs, Arthur R., 63
Students' Army Training Corps (SATC), 60
Stulen, Frank L., 221
Sullivan, William L., *71*
"Superpower," 96–105
Susskind, Alfred K., 220–221, 226, 290
Sutherland, Ivan E., 350
Swope, Gerard, 30

"Tables of Dielectric Materials," 170
Tarpley, Harold I., 216
Taylor, Norman H., *286*, 287
Taylor, Richard, 234
Teager, Herbert M., 218, 342–345
"Technology Plan," 63
Tedesco, Thomas N., *120*
Telephone, development of, 24–25
Tennessee Valley Authority, 33–34
Terman, Frederick, 63, 70, 81, 81, 206
Terwilliger, David D., 90, *91*
Therkelsen, Ernest W., 216
Thomas, Benjamin E., 23
Thomson, Elihu, 2, 29, *37*, 43, 46, 66
Thomson, James, 90
Thomson, Sir William, 90
Thornton, Richard D., 158, 174, 272
Thurston, J. N., *249*
Timbie, William H., *64*, 65, *71*, 75, 81, 82, 124–132, *125*
Time sharing, development of, 342–353
Tinlot, John H., 198
Tisza, L., 247
Tizard, Henry T., 193, 195
Tolman, Richard C., 182, 184
Townes, Charles H., 115, *346*
Transmission system simulator, *320*, 321
Traver, Harold A., 213
Troost, Laurens, 322
Troxel, Donald, 275
Trump, John G., 74, 81, 132, 160–165, *161*, 175, 200–202, 246
Truxal, James G., 265
Tucker, Carlton E., *64*, *71*, 74, 75, 107, 111, 134, *186*, 209, *312*
Tucker, John A., 128, 134–137, *135*, *347*
Tukey, John W., 287
Tuller, William G., 196, 246, 255
Tungsten, ductile, 44
Turner, L. A., 197

Tuttle, David F., Jr., 63, 157
TX-0 computer, *327*, 345
TX-2 computer, 345

Urey, Harold, 161

Vail, Theodore N., 48
Vail Library, 48
Vallarta, Manuel S., *112*, 113, 118, 253
Valley, George, 282, 294, 295
Van Atta, C. M., 123, 161
Van Atta, L. C., 123, 161
Van de Graaff, Robert J., 81, 123, 160–165
Van de Graaff generator, 123
Van Pelt, Eugene V. B., *71*
Van Rennes, Albert B., *219*
Van Valkenburg, M. E., 157, 246
Vannevar Bush Building, 175–177, *176*
Varian, Russell M., 195
Varian, Sigurd, 195
Versator II tokamak, 272
Very Large Scale Integration (VLSI) Project, 362–363
Verzuh, Frank M., 246, 293, 337
Volk, Alexander, 319
von Engel, Alfred, 175
von Hippel, Arthur R., 74, 81, 166–177, *168, 172*, 196, 204, 298
von Neumann, John, 232, 282

Wagner, Carl, 175
Wagner, K. W., 157
Walker, Francis Amasa, 12, 17
Walker, William H., 63
Walker Building, *26*
Walker Memorial, 52, *53*, 60
Wallace, William, 23
Wallman, Henry, 255, 291
Walton, Ernest T. S., 161
Wang, C., 214
Ward, Alfred G., 213–215, 226
Ward, John E., 350
Warner, Robert W., 72
Waterman, Alan T., 289
Watson, Thomas A., 24
Watson, Thomas J., 336

Watson-Watt, Robert, 193
Weaver, Warren, 184, 185, 232
Webster, Edwin S., 30, 51, *53*
Weinbach, Mendell P., 23
Weinberg, Louis, 157
Weiss, Herbert, 295
Weiss, Thomas, 275
Westinghouse, George, 2, 3
Westinghouse Electric Corporation, 99, 173, 318, 321
Westphal, William B., 170, 204
Wheeler, David, 339
Whirlwind, Project, 220, 228–235, 280–301, *292*, 336, 338–340
White, Anthony C., 29
White, David C., 173, *174*, 314, 316, *317*, 318
White, James G., 23
Whitehouse, David R., *219*
Whitney, Willis R., 30, 44, 128
Wickenden, William E., 46–47, 62, 129
Wien, Max, 166
Wiener, Norbert, 85, 92, 157, 186–188, 234, 252–254, *254, 261*, 266–267, 332
Wieser, Bob, 340
Wiesner, Jerome B., 77, 82, 115, *245*, 248, *261*, 262, 265–269, 295, *312*, 322, 323, 344, 360, *373*
Wilcox, Robert B., 216
Wildes, Karl L., *71*, 129–134, *131, 183*, 186, 187, 253, *312*
Wilkes, Maurice V., 339
Wilkins, Harold S., 148
Willems, Jan C., 226
Williams, Charles, 24
Willis, H. Hugh, 196, 214
Wilson, Carroll L., 184, 185, 195
Wilson, Gerald L., 319, *320*, 321, 361, 363–365, *364*, *373*
Wilson, Myron S., 100
Winter, David F., 246, 256
Wirephoto machine, laser-based, 275
Wiseman, Robert J., 48

Wolf, William M., 338
Wolff, Peter, *270,* 271
Woodruff, Louis F., *71, 183,*
    188, 206–207
Woodson, Herbert H., 134,
    174, 317–321
Wozencraft, John M., 258
Wurster, William W., 304

X rays, hard, 44

Yates, John E., 345
"Yellow Peril," 187
Youtz, Pat, *283,* 285–286

Zacharias, J. R., 246, 268,
    295
Zaffarano, F. P., 247
Zak, Frederick J., 145
Zames, George, 226
Zenneck, J., 121
Zierlen, Neal, 339
Zimmermann, Henry J., 134,
    196, 209, 248, *249,* 252,
    269, 316
Zobel, Otto J., 156
Zraket, Charles, 340